Lecture Notes in Artificial Intel

Edited by J. G. Carbonell and J. Siekmann

Subseries of Lecture Notes in Computer Science

Christian Müller (Ed.)

Speaker Classification II

Selected Projects

 Springer

Series Editors

Jaime G. Carbonell, Carnegie Mellon University, Pittsburgh, PA, USA
Jörg Siekmann, University of Saarland, Saarbrücken, Germany

Volume Editor

Christian Müller
International Computer Science Institute
1947 Center Street, Berkeley, CA 94704, USA
E-mail: cmueller@icsi.berkeley.edu

Library of Congress Control Number: 2007932403

CR Subject Classification (1998): I.2.7, I.2.6, H.5.2, H.5, I.4-5

LNCS Sublibrary: SL 7 – Artificial Intelligence

ISSN   0302-9743
ISBN-10   3-540-74121-6 Springer Berlin Heidelberg New York
ISBN-13   978-3-540-74121-3 Springer Berlin Heidelberg New York

Springer is a part of Springer Science+Business Media

springer.com

© Springer-Verlag Berlin Heidelberg 2007

Typesetting: Camera-ready by author, data conversion by Scientific Publishing Services, Chennai, India
Printed on acid-free paper   SPIN: 12104529   06/3180   5 4 3 2 1 0

Preface

"As well as conveying a message in words and sounds, the speech signal carries information about the speaker's own anatomy, physiology, linguistic experience and mental state. These speaker characteristics are found in speech at all levels of description: from the spectral information in the sounds to the choice of words and utterances themselves."

The best way to introduce this textbook is by using the words Volker Dellwo and his colleagues had chosen to begin their chapter "How Is Individuality Expressed in Voice?" While they use this statement to motivate the introductory chapter on speech production and the phonetic description of speech, it constitutes a framework of the entire book as well: What characteristics of the speaker become manifest in his or her voice and speaking behavior? Which of them can be inferred from analyzing the acoustic realizations? What can this information be used for? Which methods are the most suitable for diversified problems in this area of research? How should the quality of the results be evaluated?

Within the scope of this book the term *speaker classification* is defined as assigning a given speech sample to a particular class of speakers. These classes could be Women vs. Men, Children vs. Adults, Natives vs. Foreigners, etc. *Speaker recognition* is considered as being a sub-field of speaker classification in which the respective class has only one member (Speaker vs. Non-Speaker). Since in the engineering community this sub-field is explored in more depth than others covered by the book, many of the articles focus on speaker recognition. Nevertheless, the findings are discussed in the context of the broader notion of speaker classification where feasible.

The book is organized in two volumes. Volume I encompasses more general and overview-like articles which contribute to answering a subset of the questions above: Besides Dellwo and coworker's introductory chapter, the "Fundamentals" part also includes a survey by David Hill, who addresses past and present speaker classification issues and outlines a potential future progression of the field.

The subsequent part is concerned with the multitude of candidate speaker "Characteristics." Tanja Schulz describes "why it is desirable to automatically derive particular speaker characteristics from speech" and focuses on language, accent, dialect, ideolect, and sociolect. Ulrike Gut investigates "how speakers can be classified into native and non-native speakers of a language on the basis of acoustic and perceptually relevant features in their speech" and compiles a list of the most salient acoustic properties of foreign accent. Susanne Schötz provides a survey about speaker age, covering the effects of ageing on the speech production mechanism, the human ability of perceiving speaker age, as well as its automatic recognition. John Hansen and Sanjay Patil "consider a range of issues associated with analysis, modeling, and recognition of speech under stress." Anton Batliner and Richard Huber address the problem of emotion classification focusing on the specific phenomenon of irregular phonation or laryngealization and thereby point out the inherent problem of speaker-dependency, which relates the problems of speaker identification and emotion recognition with each other. The juristic implications of acquiring knowledge about the speaker on the basis of his or her speech in the context of emotion recognition is addressed by Erik Eriksson and his co-authors, discussing, "inter alia, assessment of emotion in others, witness credibility, forensic investigation, and training of law enforcement officers."

The "Applications" of speaker classification are addressed in the following part: Felix Burckhardt et al. outline scenarios from the area of telephone-based dialog systems. Michael Jessen provides an overview of practical tasks of speaker classification in forensic phonetics and acoustics covering dialect, foreign accent, sociolect, age, gender, and medical conditions. Joaquin Gonzalez-Rodriguez and Daniel Ramos point out an upcoming paradigm shift in the forensic field where the need for objective and standardized procedures is pushing forward the use of automatic speaker recognition methods. Finally, Judith Markowitz sheds some light on the role of speaker classification in the context of the deeper explored sub-fields of speaker recognition and speaker verification.

The next part is concerned with "Methods and Features" for speaker classification beginning with an introduction of the use of frame-based features by Stefan Schacht et al. Higher-level features, i.e., features that rely on either linguistic or long-range prosodic information for characterizing individual speakers are subsequently addressed by Liz Shriberg. Jacques Koreman and his co-authors introduce an approach for enhancing the between-speaker differences at the feature level by projecting the original frame-based feature space into a new feature space using multilayer perceptron networks. An overview of "the features, models, and classifiers derived from [...] the areas of speech science for speaker characterization, pattern recognition and engineering" is provided by Douglas Sturim et al., focusing on the example of modern automatic speaker recognition systems. Izhak Shafran addresses the problem of fusing multiple sources of information, examining in particular how acoustic and lexical information can be combined for affect recognition.

The final part of this volume covers contributions on the "Evaluation" of speaker classification systems. Alvin Martin reports on the last 10 years of speaker recognition evaluations organized by the National Institute for Standards and Technology (nist), discussing how this internationally recognized series of performance evaluations has developed over time as the technology itself has been improved, thereby pointing out the "key factors that have been studied for their effect on performance, including training and test durations, channel variability, and speaker variability." Finally, an evaluation measure which averages the detection performance over various application types is introduced by David van Leeuwen and Niko Brümmer, focusing on its practical applications.

Volume II compiles a number of selected self-contained papers on research projects in the field of speaker classification. The highlights include: Nobuaki Minematsu and Kyoko Sakuraba's report on applying a gender recognition system to estimate the "feminity" of a client's voice in the context of a voice therapy of a "gender identity disorder"; a paper about the effort of studying emotion recognition on the basis of a "real-life" corpus from medical emergency call centers by Laurence Devillers and Laurence Vidrascu; Charl van Heerden and Etienne Barnard's presentation of a text-dependent speaker verification using features based on the temporal duration of context-dependent phonemes; Jerome Bellegarda's description of his approach on speaker classification which leverages the analysis of both speaker and verbal content information – as well as studies on accent identification by Emmanuel Ferragne and François Pellegrino, by Mark Huckvale and others.

February 2007   Christian Müller

Table of Contents

A Study of Acoustic Correlates of Speaker Age

Susanne Schötz[1] and Christian Müller[2]

[1] Dept. of Phonetics, Centre for Languages and Literature,
Lund University, Sweden
susanne.schotz@ling.lu.se
[2] International Computer Science Institute, Berkeley, CA
cmueller@icsi.berkeley.edu

Abstract. Speaker age is a speaker characteristic which is always present in speech. Previous studies have found numerous acoustic features which correlate with speaker age. However, few attempts have been made to establish their relative importance. This study automatically extracted 161 acoustic features from six words produced by 527 speakers of both genders, and used normalised means to directly compare the features. Segment duration and sound pressure level (SPL) range were identified as the most important acoustic correlates of speaker age.

Keywords: Speaker age, Phonetics, Acoustic analysis, Acoustic correlates.

1   Introduction

Many acoustic features of speech undergo significant change with ageing. Earlier studies have found age-related variation in duration, fundamental frequency, SPL, voice quality and spectral energy distribution (both phonatory and resonance) [1,2,3,4,5,6]. Moreover, a general increase of variability and instability, for instance in F_0 and amplitude, has been observed with increasing age.

The purpose of the present acoustic study was to use mainly automatic methods to obtain normative data of a large number of acoustic features in order to learn how they are related to speaker age, and to compare the age-related variation in the different features. Specifically, the study would investigate features in isolated words, in stressed vowels, and in voiceless fricatives and plosives. The aim was to identify the most important acoustic correlates of speaker age.

2   Questions and Hypotheses

The research questions concerned acoustic feature variation with advancing speaker age: (1) What age-related differences in features can be identified in female and male speakers? and (2) Which are the most important correlates of speaker age?

C. Müller (Ed.): Speaker Classification II, LNAI 4441, pp. 1–9, 2007.

Based on the findings of earlier studies (cf. [5]), the following hypotheses were made: **Speech rate** will generally decrease with advancing age. **SPL range** will increase for both genders. **F_0** will display different patterns for female and male speakers. In females, F_0 will remain stable until around the age of 50 (menopause), when a drop occurs, followed by either an increase, decrease or no change. Male F_0 will decrease until around middle age, when an increase will follow until old age. **Jitter and shimmer** will either increase or remain stable in both women and men. **Spectral energy distribution** (spectral tilt) will generally change in some way. However, in the higher frequencies (spectral emphasis), there will be no change. **Spectral noise** will increase in women, and either increase or remain stable in men. **Resonance measures** in terms of formant frequencies will decrease in both female and male speakers.

3   Speech Material

The speech samples consisted of 810 female and 836 male versions of the six Swedish isolated words *käke* ['ɕɛ̀ːkə] (jaw), *saker* ['sàːkəʁ] (things), *själen* ['ɧɛːlən] (the soul), *sot* [suːt], *typ* [tyːp] (type (noun)) and *tack* [tak] (thanks). These words were selected because they had previously been used by the first author in a perceptual study [7] and because they contained phonemes which in a previous study had shown tendencies to contain age-related information (/p/, /t/, /k/, /s/, /ɕ/ and /ɧ/) [8]. The words were produced by 259 female and 268 male speakers, taken from the SweDia 2000 speech corpus [9] as well as from new recordings. All speakers were recorded using a Sony portable DAT recorder TCD-D8 and a Sony tie-pin type condenser microphone ECM-T140 at 48kHz/16 bit sampling frequency in a quiet home or office room. Figure 1 shows the age and gender distribution of the speakers.

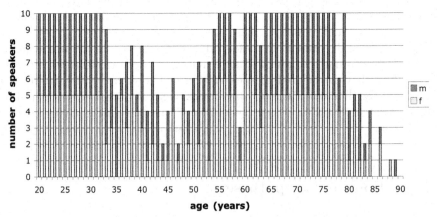

Fig. 1. Age distribution of the speakers used in this study

4   Method and Procedure

The acoustic analysis was carried out using mainly automatic methods. However, occasional manual elements were necessary in a few steps of the procedure. All words were normalised for SPL, aligned (i.e. transcribed into phoneme as well as plosive closure, VOT and aspiration segments) using several Praat [10] scripts and an automatic aligner[1]. Figure 2 shows an alignment example of the word *tack*. The alignments were checked several times using additional Praat scripts in order to detect and manually correct errors.

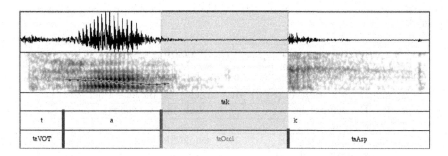

Fig. 2. Example of the word *tack*, aligned into word, phoneme, VOT, plosive closure and aspiration segments

The aligned words were concatenated; all the first word productions of a speaker were combined into one six-word sound file, all the second ones concatenated into a second file and so on until all words by all speakers had been concatenated. Figure 3 shows an example of an concatenated file.

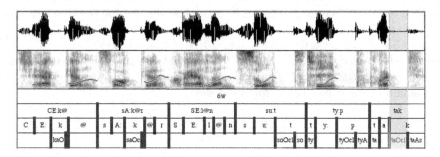

Fig. 3. Example of a concatenated file, aligned into word, phoneme, plosive closure, VOT and aspiration segments

---

[1] Originally developed by Johan Frid at the Department of Linguistics and Phonetics, Centre for Languages and Literature, Lund University, but further adapted and extended for this study by the first author.

A Praat script[1] extracted 161 acoustic features – divided into seven feature groups – from the concatenated words. Some features (e.g. syllables and phonemes per second, jitter and shimmer) were extracted only for all six words, while others (e.g. F_0, formant frequencies and segment duration) were extracted for several segments, including all six words and stressed vowels. Table 1 offers an overview of which segments were analysed in each feature group. Most features were extracted using the built-in functions in Praat. More detailed feature descriptions are given in [5].

Table 1. Segments analysed in each feature group (LTAS: long-term average spectra, HNR: harmonics-to-noise ratio, NHR: noise-to-harmonics ratio, sp.: spectral, str.: stressed)

Nr	Feature group	Segments analysed
1	syllables & phonemes per second	whole file
	segment duration (ms)	whole file, words, str. vowels,
2	sound pressure level (SPL) (dB)	fricatives, plosives (incl. VOT)
3	F_0 (Hz, semitones)	whole file, words, str. vowels
4	jitter, shimmer	
5	sp. tilt, sp. emphasis, inverse-filtered SB, LTAS	whole file
6	HNR, NHR, other voice measures	whole file, str. vowels
7	formant frequencies (F_1–F_5)	str. vowels
	sp. balance (SB)	fricatives and plosives

The analysis was performed with m3iCAT, a toolkit especially developed for corpus analysis [11]. It was used to calculate statistical measures, and to generate tables and diagrams, which displayed the variation of a certain feature as a function of age. The speakers were divided into eight overlapping "decade-based" age classes, based on the results (mean error ± 8 years) of a previous human listening test [7]. There were 14 ages in each class (except for the youngest and oldest classes): 20, aged 20–27; 30, aged 23–37; 40, aged 33–47; 50, aged 43–57; 60, aged 53–67; 70, aged 63–77; 80, aged 73–87; 90, aged 83–89.

For each feature, m3iCAT calculated actual means (μ), standard deviations (σ) and normalised means ($\bar{\mu}$) for each age class. Normalisation involved mapping the domain of the values in the following way:

$$a_i = \frac{(v_i - mean)}{stdev} \tag{1}$$

where v_i represents the actual value, *mean* represents the mean value of the data and *stdev* represents the corresponding standard deviation. Occasionally, normalisations were also carried out separately for each gender. This was done in order to see the age-related variation more distinctly when there were large differences in the real mean values between female and male speakers, e.g. in F_0 and formant frequencies. Because of the normalisation process, almost all values (except a few outliers) fall within the range between -1 and $+1$, which allows direct comparison of all features regardless of their original scaling and measurement units.

The values calculated for the eight age classes were displayed in tables, separately for female and male speakers. In addition, line graphs were generated for the age-class-related profiles or tendencies, with the age classes on the x-axis and the normalised mean values on the y-axis. The differences between the normalised mean values of all pairs of adjacent age classes are displayed as labels at the top of the diagrams (female labels above male ones). Statistical t-tests were carried out to calculate the significance of the differences; all differences except the ones within parentheses are statistically significant ($p \leq 0.01$). Figure 4 shows an example of a tendency diagram where the normalisations were carried out using all speakers (top), and the same tendencies but normalised separately for each gender (bottom).

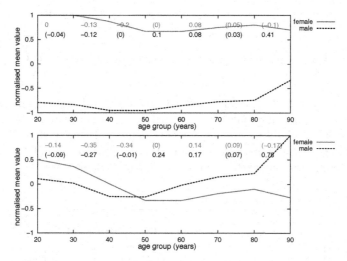

Fig. 4. Normalised tendencies for *mean F_0 (Hz) (all six words)*, 8 overlapping age classes, normalised for all speakers (top) and normalised separately for female and male speakers (bottom)

The advantage of using normalised means is that variation can be studied across features regardless of differences in the original scaling and units of the features. For instance, it allows direct comparison of the age-related variance between duration and F_0 by comparing the tendency for segment duration (in seconds) with the tendency for mean F_0 (in Hz).

5   Results

Due to the large number of features investigated, the results are presented by feature group (see Table 1). Moreover, only a few interesting results for each feature group will be described, as it would be impossible to present the results for all features within the scope of this article. A more comprehensive presentation of the results is given in [5].

The number of syllables and phonemes per second generally decreased with increased age for both genders, while segment duration for most segments increased. The tendencies were less clear for the female than the male speakers. Figure 5 shows the results for all six words.

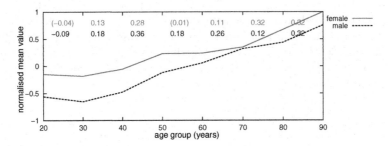

Fig. 5. Normalised tendencies for *duration (all six words)*

Average relative SPL generally either decreased slightly or remained constant with increased female and male age. The SPL range either increased or remained relatively stable with advancing age for both genders. Figure 6 shows the results for SPL range in the word *käke*. Similar tendencies were found for the other words, including the one without plosives; själen ['ɧɛːlən].

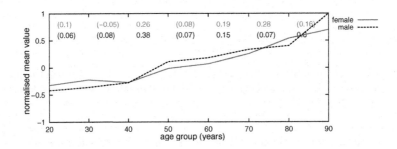

Fig. 6. Normalised tendencies for *SPL range (käke)*

Female F_0 decreased until age group 50 and then remained relatively stable. Male F_0 lowered slightly until age group 50, but then rose into old age. Due to the gender-related differences in F_0, the results for mean F_0 (Hz, all six words) are presented in Figure 7 as normalised separately for each gender to show clearer tendencies.

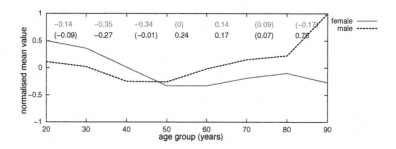

Fig. 7. Normalised (separately for each gender) tendencies for *mean F_0 (Hz, all six words)*

Although generally higher for male than female speakers, no continuous increase with age was found in either gender for jitter and shimmer. Female values remained relatively stable from young to old age. Male values generally increased slightly until age group 40, and then decreased slowly until old age, except for a considerable decrease in shimmer after age class 80. Figure 8 shows local shimmer for all six words.

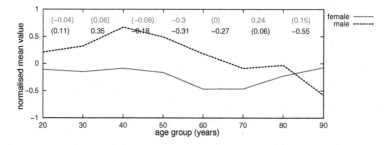

Fig. 8. Normalised tendencies for *local shimmer (all six words)*

Spectral energy distribution displayed varying results, though most measures did not change much with increased age. Figure 9 shows the LTAS amplitudes at 320 Hz, which generally increased with advancing age for both genders.

Few age-related changes were found in female NHR. Male NHR increased slightly until age class 50, where a decrease followed. Figure 10 displays the results for NHR in [ɑ:].

Resonance feature results varied with segment type in both genders. F_1 decreased in [ɛ:] (and in female [y:]), but remained stable in [a], [ɑ:] and [u:]. F_2 was stable in [y:] and increased slightly with advancing age in [ɑ:] and [ɛ:] for both genders, but decreased slightly in [a] and [u:], interrupted by increases and peaks at age group 40. In F_3 and F_4, a decrease was often observed from age class 20 to 30, followed by little change or a very slight increase. Figure 11 shows normalised tendencies for F_1 and F_2 in the vowel [ɛ:].

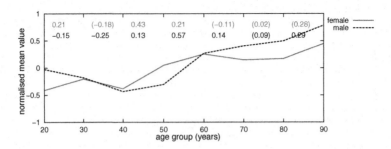

Fig. 9. Normalised tendencies for *LTAS amplitudes at 320 Hz (dB, all six words)*

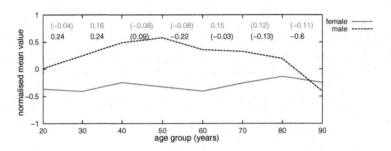

Fig. 10. Normalised tendencies for *NHR ([a:])*

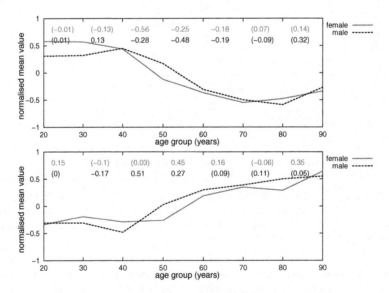

Fig. 11. Normalised (separately for each gender) tendencies for *mean F_1 (top) and F_2 (bottom)* ([ɛ:] *in the word själen* ['ɧɛ:lən])

6   Discussion and Conclusions

Many of the hypotheses were confirmed. However, some were contradicted. Possible explanations for this include differences in the speech material compared to previous studies and the fact that mainly automatic methods were used in this study. Still, the study provided some interesting results which may be used when building automatic estimators of speaker age: (1) Automatic methods can be used to analyse large speech data sets in relation to speaker age, and may yield similar results as manual studies, (2) The relatively most important correlates of adult speaker age seem to be speech rate and SPL range. F_0 also may provide consistent variation with speaker age, as may F_1, F_2 and LTAS in some segments and frequency intervals. These features may be used in combination with other features as cues to speaker age and (3) The type of speech material used in acoustic analysis of speaker age is very likely to influence the results.

These findings will be used in future studies to improve the automatic classification of speaker age.

References

1. Ryan, W.J.: Acoustic aspects of the aging voice. Journal of Gerontology 27, 256–268 (1972)
2. Amerman, J.D., Parnell, M.M.: Speech timing strategies in elderly adults. Journal of Voice 20, 65–67 (1992)
3. Xue, S.A., Deliyski, D.: Effects of aging on selected acoustic voice parameters: Preliminary normative data and educational implications. Educational Gerontology 21, 159–168 (2001)
4. Linville, S.E.: Vocal Aging. Singular Thomson Learning, San Diego (2001)
5. Schötz, S.: Perception, Analysis and Synthesis of Speaker Age. PhD thesis, Travaux de l'institut de linguistique de Lund 47. Lund: Department of Linguistics and Phonetics, Lund University (2006)
6. Schötz, S.: Acoustic analysis of adult speaker age. In: Müller, C. (ed.) Speaker Classification I. LNCS(LNAI), vol. 4343, Springer, Heidelberg (2007)
7. Schötz, S.: Stimulus duration and type in perception of female and male speaker age. In: Proceedings of Interspeech 2005, Lisbon (2005)
8. Schötz, S.: Speaker age: A first step from analysis to synthesis. In: Proceedings of ICPhS 03, Barcelona, pp. 2528–2588 (2003)
9. Bruce, G., Elert, C.C., Engstrand, O., Eriksson, A.: Phonetics and phonology of the Swedish dialects - a project presentation and a database demonstrator. In: Proceedings of ICPhS 99, San Francisco, CA, pp. 321–324 (1999)
10. Boersma, P., Weenink, D.: Praat: doing phonetics by computer (version 4.3.04) [computer program]. Retrieved (March 8, 2005) (2005), from http://www.praat.org/
11. Müller, C.: Zweistufige kontextsensitive Sprecherklassifikation am Beispiel von Alter und Geschlecht [Two-layered Context-Sensitive Speaker Classification on the Example of Age and Gender]. PhD thesis, Computer Science Institute, University of the Saarland (2005)

The Impact of Visual and Auditory Cues
in Age Estimation

Kajsa Amilon, Joost van de Weijer, and Susanne Schötz

Centre for Languages and Literature,
Box 201, 22100 Lund, Sweden
kajsa@amilon.se, vdweijer@ling.lu.se, Susanne.schotz@ling.lu.se

Abstract. Several factors determine the ease and accuracy with which we can estimate a speaker's age. The question addressed in our study is to what extent visual and auditory cues compete with each other. We investigated this question in a series of five related experiments. In the first four experiments, subjects estimated the age of 14 female speakers, either from still pictures, an audio recording, a video recording without sound, or a video recording with sound. The results from the first four experiments were used in the fifth experiment, to combine the speakers with new voices, so that there was a discrepancy in how old the speaker looked and how old she sounded. The estimated ages of these dubbed videos were not significantly different from those of the original videos, suggesting that voice has little impact on the estimation of age when visual cues are available.

Keywords: age estimation, visual cues, auditory cues, dubbed video.

1 Introduction

When meeting a person for the first time, we form an opinion of him or her in just a few seconds (Gladwell, 2006). In this process, age estimation is an important part, which provides a large amount of information about the person we meet.

When making this estimation, visual cues (such as wrinkles, hair style, posture, and clothes) probably have a large impact. Recent studies suggest that people are able to determine a persons age by approximately five years, from looking at a picture of their face (Vestlund, 2004). Auditory cues, such as word choice and voice, also play a role, but yield less exact estimates than visual cues. Studies have shown that, when visual cues are not available, people can estimate a person's age with an absolute difference of 10 years (Ptacek & Sander, 1966).

Acoustic characteristics of voice contain information about a speaker's age. Physical changes during the ageing process (decreasing lung capacity, calcification of the larynx, loss of muscular control) influences our voices to various degrees. Audible changes include a decrease in vocal intensity, lowering of vocal pitch, and narrowing of the vocal pitch range. Furthermore, voices tend to become harsher. Finally, elderly people often speak slower than younger people (Schötz, 2001). However, Brückl and

C. Müller (Ed.): Speaker Classification II, LNAI 4441, pp. 10–21, 2007.

Sendlmeier (2003) found no difference in articulation rate between young and old women during read speech.

Recently, Schötz (2006) described how these changes in voice can be synthesised using data-driven formant synthesis. In her study, acoustic parameters were extracted from the Swedish word 'själen' (the soul), spoken by four differently aged females of the same family. The information was subsequently used to synthesize voices of different ages.

The accuracy with which people can estimate a person's age, either from auditory or from visual information, has been studied into great detail. The following effects have come forward.

The expert effect
The expert effect means that training makes perfect. Studies have shown that, for instance, professionals who work with specific age groups, are more accurate at estimating the age within that age range than others (Lindstedt, 2005). Vestlund (2006) for example, showed that sales persons at the Swedish alcohol retail company (who are not allowed to sell alcohol to persons under 20) were better than a non-expert group at estimating the ages of people between 15 and 24 years old. A similar finding is reported by Nagao and Kewely-Port (2005) who compared native speakers of Japanese with native speakers of American English. Both groups were asked to estimate the age of Japanese and of American English speakers. Overall, the subjects were more correct when estimating the age of a speaker of their own language. Furthermore, the Japanese listeners were better than the English listeners at estimating the ages. This might be due to the fact that Japanese more frequently interact with elderly people (more than half of the Japanese subjects lived with at least one grandparent). In another study (Dehon & Bredart, 2001) with pictures instead of living persons, Caucasian subjects performed better at estimating the age of other Caucasians than they were on estimating the age of people of other races. A third variant of the expert effect is that it is easier to estimate the age of persons your own age (George & Hole, 1995; Vestlund, 2004). In a line-up confrontation experiment where subjects were to recognize a culprit among other persons, the subjects were more often correct when the perpetrator was their own age (Wright & Stroud, 2002).

Age
Overall, young adults tend to outperform older adults at age estimations, both from pictures (Vestlund, 2004) and sound (Linville, 1987). In a forensic context, young people have been found to be more reliable than elderly people as eyewitnesses (Wright & Stroud, 2002). Mulac and Giles (1996) added to the age effect by showing that someone's voice is not perceived as the person's chronological age, but how old the person feels. That is, if a person feels younger than his or her chronological age, this will be reflected in this person's voice and consequently, the listener will perceive him or her as younger.

Stimulus duration
The duration of the acoustic stimulus has an impact on how well subjects are able to estimate age from a speaker. Schötz (2005) found that the absolute difference

between chronological age and perceived age increases significantly from 6.5 to 9.7 years when the stimulus duration (spontaneous speech) decreases from 10 to 3 seconds. Nagao & Kewely-Port, (2005) also found that estimations improved when subjects received more contextual information.

Sex

Conflicting results exist as to whether it is easier to estimate a man's or a woman's age. Dehon and Bredart (2001) and Krauss et al. (2001), on the one hand, found that the ages of men were easier to estimate than the ages of women. Schötz (2005) and Vestlund (2006), on the other hand, found that subjects were better at estimating women's ages.

Dialect

Elderly speakers are more likely to have a strongly pronounced dialect than younger speakers, as well as men tend to be more dialectal than women (Stölten & Engstrand, 2003; Stölten, 2001). The strength of dialect consequently is an auditory cue when estimating the age of a speaker.

General effects

Several studies of both age estimation from pictures (Vestlund, 2004) and from acoustic cues (Schötz, 2005; Cerrato et al., 2000) suggest that the age of younger people often is overestimated, while the age of older people is underestimated. This indicates that subjects tend to place speakers in the middle range. Cerrato et al. (2000) received most correct estimations of people in the age group 46-52. Dehon and Bredart (2001), however, found that poor age estimations from faces are most often overestimations.

In all, age estimation is a fairly well examined research area. However, most studies only concern the impact of either voice or looks. It is very common, though, that both auditory and visual information are available at the same time. An unexplored question is whether the voice of a speaker has an impact on the estimation of a person's age when information about the person's looks is simultaneously available. Is it easier to make a correct estimation when having access to both sources of information? And if the voice does not agree with the looks, what affects the perceiver more, picture or sound?

Differences in estimations made from voices and pictures were examined by Krauss et al. (2002). They carried out two experiments in which they compared age estimation from photographs and from voices. In the first experiment, the subjects heard two sentences read by two different speakers. They were then shown two photographs which they had to match with the two voices. The subjects selected the correct photograph more than 75% of the time, which is reliably above chance level. In the second experiment a group of participants were exposed to the same sentences as in the first experiment but this time they were asked to estimate the speaker's age, height and weight. A comparison group made the same estimations from pictures of the speakers. The results showed that the estimations made from the photographs were only marginally better than the ones made from the voice samples.

It is clear that higher accuracy in age estimating is achieved when perceivers have access to a person's looks instead of only the voice (cf. an absolute difference of five respective ten years). It is unclear, however, how much simultaneously available visual and auditory cues contribute individually to the estimation of speaker age. The present study addresses this issue in a series of five related experiments. We recorded 14 differently-aged women on video tape. These video clips were subsequently presented to five groups of participants, every time in a different modality: 1) as still pictures, 2) as soundless videos, 3) as audio clips, 4) as the full video with the original sound, and, 5) as the full video but with the voice of the speaker replaced with the voice of one of the other speakers.

This design makes it possible to directly compare the estimation of age when a speaker is shown with or without voice. The goal of the Experiment 5 was to establish whether the participants' estimations could be altered by giving the speakers a voice that sounded either considerably older or younger (as was revealed by the results of Experiment 3) than their own voices. The results of this experiment were then compared with those of Experiment 4 in order to establish whether the estimated age of a speaker had changed. If so, that would indicate that the voice contributes to age estimation. Experiment 2 was included in the present study mainly to establish whether the extra information provided by soundless video yielded significantly better estimates than still pictures.

2 Method

Speaker characteristics
Since previous research (Dehon & Bredart, 2001; Krauss et al., 2001; Schötz, 2005; Vestlund 2006) had shown a difference in the ability of estimating men and women, we decided to use only female speakers for the present study. Because of the pronounced change of voice in puberty only adults participated. The age interval of approximately five years was chosen because previous studies show that participants are able to estimate a speaker's age with an accuracy of approximately five years. The women's ages were: 17, 25, 31, 35, 38, 45, 47, 54, 55, 58, 65, 73, 76, 81, representing the entire adult age range. All women spoke Southern Swedish dialect. They were non smokers and did not have any known pathological voice disorder. Nor did any of them have a cold at the time of the recording. The speakers had not received any instructions about visual appearance (clothing, make up, hair style). They all wore casual clothing, and, according to us, they did not look or sound older or younger than their chronological age.

Equipment, recording conditions
The recordings were made at some time between 10 AM and 4 PM in an isolated recording booth. The speakers sat down on a chair in front of a neutral light background where they were recorded in portrait form. Recordings were made with a DV camera (Panasonic NVGS-180) and an external wireless microphone (Sony UTX-H1), mounted on a tripod. All women were first given the opportunity to look at the test

sentence for as long as they needed. A sign with the text was placed next to the camera, so that the speakers could read it out loud and did not need to memorize the exact text. In order to facilitate dubbing we tried to obtain test sentences that were all read at approximately equal speed. The reading speed of the first speaker (age 25) was used as a reference during the subsequent recordings. The other speakers were first asked to read in their own tempo, and then, if necessary, told to increase or decrease their speaking rate to end up as close as possible to the reference time, without their speaking style becoming unnatural.

Material
The sentence used for the investigation was "En undersökning som denna är viktig för förståelsen av folks röster i olika åldrar" (A study like this is important for the understanding of people's voices in different ages). In order to minimize the effect of dubbing a new voice onto a speaker, the sentence had few visibly salient speech sounds (for example, bilabial, labiodental). Furthermore, the sentence was not too long (a duration of about five seconds). According to previous research results (Schötz, 2005) five seconds is rather short to make a fair estimation of the age. In order to compensate for this, the participants in our study were allowed to watch or listen to the stimuli as many times as they wished.

Preparation of the materials
The video recordings were transferred to a Macintosh computer for preparation for the five experiments. Preparation was done using the video-editing program iMovie, and the speech-editing software Audacity. 14 clips with similar durations that seemed most suitable for the dubbing experiment were first selected from the available material. These clips were used for Experiments 2 and 4. For Experiment 1, a neutral-looking still picture of each speaker was taken from the original clips, and for Experiment 3, the mono sound signals were extracted and saved as audio files.

Our aim for Experiment 5 was to combine speakers with a new voice that sounded between 15 and 20 years younger or older than the speaker's own voice. It turned out that this was not possible for all speakers. In spite of the effort taken to equalize reading speed and utterance duration during the recordings, there was, inevitably, considerable variation in phrasing and pausing, which made it impossible to find an appropriate combination with a new voice for some speakers. We found 10 combinations of speakers and voices that appeared good enough for participants not to notice that the voice of the speaker was dubbed. In order to mask the fact that some video clips were dubbed, five dubbed clips were combined with five original video clips, and presented in two experimental sessions. The voices were dubbed onto the recordings with iMovie. In some cases, minor changes (e.g., shortening of fricative durations, shortening or lengthening of pauses) were made to the sound signal with Audacity so that the fit between sound and image improved.

Subjects
To participate in the sound only experiment (Experiment 3), the original video (Experiment 4) experiment or the dubbed video (Experiment 5) experiment, participants could only be native Swedish speakers. This was to prevent a possible

'other-language effect' affecting the results. In the other experiments, there were no nationality restrictions for participating. All participation was voluntary.

Procedure

Most participants were tested in a computer room at Lund University. An additional number were tested at other locations. In those cases, a portable Macintosh with headphones was used.

Participants received written instructions, which were similar for all five experiments, except for the description of the stimuli. The instruction for Experiment 5 (dubbed video) was identical to that of Experiment 4 (original video).

At the start of the experiments, all participants provided information about their age, sex, native language and dialect. They were not allowed to ask any questions during the experiment.

The video clips and still pictures were displayed on the computer screen. The picture was approximately 20 x 20 centimeters. Sound was presented via headphones at a comfortable listening level, but participants were allowed to adjust the volume if they wanted to. For every stimulus, the participants typed the speaker's estimated age in exact whole years (e.g., 24, 82, 57), and rated their judgment on a scale from 1 to 5, indicating whether they were very confident (5) about their answer or not at all (1). The stimuli were presented in random order. The results were saved automatically on the computer's hard disk.

At the end of the experiment, the subjects filled in a questionnaire with the following six questions: What information did you use for your estimation? What was your main cue? Did you find anything hard in particular? Did you find any of the persons harder to estimate than the others? Did you react to anything in particular? Did you recognize any of the women in the clips? Responses to speakers that the participant had recognized were excluded from further analysis. The answers to these questions were used for evaluating the results. At the end of Experiment 5, the participants were additionally asked if they had noticed that the videos were dubbed. If so, the participant was excluded from the analysis. All experiments took approximately 15 minutes to complete.

Analysis

Two main comparisons were carried out. The first comparison concerns the accuracy of the participants' estimates. As in several previous studies (cf. Vestlund, 2006; Schötz, 2005; Braun & Cerrato, 1999; Nagao & Kewley-Port, 2005), the error is calculated as the absolute difference between the speaker's chronological age and the perceived age. If this difference is not lower on average in Experiment 4 (original video) than in Experiment 2 (soundless video), then it seems that auditory cues are more or less neglected when visual cues are available. The second comparison concerns the difference in estimated age of the speakers in Experiment 4 (original video) and those in Experiment 5 (dubbed video). If the estimated age of a speaker does not change when she is given a voice that is considerably older or younger than her own voice, that would additionally suggest that auditory information is neglected when visual information is available.

3 Results

In all, 141 persons participated. The majority were students or staff at the Centre for Languages and Literature at Lund University. None of them participated in more than one experiment. None of them reported any hearing disorder.

Experiment 1: Still pictures
14 women and 9 men participated in Experiment 1. Their average age was 33 years. 28 were Swedish, and 5 had a different nationality. In total there were 322 responses, of which 13 (4.0%) were excluded, either because the participant had recognized a speaker or because of equipment failure. The results (given in Table 1) show that there was an average absolute difference of 5.7 years. Furthermore, the average difference between chronological age and perceived age was –2.7, indicating that the participants somewhat underestimated the age of the speakers.

Table 1. Results of Experiment 1 (Still picture)

chronological age	perceived age	difference	absolute difference	confidence rating
17	21.6	4.6	4.7	3.30
25	24.2	–0.8	2.5	3.18
31	25.9	–5.1	5.1	3.58
35	36.2	1.2	3.5	3.26
38	37.0	1.0	2.4	3.26
45	41.2	–3.8	4.6	3.23
47	47.6	0.6	4.6	3.17
54	43.8	–10.2	10.9	2.65
55	50.5	–4.5	5.4	3.23
58	50.9	–7.1	8.0	2.95
65	69.4	4.4	6.1	3.13
73	68.3	4.7	6.0	3.22
76	70.7	–5.3	7.7	3.05
81	74.0	–7.0	8.0	3.35
Average		**–2.7**	**5.7**	**3.18**

Experiment 2: Soundless video
14 women and 14 men participated in experiment 2. Their average age was 27. 19 were Swedish, and 9 had a different nationality. 380 responses were analyzed. 12 responses were excluded (3.1%) because the participant had recognized a speaker or because of equipment failure. The overall results of Experiment 2 were similar to those of Experiment 1. As in Experiment 1, there was a tendency to underestimate the age of the speakers and the absolute error was somewhat over five years. The average confidence ratings were similar in both experiments, suggesting that the participants did not find the videos more difficult to assess than the still pictures.

Table 2. Results of Experiment 2 (Soundless video)

chronological age	perceived age	difference	absolute difference	confidence rating
17	20.0	3.0	3.0	3.29
25	24.5	−0.5	1.9	3.57
31	27.2	−3.8	4.6	3.08
35	33.1	−1.9	3.6	3.21
38	35.9	−2.1	4.1	3.13
45	40.8	−4.2	5.4	3.22
47	48.4	1.4	4.4	3.30
54	47.9	−6.1	7.1	2.93
55	51.0	−4.0	6.0	3.14
58	51.0	−7.0	8.0	2.81
65	67.4	2.4	5.3	2.89
73	67.6	−5.4	6.6	3.07
76	73.0	3.0	4.9	3.18
81	77.2	−3.8	5.6	3.14
average		**−2.5**	**5.0**	**3.14**

Experiment 3: Sound only

16 women and 12 men participated in Experiment 3. Their average age was 29. There were 392 responses, of which 11 (2.8%) were excluded, either because the participant had recognized a speaker or because of equipment failure.

Table 3. Results of Experiment 3 (Sound only)

Chronological age	Perceived age	Difference	Absolute difference	Confidence rating
17	18.5	1.5	3.4	3.64
25	24.0	−1.0	2.1	3.74
31	29.3	−1.7	5.6	3.11
35	35.2	0.2	7.0	2.96
38	26.3	−11.7	12.2	3.27
45	40.8	−4.2	6.6	2.86
47	42.0	−5.0	6.8	2.85
54	47.2	−6.8	8.9	3.11
55	46.2	−8.8	9.9	3.00
58	50.7	−7.3	10.0	2.67
65	42.9	−22.1	22.1	2.78
73	76.2	3.2	6.7	3.64
76	66.6	−9.4	10.6	3.18
81	55.7	−25.3	25.3	2.93
average		**−7.0**	**9.7**	**3.13**

The sound only experiment appeared, like previous research had shown, to constitute the most difficult assignment for the participants. With an average absolute difference of 9.7 years years, the error was twice as large compared to the results of Experiments 1 and 2. This difficulty was nevertheless not reflected in particularly lower confidence ratings in Experiment 3 compared to those in Experiments 1 and 2. Remarkably, the tendency to underestimate the speakers' ages was even more pronounced in Experiment 3. The average estimated age was even further below the speakers' chronological age in Experiment 3. This was especially the case for the speakers aged 65 and 81, whose voices were estimated more than 20 years below their chronological ages.

Experiment 4: Original video
9 women and 12 men participated in this experiment. Their average age was 25. A total of 289 responses were analyzed. 5 responses (1.7%) were excluded either because the participant had recognized a speaker, or because of equipment failure. The results are given in Table 4.

Table 4. Results of Experiment 4 (original video)

chronological age	perceived age	difference	absolute difference	confidence rating
17	18.3	1.3	2.0	3.86
25	24.6	−0.4	2.0	3.29
31	27.1	−3.9	4.3	3.20
35	32.7	−2.3	3.0	3.35
38	34.7	−3.3	5.0	3.05
45	39.9	−5.1	6.1	3.19
47	45.9	−1.1	2.9	3.00
54	47.0	−7.0	7.65	2.95
55	50.1	−4.9	5.1	3.48
58	51.4	−6.6	7.3	2.86
65	62.6	−2.4	4.0	3.20
73	70.0	−3.0	5.9	2.95
76	69.9	−6.1	8.3	3.14
81	74.0	−7.0	7.7	3.19
average		**−3.7**	**5.1**	**3.19**

The average absolute difference between estimated age and chronological age was 5.1 years, which is comparable to the values found in Experiments 1 and 2. Interestingly, the two speakers aged 65 and 81 were not estimated considerably younger in Experiment 4 than in Experiment 1, in spite of the fact that their voices sounded more than 20 years younger in Experiment 3. This is a first indication that voice did not play a significant role in the judgment of age.

Comparison of the results of Experiments 1 to 4
The absolute differences between chronological age and perceived age in Experiments 1 (soundless video), 2 (still picture) and 4 (original video) were tested in an ANOVA with 'experiment' as fixed factor, and 'participants' and 'speaker age' as random factors. Overall, the differences between the experiments were significant (F[3, 52.048] = 5.308, p < .05). Post-hoc comparisons using Tukey's HSD procedure showed that the differences between Experiments 1, 2, and 4 were not significant. Stated otherwise, soundless video did not yield more accurate responses than still pictures, and video with sound did not yield more accurate responses than video without sound.

The average confidence ratings, on the contrary, were not significantly different across the first four experiments (F[3, 52.340] = 0.229, p > .05).
    The results of the first four experiments do not suggest that speakers' voices contributed substantially to the estimation of age when visual information was available. This suggestion is supported by the answers that the participants gave the questionnaires. Only four participants of Experiment 4 wrote that they also had paid attention to the voices of the speakers while making their judgments. Most participants, though, only mentioned visual cues, such as wrinkles, hair color, hair style, and so on.

Experiment 5: Dubbed video
24 women and 17 men participated in the final experiment. They had an average age of 35.2. years. A total of 200 responses were analyzed. 5 responses were excluded because the participant had recognized the speaker, or because of equipment failure. Table 5 gives an overview of the results.

Table 5. Results of Experiment 5 (Dubbed video)

chronological age	combined with voice perceived as	perceived age in original video	perceived age in dubbed version	confidence rating
25	40.8	24.6	26.6	3.33
35	50.7	32.7	34.8	3.14
38	24.0	34.7	33.1	3.35
38	47.2	34.7	40.3	2.95
45	29.3	39.9	40.1	3.05
47	29.3	45.9	48.4	3.25
55	76.2	50.1	58.1	2.70
58	66.6	51.4	55.0	2.70
65	66.6	62.6	65.4	3.19
76	50.7	69.9	70.2	2.75
average				**3.04**

The main comparison of interest is how the speaker's perceived age changes as a consequence of the new voice. This change is shown in the third and fourth column of the table. For all but two of the speakers (aged 38 and 55), the average perceived age in Experiment 5 was more than five years higher than the perceived age in Experiment 4. For the other speakers the differences in perceived age were smaller than five years.

Comparison of experiments 4 and 5

The estimated ages of the speakers in Experiment 4 were compared with those in Experiment 5 in order to establish whether the new voices had a significant effect on the participants' judgments. This turned out not to be the case. The differences in perceived age were not significant ($F[1,17.018] = 0.046$, $p > .05$). Overall, the confidence ratings were a bit lower in Experiment 5 than in Experiment 4. This difference was not significant either ($F[1, 16.987] = 1.602$, $p > .05$).

4   Discussion

The purpose of the present study was to investigate the impact of voice in age estimation. To what extent is age estimation influenced by voice when we see and hear a person simultaneously?

The results of our experiments are generally consistent with those of previous studies. Participants were more accurate at estimating the age of the speaker based on visual information compared to auditory information. Soundless video did not yield more accurate estimations than still pictures.

It was found that the accuracy with which participants estimated the ages of the fourteen speakers did not improve significantly when sound was added to the videos. Furthermore, deliberately manipulating the voices in order to make the speaker appear younger or older, did not have a significant impact on the participants' estimations. This was corroborated by the qualitative evaluations given by the participants at the end of the experiment. They reported that they paid attention to visual details, and hardly ever did they mention voice as an important characteristic. Taken together, these results suggest that voice does not contribute substantially to the estimation of age, and that estimations are based on visual appearance mainly, if not to say only.

This conclusion is in a sense limited, since voice is not the only auditory cue that contains information about speaker age. Other cues, such as speaking rate, word choice, dialect, were kept constant to a maximal extent in the present study. Nor do the present results show whether the same conclusion can be drawn for male speakers.

Even if voice does not appear to contribute significantly to the estimation of speaker age, it will be interesting to establish whether the impact of voice is larger for the estimation of other speaker characteristics, such as personality or mood. The present methodology of dubbing voices onto different speakers appears to be a promising way of investigating speaker judgments. Nevertheless, the number of dubbing possibilities was limited with the available material. In order to overcome these limitations, a tool such as the speaker age synthesizer (Schötz, 2006) can be of great value in the future for carrying out new experiments with more speakers and naturalistic material.

References

Braun, A., Cerrato, L.: Estimating speaker age across languages. In: Proceedings of ICPhS 99, San Francisco, CA, pp. 1369–1372 (1999)

Brückl, M., Sendlmeier, W.: Aging female voices: An acoustic and perceptive analysis. Institut für Sprache und Kommunikation, Technische Universität, Berlin, Germany. Voqual '03, Geneva (August 27-29, 2003)

Cerrato, L., Falcone, M., Paoloni, A.: Subjective age estimation of telephonic voices. Speech Communication 31, 107–112 (2000)

Dehon, H., Bredart, S.: An 'other-race' effect in age estimation from faces. Perception 26(9), 1107–1113 (2001)

George, P., Hole, G.: Factors influencing the accuracy of age estimations of unfamiliar faces. Perception 24(9), 1059–1073 (1995)

Gladwell, M.: Blink: The Power of Thinking without Thinking. Time Warner Book Group, New York (2006)

Krauss, R., Freyberg, R., Morsella, E.: Inferring speakers' physical attributes from their voices. Journal of Experimental Social Psychology 38, 618–625 (2002)

Lindstedt, R.: Finns det experter på åldersbedömning? (Are there experts on the estimation of age?) Unpublished Paper. Department of Education and Psychology, University of Gävle (2005)

Linville, S.: Acoustic-perceptual studies of an aging voice in women. Journal of Voice 1, 44–48 (1987)

Mulac, A., Giles, H.: You're only as old as you sound: Perceived vocal age and social meanings. Health Communication 3, 199–215 (1996)

Nagao, K., Kewley-Port, D.: The effect of language familiarity on age perception. In: International Research Conference on Aging and Speech Communication, Bloomington, Indiana (October 10, 2005)

Ptacek, P., Sander, E.: Age recognition from voice. Journal of Speech and Hearing Research 9, 273–277 (1966)

Schötz, S.: Röstens ålder – en perceptionsstudie (The age of the voice – a perceptual study). Unpublished Paper. Department of Linguistics, Lund University (2001)

Schötz, S.: Effects of stimulus duration and type on perception of female and male speaker age. In: Proceedings, FONETIK, 2002. Department of Linguistics, Gothenburg University (2005)

Schötz, S.: Data-driven formant synthesis of speaker age. Working Papers. Lund University. 52, 105–108 (2006)

Stölten, K.: Dialektalitet som ålders- och könsmarkör i arjeplogsdialekten: ett auditivt test och mätningar på preaspiration och VOT (Dialect as a marker of age and sex in the Arjeplog dialect: an auditory experiment and measurements of pre-aspiration and VOT.) Unpublished paper. Department of Linguistics Stockholm University (2001)

Stölten, K., Engstrand, O.: Effects of perceived age on perceived dialect strength: A listening test using manipulations of speaking rate and F0. Phonum 9, 29–32 (2003)

Vestlund, J.: Åldersbedömning av ansikten – precision och ålderseffekter (Estimation of age from faces – Precision and age effects). Unpublished paper. Department of Education and Psychology, University of Gävle (2004)

Vestlund, J.: Åldersbedömning av ansikten – expertkunskaper, könseffekter och jämnårighetseffekter (Estimation of age from faces – expert effects, sex effects, and same-age effects). Unpublished paper. Department of Education and Psychology, University of Gävle (2006)

Wright, D., Stroud, J.: Age differences in line-up identification accuracy: people are better with their own age. Law and Human Behaviour 26, 641–654 (2002)

Development of a Femininity Estimator for Voice Therapy of Gender Identity Disorder Clients

Nobuaki Minematsu[1] and Kyoko Sakuraba[2]

[1] The University of Tokyo
[2] Kiyose-shi Welfare Center for the Handicapped
mine@k.u-tokyo.ac.jp, sakuraba@mtd.biglobe.ne.jp

Abstract. This work describes the development of an automatic estimator of perceptual femininity (PF) of an input utterance using speaker
verification techniques. The estimator was designed for its clinical use
and the target speakers are Gender Identity Disorder (GID) clients, especially MtF (Male to Female) transsexuals. The voice therapy for MtFs,
which is conducted by the second author, comprises three kinds of training; 1) raising the baseline F_0 range, 2) changing the baseline voice quality, and 3) enhancing F_0 dynamics to produce an exaggerated intonation
pattern. The first two focus on static acoustic properties of speech and
the voice quality is mainly controlled by size and shape of the articulators, which can be acoustically characterized by the spectral envelope.
Gaussian Mixture Models (GMM) of F_0 values and spectrums were built
separately for biologically male speakers and female ones. Using the four
models, PF was estimated automatically for each of 142 utterances of
111 MtFs. The estimated values were compared with the PF values obtained through listening tests with 3 female and 6 male novice raters.
Results showed very high correlation (R=0.86) between the two, which
is comparable to the intra- and inter-rater correlation.

Keywords: Gender identity disorder, femininity, voice therapy, vocal
tract shape, fundamental frequency, speaker verification, GMM.

1 Introduction

Advanced speech technologies are applied not only for man-machine interface
and entertainment but also for medical treatment [1] and pronunciation training
of foreign language education [2]. Many works were done for developing cochlea
implants [3,4,5] and artificial larynxes [6,7,8] and, recently, the technologies have
been applied to realize an on-line screening test of laryngeal cancer [9] as well as
an on-line test of pronunciation proficiency of foreign languages [10]. The present
work examines the use of the technologies for another medical treatment; voice
therapy for GID clients.

A GID individual is one who strongly believes that his or her true psychological gender identity is not his or her biological or physical gender, i.e., sex.
In most of the cases, GID individuals live for years trying to conform to the
social role required by their biological gender, but eventually seek medical and

C. Müller (Ed.): Speaker Classification II, LNAI 4441, pp. 22–33, 2007.

surgical help as well as other forms of therapy in order to achieve the physical characteristics and the social role of the gender which they feel to be their true one. In both cases of FtMs (Female-to-Male) and MtFs, many of them take hormone treatment in order to make physical change of their bodies and the treatment is certainly effective for both cases. However, it is known that the hormone treatment brings about sufficient change of the voice quality only for FtMs [11,12]. Considering that the voice quality is controlled by physical conditions of the articulators, the vocal folds and the vocal tract are presumed to retain their pretreatment size and shape in the case of MtFs. To overcome this hardship and mainly to shift up the baseline F_0 range, some MtFs take surgical treatment. Although the F_0 range is certainly raised in the new voice, as far as the second author knows, it is a pity that the naturalness is decreased in the new voice instead. Further, many clinical papers and engineering papers on speech synthesis claim that raising the F_0 range alone does not produce good femininity [13,14,15]. Since shape of the vocal tract has a strong effect on the voice quality, good control of the articulators has to be learned to achieve good femininity. Considering small effects of the hormone treatment and the surgical treatment on MtF clients, we can say that the most effective and least risky method to obtain good femininity is taking voice training from speech therapists or pathologists with good knowledge of GID.

2   Why Femininity Estimator?

In the typical therapy conducted by the second author, the following three methods are used based on [16]. 1) raising the baseline F_0 range, 2) changing the baseline voice quality, and 3) enhancing F_0 dynamics to produce an exaggerated intonation pattern. One of the most difficult things in the voice therapy lies not on a client's side but on a therapist's side, i.e., accurate and objective evaluation of the client's voice. It is often said that as synthetic speech samples are presented repeatedly, even expert speech engineers tend to perceive better naturalness in the samples, known as habituation effect. This is the case with good therapists. To avoid this effect and evaluate the femininity unbiasedly, listening tests with novice listeners are desirable. But the tests take a long time and a large cost because a new test has to be done whenever some acoustic change happens in the client's voice through the therapy.

Further, in most of the cases, GID clients are very eager to know how they sound to novice listeners, not experts. Some clients, not so many, claim that they sound feminine enough although they sound less feminine to anybody else. The objective evaluation of their voices is very effective to let these clients know the truth. For these two reasons, in this study, a listening test simulator was developed by automatically estimating the femininity which novice listeners would perceive if they heard the samples.

Among the above three methods, the first two ones focus on static acoustic properties and the last one deals with dynamic F_0 control. The dynamic control of F_0 for various speaking styles is a very challenging task in speech synthesis research and, therefore, we only focused on the femininity controlled by the F_0 range and the voice quality. In medical and educational applications, unlike entertainment applications, technologies should not be used easily if they are not mature enough. In this work, good discussion was done in advance about what should be done by machines and what should be done by humans in the therapy. Only when there are some things difficult for humans and easy for machines, then, those things should be treated by machines.

GMM modeling of F_0 values and that of spectrums were done separately for biologically male speakers and female ones. By using the four models, the estimator was developed. In addition to the experimental results of the femininity estimation, some merits and demerits of using the estimator in actual voice therapy are described.

3   GMM-Based Modeling of Femininity

3.1   Modeling Femininity with Isolated Vowel Utterances

Questions of acoustic cues of good femininity were often raised in previous studies [17,18,19,20,21]. Acoustic and perceptual analysis of speech samples of biologically male and female speakers and those of MtF ones were done and the findings lead to the three kinds of methods in the previous section. About the voice quality, as far as the authors know, all the studies focused on isolated vowel utterances and formant frequencies were extracted to estimate the femininity. It is true that, even from a single /a/ utterance, it is possible to estimate vocal tract length [22] and then, the femininity. It is also true, however, that even if a client can produce very feminine isolated vowels with careful articulation, it does not necessarily mean that the client can produce continuous speech with good femininity. This is the case with foreign language pronunciation. Even if a learner can produce very good isolated vowels, the learner is not always a good speaker of the target language in normal speech communication. This is partly because good control of speech dynamics including prosody is required in continuous speech. We can consider another reason that so much attention cannot be paid to every step of producing vowels in a sentence. We can say that the desired tool for MtF voice therapy is an estimator of the femininity from naturally-spoken continuous speech. In this case, we have a fundamental problem. With the analysis methods used in the previous studies, it is difficult to estimate the femininity from continuous speech because formant frequencies change not only due to the femininity but also due to phonemic contexts of the target vowels.

3.2   Modeling Femininity with Continuous Speech

This problem can be solved by using GMM-based speaker recognition/verification techniques. In continuous speech, various phonemes are found and the phonemes naturally cause spectral changes. If the utterance has sufficient spectral variations, averaging the spectrum slices over time can effectively cancel the spectral changes caused by the phonemic variation. The resulting average pattern of spectrum comes to have a statically biased form of spectrum, which is considered to characterize the speaker identity and the stationary channel. In GMM-based speaker recognition/verification, the average pattern is modeled not as a single spectrum slice but as a mixture of Gaussian distributions, where the spectrum is often represented as cepstrum vector.

What kind of phone does the averaged spectrum correspond to? If a continuous speech includes vowels only, it is possible to give a clear phonetic interpretation of the averaged spectrum [23]. Figure 1 shows the vowel chart of American English, where the 10 monophthongs are plotted. This figure clearly shows the articulatory center (average) of the vowels corresponds to /ə/, schwa sound. The other figure in Figure 1 is a result of MDS (Multi-Dimensional Scaling) analysis of vowel samples from an American female speaker. Here, a single Gaussian distribution was used to model each vowel acoustically and a distance matrix was calculated from the vowel examples using Bhattacharyya distance measure. In the figure, we can say that the acoustic center of the vowels also corresponds to /ə/. It is known that schwa is generated with a sound tube of a uniform cross-sectional area, which implies that schwa is produced with the least articulatory effort. In continuous speech, most of the unstressed vowels are reduced to be schwa sounds, meaning that the schwa is the most frequent sound observed in naturally-spoken sentences. The averaging operation not only can cancel the spectral changes caused by the phonemic variation but also can represent the acoustic quality of the most frequent phone (vowel) of that speaker.

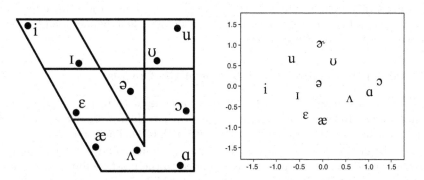

Fig. 1. The vowel chart of American English and an MDS chart of its vowel examples

With speech samples of any text spoken by a large number of female speakers, a GMM was trained to characterize the spectrum-based femininity, M_F^s. With male speakers, a GMM for the masculinity was trained, M_M^s. Using both models, the eventual spectrum-based femininity for a given cepstrum vector o, $F^s(o)$, was defined as the following formula [24,25];

$$F^s(o) = \log \mathcal{L}(o|M_F^s) - \log \mathcal{L}(o|M_M^s). \tag{1}$$

Similar models are trained for the F_0-based femininity and masculinity;

$$F^f(o) = \log \mathcal{L}(o|M_F^f) - \log \mathcal{L}(o|M_M^f). \tag{2}$$

Integration of the four models, M_F^s, M_M^s, M_F^f, and M_M^f can be done through generalizing the above formulae by linear regression analysis.

$$F(o) = \alpha \log \mathcal{L}(o|M_F^s) + \beta \log \mathcal{L}(o|M_M^s) +$$
$$\gamma \log \mathcal{L}(o|M_F^f) + \varepsilon \log \mathcal{L}(o|M_M^f) + C, \tag{3}$$

where α, β, γ, ε, and C are calculated so that the $F(o)$ can predict perceptual femininity (PF) of o the best. The PF scores were obtained in advance through listening tests with novice listeners.

4   Femininity Labeling of MtF Speech Corpus

4.1   MtF Speech Corpus

A speech corpus of 111 Japanese MtF speakers was built, some of whom sounded very feminine and others sounded less feminine and needed additional therapy. Each speaker read the beginning two sentences of "Jack and the beanstalk" with natural speaking rate and produced isolated Japanese vowels of /a, i, u, e, o/. The two sentences had 39 words. All the speech samples were recorded and digitized with 16 bit and 16 kHz AD conversion. Some clients joined the recording twice; before and after the voice therapy. Then, the total number of recordings was 142. For reference, 17 biologically female Japanese read the same sentences and produced the vowels.

4.2   Perceptual Femininity Labeling of the Corpus

All the sentence utterances were randomly presented to 6 male and 3 female adult Japanese listeners through headphones. All of them were in their 20s with normal hearing and they were very unfamiliar with GID. The listeners were asked to rate subjectively how feminine each utterance sounded and write down a score using a 7-degree scale, where +3 corresponded to the most feminine and −3 did to the most masculine. Some speech samples of biological female speakers were used as dummy samples. Every rater joined the test twice and 18 femininity judgments were obtained for each utterance. Figure 2 shows histogram of the averaged PF scores for the individual MtF utterances. Although some utterances still sounded rather masculine, a good variance of PF was found in the corpus. The averaged PF of biological female speakers was 2.74. While, in Figure 2, the averaging operation was done over all the raters, in the following section, it will be done dependently on the rater's biological gender.

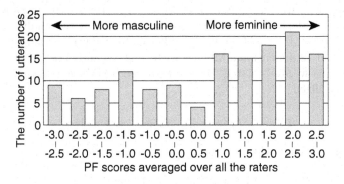

Fig. 2. Histogram of the averaged PF scores of the 142 utterances

4.3   Intra-rater and Inter-rater Judgment Agreement

Agreement of the judgments within a rater was examined. Every rater joined the test twice and the correlation between the two sessions was calculated for each. The averaged correlation over the raters was 0.80, ranging from 0.48 to 0.91. If the rater with the lowest correlation can be ignored, the average was recalculated as 0.84 (0.79 to 0.91).

PF scores by a rater were defined as the scores averaged over the two sessions. Using these scores, the judgement agreement between two raters was analyzed. The agreement between a female and another female was averaged to be 0.76, ranging from 0.71 to 0.83. In the case of the male raters, the agreement was averaged to be 0.75, ranging from 0.59 to 0.89. The agreement between a female and a male was 0.71 on average, ranging from 0.60 to 0.79. Some strategic differences in the judgment may be found between the two sexes.

PF scores by the female were defined as the averaged scores over the three female raters. Similarly, *PF scores by the male* were defined. The correlation between the two sexes was 0.87, which is very high compared to the averaged inter-rater correlation between the two sexes (0.71). This is because of the double averaging operations, which could reduce inevitable variations in the judgments effectively.

Now, we have 12 different kinds of PF scores; 9 from the 9 raters, 2 as the scores by the male and the female, and the other one obtained by averaging the male score and the female one. In the following sections, the correlations between the original PF scores and the automatically estimated PF scores, defined in Section 3.2, are investigated.

5   Training of M_F^s, M_M^s, M_F^f and M_M^f

As described in Section 3.2, automatic estimation of the femininity is examined based on GMM-based modeling. As speech samples for training, JNAS (Japanese Newspaper Article Sentences) speech database, 114 biological males and 114 biological females, was used [26]. The number of sentence utterances was 3,420 for each sex. Table 1 shows acoustic conditions used in the analysis. For the spectrum-based GMMs of M_F^s and M_M^s, silence removal was carried out from the speech files and 12 dimensional MFCCs with their Δs and $\Delta\Delta$s of the remaining speech segments were calculated. For the F_0-based GMMs of M_F^f and M_M^f, $\log F_0$ values were utilized.

Table 1. Acoustic conditions of the analysis

sampling	16bit / 16kHz
window	25 ms length and 10 ms shift
parameters	MFCC with its Δ and $\Delta\Delta$ for M_F^s and M_M^s
	$\log F_0$ for M_F^f and M_M^f
GMM	mixture of 16 Gaussian distributions

6   Automatic Estimation of Femininity

6.1   Simple Estimation Based on F^s and F^f

For each of the 142 MtF utterances, their femininity scores were estimated using F^s and F^f. The estimated scores were compared with the 12 different PF scores and the correlation was calculated separately. The averaged correlation over the first 9 PF scores is 0.64 for F^s and 0.66 for F^f. Table 2 shows the correlation of F^s and F^f with the other three PF scores; the male and the female scores and their averaged one. While the female PF is more highly correlated with F^s than F^f, the male PF is more highly correlated with F^f than F^s. This may imply different strategies of judging the femininity between the male raters and the female ones. It seems that the male tend to perceive the femininity more in high pitch of the voice. This finding is accordant with the results obtained in a previous study done by the second author [27]. In the study, it was shown that male listeners tend to assign higher femininity scores to speech samples with higher pitch.

Table 2. Correlation of F^s and F^f with the three PF scores

	female PF	male PF	averaged PF
F^s	0.71	0.70	0.73
F^f	0.67	0.76	0.74

6.2   Integrated Estimation with Weighting Factors

Linear regression analysis was done to predict the 12 PF scores using the four models, where the PF scores were converted to have a range from 0 (the most masculine) to 100 (the most feminine). As shown in Equation 3, four weighting factors and one constant term were calculated to minimize the prediction error.

The averaged multiple correlation over the first 9 PF scores was increased up to 0.76. Table 3 shows the multiple correlation coefficients with the other three PF scores. Figure 3 graphically shows the correlation with the female and male scores. Here, for utterance(s) of an MtF speaker, the weighting factors were calculated using utterances of the other MtF speakers and then, the femininity score(s) of utterance(s) of that MtF speaker were estimated. Namely, the estimation was done in a speaker-open mode. Considering magnitude of the intra- and inter-rater correlations of PF, we can say that $F(o)$ is a very good estimator of PF.

Table 3. Correlation of F with the three PF scores

	female PF	male PF	averaged PF
F	0.78	0.86	0.84

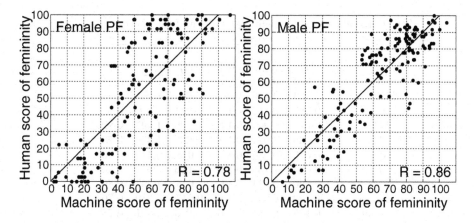

Fig. 3. Correlation between the original and the estimated PF scores

6.3 Discussions

The four weighting factors and the constant term in Equation 3 show different values for the 12 PF scores. The difference in values between two raters characterizes the difference in judging strategies between them. Table 4 shows 12 patterns of α, β, γ, ε, and C with the multiple correlation coefficient (R). As was found in Section 6.1, clear difference was found between the female and the male raters. The female tend to focus on spectral properties (α=0.125 and 0.068 for female PF and male PF), while the male tend to focus on pitch (γ=0.107 and 0.144 for female PF and male PF). In this sense, F3's judgment is very male because she emphasized pitch (γ=0.167) and de-emphasized spectral properties (α=0.085). In Table 3 and Figure 3, the multiple correlation was not so high for the female PF scores (R=0.78) and this can be considered probably because of F3.

For every PF score, the absolute value of α and that of β are similar. α and β of the female PF are 0.125 and -0.127. Those of the male PF are 0.068 and -0.064. This directly means that, with spectral properties of the voice, it is as important to shift the client's voice closer to the female region as to shift the voice away from the male region. On the contrary, the absolute value of γ and that of ε show a large difference. γ and ε of the female PF are 0.107 and -0.013. Those of the male PF are 0.144 and 0.009. In every case, ε takes a very small value, near to zero, compared to γ. This indicates that, as for F_0, although it is important to shift the voice into the female region, it matters very little if the voice is still located in the male region. This asymmetric effects of spectrum and F_0 can be summarized as follows by using terms of bonus and penalty. If the voice is closer to the female region, larger bonus is given and if the voice is closer to the male region, larger penalty is given. Although both bonus and penalty should be considered with spectral properties of the voice, only bonus is good enough with its F_0 properties.

Table 4. Values of the weighting factors and the constant term

	α	β	γ	ε	C	R
	M_F^s	M_M^s	M_F^f	M_M^f		
F1	0.154	-0.143	0.068	-0.017	1.301	0.77
F2	0.133	-0.126	0.086	-0.012	1.005	0.68
F3	0.087	-0.112	0.167	-0.011	-0.742	0.72
M1	0.074	-0.069	0.120	-0.009	1.091	0.82
M2	0.035	-0.024	0.176	-0.030	1.479	0.73
M3	0.076	-0.085	0.150	0.028	0.446	0.80
M4	0.034	-0.035	0.172	-0.005	0.941	0.76
M5	0.090	-0.084	0.141	0.023	1.237	0.86
M6	0.100	-0.090	0.107	0.047	1.389	0.70
female PF	0.125	-0.127	0.107	-0.013	0.521	0.78
male PF	0.068	-0.064	0.144	0.009	1.097	0.86
average PF	0.097	-0.096	0.126	-0.002	0.809	0.84

7   Actual Use of the Estimator in Voice Therapy – Merits and Demerits –

The second author has used the estimator in her voice therapy for MtF clients since Feb. 2006. Figure 4 shows an actual scene of using the estimator in the therapy and Figure 5 is the interface of the estimator. It was found that, when biologically male speakers without any special training pretended to be female, it was very difficult to get a score higher than 80. However, it is very interesting that good MtF speakers, who can change their speaking mode voluntarily from male to female, could have the estimator show a very low score (very masculine) and a very high score (very feminine) at their will. Since the estimator is focusing on only static acoustic properties, we consider that these MtFs have two baseline shapes of the vocal tract, which may be realized by different positioning of the tongue, and two baseline ranges of F_0. In this sense, the estimator helped the clients a great deal who were seeking for another baseline of the vocal tract shape and/or that of F_0 range through try-and-errors. Needless to say, quantitative and objective evaluation of their trials motivated the clients very well.

Fig. 4. An actual scene of the voice therapy with the estimator

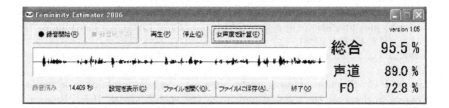

Fig. 5. Interface of the estimator

Only the focus on static acoustic properties naturally caused some problems. As described in Section 1, by producing a rather exaggerated pattern of intonation, listeners tend to perceive higher femininity. Although this exaggeration is a good technique to obtain high PF in the voice, the estimator completely ignores this aspect and then, some clients received unexpectedly high scores or low scores. In actual therapies, the therapist has to use the machine by carefully observing what kind of strategy the client is trying to use. If speech dynamics is effectively controlled, then, the therapist should not use the machine but give adequate instructions only based on the therapist's ears. Further, especially when the clients evaluate their voices by themselves, what is possible and what is impossible by the machine should be correctly instructed to them.

8 Conclusions

This work described the development of an automatic estimator of the perceptual femininity from continuous speech using speaker verification techniques. Spectrum-based and F_0-based GMMs were separately trained with biological male and female speakers. By integrating these models, the estimator was built. The correlation of the estimated values and the perceptual femininity scores obtained through listening tests was 0.86, comparable to the intra- and inter-rater correlation. Some analyses were done about sexual differences of the femininity judgment and some strategic differences in the use of spectrum-based cues and F_0-based cues were shown. It was indicated that the male tend to give higher scores to the voices with higher pitch. Further, it was shown independently of the rater's sex that the penalty of the F_0 range being still in the male region is remarkably small. As future work, we are planning to take MRI pictures of a good MtF speaker's control of the articulators when producing feminine vowels and masculine ones. We hope that the estimator will help many MtF clients improve the quality of their lives.

References

1. Levitt, H.: Processing of speech signals for physical and sensory disabilities. National Academy of Sciences 92(22), 9999–10006 (1995)
2. Ehsani, F., Knodt, E.: Speech technology in computer-aided language learning: strategies and limitations of a new CALL paradigm. Language Learning & Technology 2(1), 45–60 (1998)
3. Loizou, P.C.: Mimicking the human ear. IEEE Signal Process Magazine 15, 101–130 (1998)
4. Loizou, P.C.: Signal-processing techniques for cochlear implants – a review of progress in deriving electrical stimuli from the speech signal. IEEE Engineering in Medicine and Biology Magazine 18(3), 34–46 (1999)
5. Suhail, Y., Oweiss, K.G.: Augmenting information channels in hearing aids and cochlear implants under adverse conditions. In: Proc. IEEE Int. Conf. Acoustics, Speech, and Signal Processing, pp. 889–892 (May 2006)
6. Barney, H.L., Haworth, F.E., Dunn, H.K.: An experimental transistorized artificial larynx. Bell System Technical Journal 38, 1337–1356 (1959)
7. Houston, K.M., Hillman, R.E., Kobler, J.B., Meltzner, G.S.: Development of sound source components for a new electrolarynx speech prosthesis. In: Proc. IEEE Int. Conf. Acoustics, Speech, and Signal Processing, pp. 2347–2350 (March 1999)
8. Meltzner, G.S., Kobler, J.B., Hillman, R.E.: Measuring the neck frequency response function of laryngectomy patients: implications for the design of electrolarynx devices. Acoustic Society of America 114(2), 1035–1047 (2003)
9. Mori, H., Otawa, H., Ono, T., Ito, Y., Kasuya, H.: Internet-based acoustic voice evaluation system for screening of laryngeal cancer, Acoustic Society of Japan 62(3), 193–198 (2006) (in Japanese)

10. Jong, J.D., Bernstein, J.: Relating phonepass scores overall scores to the council of europe framework level descriptors. In: Proc. European Conf. Speech Communication and Technology, pp. 2803–2806 (September 2001)
11. Edgerton, M.T.: The surgical treatment of male transsexuals. Clinics in Plastic Surgery 1(2), 285–323 (1974)
12. Wolfort, F.G., Parry, R.G.: Laryngeal chondroplasty for appearance. Plastic and reconstructive surgery 56(4), 371–374 (1975)
13. Bralley, R.C., Bull, G.L., Gore, C.H., Edgerton, M.T.: Evaluation of vocal pitch in male transsexuals. Communication Disorder 11, 443–449 (1978)
14. Spencer, L.E.: Speech characteristics of MtF transsexuals: a perceptual and acoustic study. Folia phoniat. 40, 31–42 (1988)
15. Mount, K.H., Salmon, S.J.: Changing the vocal characteristics of a postoperative transsexual patient: a longitudinal study. Communication Disorder 21, 229–238 (1988)
16. Gelfer, M.P.: Voice therapy for the male-to-female transgendered client. American J. Speech-Language Pathology 8, 201–2008 (1999)
17. Sato, H.: Acoustic cues of female voice quality. IECE Transaction 57(1), 23–30 (1974) (in Japanese)
18. Bennett, S.: Acoustic correlates of perceived sexual identity in preadolescent children's voices. Acoustic Society of America 66(4), 989–1000 (1979)
19. Andrews, M.L., Schmidt, C.P.: Gender presentation: perceptual and acoustical analyses of voice. J. Voice 11(3), 307–313 (1997)
20. Gelfer, M.P., Schofield, K.J.: Comparison of acoustic and perceptual measures of voice in MtF transsexuals perceived as female vs. those perceived as male. J. Voice 14(1), 22–33 (2000)
21. Wolfe, V.I.: Intonation and fundamental frequency in MtF TS. J. Speech Hearing Disorders 55, 43–50 (1990)
22. Paige, A., Zue, V.W.: Calculation of vocal tract length, AU-18(3), 268–270 (1970)
23. Minematsu, N., Asakawa, S., Hirose, K.: Para-linguistic information represented as distortion of the acoustic universal structure in speech. In: Proc. IEEE Int. Conf. Acousitcs, Speech, and Signal Processing. vol. 1, pp. 261–264 (May 2006)
24. Rosenberg, A.E., Parthasarathy, S.: Speaker background models for connected digit password speaker verification. In: Proc. IEEE Int. Conf. Acoustics, Speech, and Signal Processing, pp. 81–84 (May 1996)
25. Heck, L.P., Weintraub, M.: Handset-dependent background models for robust text-independent speaker recognition. In: Proc. IEEE Int. Conf. Acoustics, Speech, and Signal Processing, pp. 1071–1074 (April 1997)
26. http://www.mibel.cs.tsukuba.ac.jp/jnas/
27. Sakuraba, K., Imaizumi, S., Hirose, K., Kakehi, K.: Sexual difference between male and female listeners in the perceptual test with the voice produced by MtF transsexuals. In: Proc. Spring Meeting of Acoustic Society of Japan, 2-P-4, pp. 337–338 (March 2004) (in Japanese)

Real-Life Emotion Recognition
in Speech

Laurence Devillers and Laurence Vidrascu

LIMSI-CNRS, BP133, 91403 Orsay Cedex
{devil,vidrascu}@limsi.fr
http://www.limsi.fr/Individu/devil

Abstract. This article is dedicated to Real-life emotion detection using
a corpus of real agent-client spoken dialogs from a medical emergency
call center. Emotion annotations have been done by two experts with
twenty verbal classes organized in eight macro-classes. Two studies are
reported in this paper with the four macro classes: Relief, Anger, Fear
and Sadness: the first investigates automatic emotion detection using
linguistic information whith a detection score of about 78% and a very
good detection of Relief, whereas the second investigates emotion detection with paralinguistic cues with 60% of good detection, Fear being best
detected.

Keywords: emotion detection, real-life data, linguistic and paralinguistic cues.

1   Introduction

The emotion detection work reported here is part of a larger study aiming to
model user behaviour in real interactions. We have already worked on other real
life data: financial call centers [1] and EmoTV clips [2]. In this paper, we make
use of a corpus of real agent-client spoken dialogs in which the manifestation
of emotion is stronger [1][3]. The context of emergency gives a larger palette
of complex and mixed emotions. About 30% of the utterances are annotated
with non-neutral emotion labels in the medical corpus compared to 11% for
the financial data. Emotions are less shaded than in the financial corpus where
the interlocutors attempt to control the expression of their internal attitude. In
the context of emergency, emotions are not played but really felt in a natural way.
In contrast to research carried out with artificial data with simulated emotions
or with acted data, for real-life corpora the emotions are linked to internal or
external emotional event(s). We might think that natural and complex emotion
behaviour could be found in movies data. Yet, emotions are still played and in
most cases, except for marvellous actors, they are not really felt. However, it is
also of great interest to study professional movie actors in order to portray a
recognisable emotion and to define a scale of naturalness [4]. The difference is
mainly due to the context. The context is the set of events that are at the origin
of a person's emotions. These events can be external or internal. Different events

C. Müller (Ed.): Speaker Classification II, LNAI 4441, pp. 34–42, 2007.

might trigger different emotions at the same time: for instance a physical internal event as a stomachache triggering pain with an external event as someone helping the sick person triggering relief. In the artificial data, this context is rubbed out or simulated so that we can expect to have much more simple full-blown affect states which are far away from real affective states.

In contrast to research carried out with artificial data and simulated emotions, for real-life corpora the set of appropriate emotion labels must be determined. There are many reviews on the representation of emotions. For a recent review, the reader is referred to [5]. We have defined in the context of Humaine NoE, an annotation scheme Multi-level Emotion and Context Annotation Scheme [1][2] to represent the complex real-life emotions in audio and audiovisual natural data. It is a hierarchical framework allowing emotion representation with several layers of granularity, including both dominant (Major) and secondary (Minor) labels as well as the context representation. This scheme includes verbal, dimensional and appraisal labels. Our aim in this study is to find robust lexical and paralinguistic cues for emotion detection.

One of the challenges when studying real-life speech call center data is to identify relevant cues that can be attributed to an emotional behavior and separate them from those that are simply characteristic of spontaneous conversational speech. A large number of linguistic and paralinguistic features indicating emotional states are present in the speech signal. Among the features mentioned in the literature as relevant for characterizing the manifestations of emotions in speech, prosodic features are the most widely employed, because as mentioned above, the first studies on emotion detection were carried out with acted speech where the linguistic content was controlled. At the acoustic level, the different features which have been proposed are prosodic (fundamental frequency, duration, energy), and voice-quality features [6].

Additionally, lexical and dialogic cues can also help to distinguish between emotion classes [1], [7], [8], [9], [10]. Speech disfluencies have also been shown as relevant cues for emotion characterization [11] and can be automatically extracted. Non-verbal speech cues such as laughter or mouth noise are also helpful for emotion detection. The most widely used strategy is to compute as many features as possible. All the features are, more or less, correlated with each other. Optimization algorithms are then often applied to select the most efficient features and reduce their number, thereby avoiding making hard a priori decisions about the relevant features. Trying to combine the information of different natures, paralinguistic features (prosodic, spectral, disfluences, etc) with linguistic features (lexical, dialogic), to improve emotion detection or prediction is also a research challenge. Due to the difficulty of categorization and annotation, most of the studies [7], [8], [9], [10], [11] have only focused on a minimal set of emotions.

Two studies are reported in this paper: the first investigates automatic emotion detection using linguistic information, whereas the second investigates emotion detection through paralinguistic cues. Sections 2 and 3 describe the corpus and the adopted annotation protocol. Section 4 relates experiments with respectively lexical and paralinguistic features. Finally, in the discussion and conclusion (section 5), the results obtained with lexical and paralinguistic are compared and future research is discussed.

2   The CEMO Corpus

The studies reported in this paper make use of a corpus of naturally-occurring dialogs recorded in a real-life medical call center. The dialog corpus contains real agent-client recordings obtained from a convention between a medical emergency call center and the LIMSI-CNRS. The use of these data carefully respected ethical conventions and agreements ensuring the anonymity of the callers, the privacy of personal information and the non-diffusion of the corpus and annotations.

The service center can be reached 24 hours a day, 7 days a week. The aim of this service is to offer medical advice. The agent follows a precise, predefined strategy during the interaction to efficiently acquire important information. The role of the agent is to determine the call topic, the caller location, and to obtain sufficient details about this situation so as to be able to evaluate the call emergency and to take a decision. In the case of emergency calls, the patients often express stress, pain, fear of being sick or even real panic. In many cases, two or three persons speak during a conversation. The caller may be the patient or a third person (a family member, friend, colleague, caregiver). Table 1 gives the caracteristics of the CEMO corpus.

Table 1. CEMO corpus characteristics: 688 agent-client dialogs of around 20 hours (M: male, F: female)

#agents	7 (3M, 4F)
#clients	688 dialogs (271M, 513F)
#turns/dialog	Average: 48
#distinct words	9.2 k
#total words	262 k

The transcription guidelines are similar to those used for spoken dialogs in previous work [1]. Some additional markers have been added to denote named-entities, breath, silence, intelligible speech, laugh, tears, clearing throat and other noises (mouth noise). The transcribed corpus contains about 20 hours of data About 10% of speech data is not transcribed since there is heavily overlapping speech.

3   Emotion Annotation

We use categorical labels and abstract dimensions in our annotation scheme of emotions. The valence and activation dimensions do not allow to separate Anger from Fear which are the main classes observed in our corpus and that we wanted to detect. For detection purpose, the use of categorical labels cannot be overlooked. How to choose these labels is a real challenge.

The labels are task-dependent. There are many different possible strategies for finding the best N category labels. Two extreme strategies are the direct selection of a minimal number of labels (typically from 2 to 5) or a free annotation which leads to a high number of verbal labels that must be reduced to be tractable, for instance from 176 (after normalization) to 14 classes in experiments by [2]. The mapping from fine-grained to coarse-grained emotion labels is not straightforward when free annotations are used. In the previous experiment, the mapping from 176 to 14 was done manually by the same annotators after a consensus was made on a shorter list of 14 emotion labels. An alternative strategy, which seems to us to be more powerful, is to select by majority vote a set of labels before annotating the corpus. However to adopt this strategy, a group of (at least 3) persons who have already worked with the corpus, need to select emotions with high appropriateness, appropriateness, moderate appropriateness from a list of reference emotions. Then a majority voting procedure allows a sub-list of verbal categories, the best 20 for instance, to be selected. Several different reference lists can be found in the literature (see http://www.emotion-research.fr). We have adopted the last strategy described to select labels in this work. 5 persons were involved in this task.

The set of 20 labels is hierarchically organized (see Table 2) from coarse-grained to fine-grained labels in order to deal with the lack of occurrences of fine-grained emotions and to allow for different annotator judgments.

There are a lot of different manifestations of Fear in the corpus. Emotions like Stress which, in other conditions could be linked to the coarse label Anger for instance, were judged as fitting in the Fear class. A perceptive test will be carried out to confirm this hierarchy.

Table 2. Emotion classes hierarchy: multi-level of granularity

Coarse level (8 classes)	Fine-grained level (20 classes + Neutral)
Fear	Fear, Anxiety, Stress, Panic, Embarrassment, Dismay
Anger	Anger, Annoyance, Impatience, ColdAnger, HotAnger
Sadness	Sadness, Disappointment, Resignation, Despair
Hurt	Hurt
Surprise	Surprise
Relief	Relief
Other Positive	Interest, Compassion, Amusement
Neutral	Neutral

Representing complex real-life emotion and computing inter-labeler agreement and annotation label confidences are important issues to address. A soft emotion vector is used to combine the decisions of the two annotators and represent emotion mixtures [1], [2]. This representation allows to obtain a much more reliable and rich annotation and to select the part of the corpus without conflictual blended emotions for training models. Sets of pure emotions or blended emotions can then be used for testing models. In this experiment utterances without emotion mixtures were considered.

The annotation level used to train emotion detection system can be chosen based on the number of segments available. The repartition of fine labels (5 best classes) only using the emotion with the highest coefficient in the vector [1] is given Table 3.

Table 3. Repartition of fine labels (688 dialogues). Other gives the percentage of the 15 other labels. Neu: Neutral, Anx: Anxiety, Str: Stress, Hur: Hurt, Int: Interest, Com: Compassion, Sur: Surprise, Oth: Other.

Caller	Neu.	Anx.	Str.	Rel.	Hur.	Oth.
10810	67.6	17,7	6.5	2.7	1.1	4.5
Agent	Neu.	Int.	Com.	Ann.	Sur.	Oth.
11207	89.2	6.1	1.9	1,7	0.6	0.6

The Kappa coefficient was computed for agents (0.35) and clients (0.57). Most confusion is between a so-called neutral state and an emotional set. Because we believe there can be different perceptions for a same utterance, we considered an annotator as coherent if he chooses the same labels for the same utterance at any time. We have thus adopted a self re-annotation procedure of small sets of dialogs at different time (for instance once a month) in order to judge the intra-annotator coherence over time. About 85%.of the utterances are similarly re-annotated [1]. A perceptive test was carried out [13]. Subjects have perceived complex mixtures of emotions within different classes both of the same and of different valence. The results validate our annotation protocol, the choice of labels and the use of a soft vector to represent emotions.

4   Classification

Our goal is to analyze the emotional behaviors observed in the linguistic and paralinguistic material of the human-human interactions present in the dialog corpus in order to detect what, if any, lexical information or paralinguistic is particularly salient to characterize each of the four emotion selected. Several classifiers and classification strategies well described in the machine learning literature are used.

For this study, four classes at the coarse level have been considered: Anger, Fear, Relief and Sadness (see Table 4). We only selected utterances of callers and non-mixed emotions for this first experiment.

Table 4. Train and test corpus characteristics

Corpus	Train	Test
#Speaker turn	1618	640
#Speakers	501(182 M, 319F)	179(60M, 119F)
Anger	179	49
Fear	1084	384
Relief	160	107
Sadness	195	100

4.1   Lexical Cues

Our emotion detection system is based on a unigram model, as used in the LIMSI Topic Detection and Tracking system. The lexical model is a unigram model, where the similarity between an utterance and an emotion is the normalized log likelihood ratio between an emotion model and a general task-specific model *eq. (1)*. Four unigram emotion models were trained, one for each annotated emotion, using the set of on-emotion training utterances. Due to the sparseness of the on-emotion training data, the probability of the sentence given the emotion is obtained by interpolating its maximum likelihood unigram estimate with the general task-specific model probability. The general model was estimated on the entire training corpus. An interpolation coefficient of $\lambda = 0.75$ was found to optimize the results of CL and RR. The emotion of an unknown sentence is determined by the model yielding the highest score for the utterance u, given the emotion model E.

$$\log P(u/E) = \frac{1}{L_u} \sum_{w \epsilon u} tf(w,u) \log \frac{\lambda P(W/E) + (1-\lambda)P(w)}{P(w)} \tag{1}$$

where P(w/E) is the maximum likelihood estimate of the probability of word w given the emotion model, P(w) is the general task-specific probability of w in the training corpus, tf(w,u) are the term frequencies in the incoming utterance u, and L_u is the utterance length in words. Stemming procedures are commonly used in information retrieval tasks for normalizing words in order to increase the likelihood that the resulting terms are relevant. We have adopted this technique for emotion detection. The training is done on 501 speakers and the test corresponds to 179 other speakers. Table 5 relates experiments with a stemming procedure and without a normalization procedure (the baseline).

Table 6 shows the emotion detection results for the baseline unigram system, and with the normalization procedure of stemming. Since the normalization procedures change the lexical forms, the number of words in the lexicon is also given. Results are given for the complete test set and for different λ. Using the baseline system, emotion can be detected with about 67% precision. Stemming is seen to improve the detection rate, we obtained around 78% of recognition rate (67.2% for class-wise averaged recognition rate). The results in Table 6 show that some emotions are better detected than others, the best being the Fear class

Table 5. Emotion detection with lexical cues

	Baseline		Stemming	
Size of lexicon	2856		1305	
lambda	RR	CL	RR	CL
0.65	62.7	47.5	75.9	67.1
0.75	66.9	47.5	78.0	67.2
0.85	67.5	44.4	80.3	64.6

Table 6. Repartition for the 4 classes for stemming condition and lambda = 0.75. Utt: Utterances, A: Anger, F: Fear, R: Relief, S: Sadness.

Stemming	Total	A	F	R	S
#Utt.	640	49	384	107	100
% rec.	78	59	90	86	34

and the worst Sadness. Anxiety is the main emotion for the callers. The high detection of Relief can be attributed to strong lexical markers which are very specific to this emotion (thanks, I agree). In contrast, the expression of Sadness is more prosodic or syntactic than lexical in this corpus. The main confusions are between Fear and Sadness, and Fear and Anger.

4.2 Paralinguistic Cues

A crucial problem for all emotion recognition systems is the selection of the set of relevant features to be used with the most efficient machine learning algorithm. In the experiments reported in this paper, we have focused on the extraction of prosodic, spectral, disfluency and non-verbal events cues, The Praat program [14] was used for prosodic (F0 and energy) and spectral cue extraction. About a hundred features are input to a classifier which selects the most relevant ones:

- F0 and Spectral features (Log-normalized per speaker): min, median, first and third quartile, max, mean, standard deviation, range at the turn level, slope (mean and max) in the voiced segments, regression coefficient and its mean square error (performed on the voiced parts as well), maximum cross-variation of F0 between two adjoining voiced segments (inter-segment) and with each voiced segment(intra-segment), position on the time axis when F0 is maximum (resp. minimum), ratio of the number of voiced and non-voiced segments, formants and their bandwidth, difference between third and second formant, difference between second and first formant: min, max, mean, standard deviation, range.
- Microprosody : jitter, shimmer, NHR, HNR
- Energy features (normalized): min, max, mean, standard deviation and range at the segment level, position on the time axis when the energy is maximum (resp. minimum).

- Duration features: speaking rate (inverse of the average length of the speech voiced parts), number and length of silences (unvoiced portions between 200-800 ms).
- Disfluency features: number of pauses and filled pauses ("euh" in French) per utterance annotated with time-stamps during transcription.
- Non linguistic event features: inspiration, expiration, mouth noise laughter, crying, and unintelligible voice. These features are marked during the transcription phase.

The above set of features are computed for all emotion segments and fed into a classifier. The same train and test are used as for the classifier based on the lexical features. Table 7 shows the emotion detection results using a SVM classifier.

Table 7. Repartition for the 4 with a SVM classifier. A: Anger, F: Fear, R: Relief, S: Sadness.

	Total	A	F	R	S
#Utterances	640	49	384	107	100
% rec.	59,8	39	64	58	57

As for lexical results, the Fear is best detected (64%). The Anger is worst detected (39%) while still above chance. It is mostly confused with Fear (37%). This might be due to the fact that Fear is often in the background.

5 Discussion and Conclusion

We have obtained about 78% and 60% of good detection for respectively lexical and paralinguistic cues on four real-life emotion classes. Both results were better for FearAnxiety detection, which is the most frequent emotion in the corpus and occurs with different intensity (anxiety, stress, fear, panic). Because Anger recognition is very low with the paralinguistic model and Sadness is low with the lexical model, we believe there might be a way to combine the two models and yield better results. Thus, future work will be to combine information of different natures: paralinguistic features (prosodic, spectral, disfluences, etc) with linguistic features (lexical), to improve emotion detection or prediction. Comparison with our previous results on lexical, paralinguistic and combined cues on other call center data will be done in a next future.

Acknowledgements

The work is conducted in the framework of a convention between the APHP France and the LIMSI-CNRS. The authors would like to thank the Professor P. Carli, the Doctor P. Sauval and their colleagues N. Borgne, A. Rouillard and G. Benezit. This work was partially financed by NoE HUMAINE.

References

1. Devillers, L., Vidrascu, L., Lamel, L.: Challenges in real-life emotion annotation and machine learning based detection, Journal of Neural Networks 2005. special issue: Emotion and Brain 18(4), 407–422 (2005)
2. Devillers, L., Abrilian, S., Martin, J.-C.: Representing real life emotions in audiovisual data with non basic emotional patterns and context features. In: Tao, J., Tan, T., Picard, R.W. (eds.) ACII 2005. LNCS, vol. 3784, Springer, Heidelberg (2005)
3. Vidrascu, L., Devillers, L.: Real-life Emotions Representation and Detection in Call Centers. In: Tao, J., Tan, T., Picard, R.W. (eds.) ACII 2005. LNCS, vol. 3784, Springer, Heidelberg (2005)
4. Clavel, C., Vasilescu, I., Devillers, L., Ehrette, T.: Fiction database for emotion detection in abnormal situation, ICSLP (2004)
5. Cowie, R., Cornelius, R.R.: Describing the emotional states expressed in speech. Speech Communication 40(1-2), 5–32 (2003)
6. Campbell, N.: Accounting for Voice Quality Variation, Speech Prosody, 217–220 (2004)
7. Batliner, A., Fisher, K., Huber, R., Spilker, J., Noth, E.: How to Find Trouble in Communication. Journal of Speech Communication 40, 117–143 (2003)
8. Lee, C.M., Narayanan, S., Pieraccini, R.: Combining acoustic and language information for emotion recognition, ICSLP (2002)
9. Forbes-Riley, K., Litman, D.: Predicting Emotion in Spoken Dialogue from Multiple Knowledge Sources. In: Proceedings of HLT/NAACL (2004)
10. Devillers, L., Vasilescu, I., Lamel, L.: Emotion detection in task-oriented dialog corpus. In: Proceedings of IEEE International Conference on Multimedia (2003)
11. Devillers, L., Vasilescu, I., Vidrascu, L.: Anger versus Fear detection in recorded conversations. In: Proceedings of Speech Prosody, pp. 205–208 (2004)
12. Steidl, S., Levit, M., Batliner, A., Nth, E., Niemann, E.: Off all things the measure is man Automatic classification of emotions and inter-labeler consistency. In: Proceeding of the IEEE ICASSP (2005)
13. Vidrascu, L., Devillers, L.: Real-life emotions in naturalistic data recorded in a medical call center. In: Workshop on Emotion, LREC (2006)
14. Boersma, P.: Accurate short-term analysis of the fundamental frequency and the harmonics-to-noise ratio of a sampled sound. In: Proceedings of the Institute of Phonetic Sciences, pp. 97–110 (1993)

Automatic Classification of Expressiveness in Speech: A Multi-corpus Study*

Mohammad Shami and Werner Verhelst

Vrije Universiteit Brussel, Interdisciplinary Institute for Broadband Technology - IBBT,
department ETRO-DSSP, Pleinlaan 2, B-1050 Brussels, Belgium.
wverhels@etro.vub.ac.be

Abstract. We present a study on the automatic classification of expressiveness in speech using four databases that belong to two distinct groups: the first group of two databases contains adult speech directed to infants, while the second group contains adult speech directed to adults. We performed experiments with two approaches for feature extraction, the approach developed for Sony's robotic dog AIBO (AIBO) and a segment based approach (SBA), and three machine learning algorithms for training the classifiers. In mono corpus experiments, the classifiers were trained and tested on each database individually. The results show that AIBO and SBA are competitive on the four databases considered, although the AIBO approach works better with long utterances whereas the SBA seems to be better suited for classification of short utterances. When training was performed on one database and testing on another database of the same group, little generalization across the databases happened because emotions with the same label occupy different regions of the feature space for the different databases. Fortunately, when the databases are merged, classification results are comparable to within-database experiments, indicating that the existing approaches for the classification of emotions in speech are efficient enough to handle larger amounts of training data without any reduction in classification accuracy, which should lead to classifiers that are more robust to varying styles of expressiveness in speech.

Keywords: Affective computing, emotion, expressiveness, intent.

1 Introduction

Affective computing aims at the automatic recognition and synthesis of emotions in speech, facial expressions, or any other biological communication channel [1]. Within the field of affective computing, this paper addresses the problem of the automatic recognition of emotions in speech. Data-driven approaches to the classification of emotions in speech use supervised machine learning algorithms (such as neural

---

* This paper summarizes research that was reported in the manuscript "An evaluation of the robustness of existing supervised machine learning approaches to the classification of emotions in speech", which was accepted for publication in Speech Communication.

C. Müller (Ed.): Speaker Classification II, LNAI 4441, pp. 43–56, 2007.
© Springer-Verlag Berlin Heidelberg 2007

networks or support vector machines, etc.) that are trained on patterns of speech prosody. Typically, statistical measures of speech pitch and intensity contours are used as features of the expression of emotions in speech. These features are provided as input to a machine learning algorithm along with the known emotional labels of a training set of emotional utterances. The output of the supervised learning phase is a classifier capable of distinguishing between the emotional classes it was trained with.

Previous studies have focused on different aspects of the emotion recognition problem. Some studies focus on finding the most relevant acoustic features of emotions in speech [2], [3], [4]. Other studies search for the best machine learning algorithm for constructing the classifier [5] or investigate different classifier architectures [6]. Lately, segment based approaches that try to model the acoustic contours more closely are becoming popular [7], [8], [9]. In all of these studies, a single corpus is used for training and testing a machine learned classifier. To our knowledge, emotion recognition using parallel emotional corpora has not been attempted. In this study, we used four emotional speech databases. Our aims were twofold; an estimation of the accuracy of the classification approaches in single corpus experiments and the assessment of their robustness in multi-corpus experiments utilizing parallel emotional speech corpora.

2   Description of the Speech Corpora

We used four different databases that each belong to one of two different groups. One group of two databases contains adult speech directed to infants (Kismet and BabyEars), while the second group contains adult speech directed to adults (Berlin and Danish).

2.1   The Databases with Adult-to-Infant Expressive Speech

The two adult-to-infant corpora used in this study are Kismet and BabyEars. They contain expressions of non-linguistic communication (affective intent) conveyed by a parent to a pre-verbal child.[1] The breakdown of the expressive class distribution in the Kismet and the BabyEars databases is given in Table 1.

Table 1. Emotional Classes in the Kismet and BabyEars database pairs and number of utterances per class

Kismet		BabyEars	
Approval	185	Approval	212
Attention	166	Attention	149
Prohibition	188	Prohibition	148
Soothing	143		
Neutral	320		

---

[1] Actually, Kismet is a small robotic creature with infantile looks; the speech type utilized when addressing the robot is essentially of a same type as that addressed to human infants.

Since the Kismet corpus contains two extra emotional classes that are not available in the BabyEars corpus, it was necessary to remove those two classes of emotions when performing multi-corpus experiments. Assuming that the recorders of the two corpora intended to have the same color of emotion under the same emotional label, the removal of the extra classes makes the two corpora compatible for machine learning experiments.

Kismet

The first corpus is a superset of the Kismet speech corpus that has been initially used in [6]. The corpus used in this work contains a total of 1002 American English utterances of varying linguistic content produced by three female speakers in five classes of affective communicative intents. These classes are Approval, Attention, Prohibition Weak, Soothing, and Neutral utterances.

The affective communication intents sound acted and are generally expressed rather strongly. Recording is performed with 16-bit per sample, with occurrences of 8 and 22 kHz sampled speech, and under varying amounts of noise. The speech recordings are of variable length, mostly in the range of 1.8 to 3.25 seconds.

BabyEars

The second speech corpus is the BabyEars speech corpus that has also been used in [10], [11]. The corpus consists of 509 recordings in American English of six mothers and six fathers as they addressed their infants while naturally interacting with them. Three emotional classes are included in the corpus, namely: Approval, Attention, and Prohibition.

The emotions expressed in the recordings sound natural and unexaggerated. The utterances are typically between 0.53 to 8.9 seconds in length.

2.2 The Databases with Adult-to-Adult Expressive Speech

The two databases with adult-to-adult expressive speech used in this study are the Berlin and Danish databases. Table 2 shows their emotional class distributions.

Table 2. Emotional Classes in the Berlin and Danish databases and nr. of utterances per class

Berlin		Danish	
Anger	127	Angry	52
Sadness	52	Sad	52
Happiness	64	Happy	51
Neutral	78	Neutral	133
Fear	55	Surprised	52
Boredom	79		
Disgust	38		

Similarly to the Kismet and BabyEars database pair, the Berlin and Danish databases share a number of emotional classes. Only these common emotional classes were used in multi-corpus experiments.

Berlin
This emotional speech database was recorded at the Technical University of Berlin [12]. Five female and five male actors uttered ten sentences in German that have little textual emotional content. The database contains a total of 493 utterances, divided among seven emotional classes: Neutral, Anger, Fear, Joy, Sadness, Disgust, and Boredom.

The recordings were made using high-quality recording equipment with 16-bit precision at a sampling rate of 22 kHz.

Danish
This database is described in detail in [13]. It consists of a combination of short and long utterances in Danish spoken by two male and two female speakers in five emotions. These emotions are Neutral, Surprised, Happy, Sad, and Angry.

The recordings were made with 16-bit precision at a sampling rate of 20 kHz in a recording studio.

3   Approaches Used for the Classification of Expressive Speech

Two different approaches for the classification of expressive speech have been used in this study; the AIBO approach uses feature vectors that are composed of statistical measures of acoustical parameters at the utterance level, while SBA uses a combination of such statistical measures at the utterance level and the level of individual voiced segments in the utterance. More details on these two feature sets will be given in subsections 3.1 and 3.2, respectively.

Based on these feature vectors, several supervised machine learning techniques were used to construct the classifiers. These were K-nearest neighbours (KNN) Support Vector Machines (SVM) and Ada-boosted C4.5. The method used to evaluate the resulting classifiers is stratified 10-fold cross validation. In all classification experiments, we used the implementation of the machine learning algorithms as available in the data mining toolkit Weka [14] and Milk [15].

3.1   Utterance Based Classification of Emotions: The AIBO Feature Space

The AIBO approach is a bottom up approach that relies on using an extensive feature set of low level statistics of prosodic parameters that are used in conjunction with a supervised machine learning algorithm to construct a classifier from the labeled database [5].

First the pitch, intensity, lowpass intensity, highpass intensity, and the norm of the absolute vector derivative of the first 10 Mel Frequency Cepstral Coefficients (MFCCs) are extracted from the speech signal. Next, out of each one of these five time series, four derived series are further extracted: the series of minima, the series of maxima, the series of the durations between local extrema of the 10 Hz smoothed curve, and the series itself. In the last step 10 statistics of each of the resulting 20 series are calculated, as shown in Table 3.

Table 3. Statistical measures used in the AIBO approach

Acoustic features measured	Derived time series	Statistics used
-intensity -lowpass intensity -highpass intensity -norm absolute vector derivative of first 10 MFCCs	-minima -maxima -durations between local extrema -the feature series itself	-Mean -maximum -minimum -range -variance -median -first quartile -third quartile -inter-quartile range -Mean absolute value of the local derivative

3.2 Segment Based Classification of Emotions: The SBA Feature Space

As the flowchart in Fig. 1 shows, the speech utterance as a whole is first summarized using statistical measures of spectral shape, intensity, and pitch contours. As a by-product of the pitch extraction process, the utterance is segmented into a sequence of N voiced segments.

Using that segmentation information, the same statistical measures that were also used at the whole utterance level are now calculated for each of the detected segments. Next, a feature vector consisting of both the utterance level information and the information local to the voiced segment is formed for each of the voiced segments.

Since class labels in the database are provided for the utterances as a whole, each of the voiced segments of the utterance is given the same label as the whole utterance it belongs to. A segment classifier is trained using the segment feature vectors and these segment labels.

For the classification of the whole utterance, the decisions made by the segment classifier for each of the utterances voiced segments, expressed as a posteriori class probabilities, are aggregated to obtain a single utterance level classification decision.

The feature set is made up of 12 statistical measures of pitch, intensity, and spectral shape variation (Delta-MFCCs). For pitch extraction PRAAT [16] is employed, which uses a pitch extraction algorithm that is based on an autocorrelation method [17]. As shown in Table 4, six statistical measures are used to describe the pitch contour, three for the intensity contour, and three for the spectral rate of change. More details on the specifics of the algorithm are in [11], [18].

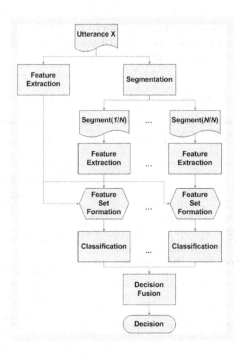

Fig. 1. Flowchart of the segment based approach

Table 4. Statistical measures used in the SBA approach

Pitch	Intensity	Speech Rate
-Variance - Slope - Mean - Range - Max - Sum of Abs Delta	-Variance - Mean - Max	- Sum of Absolute Delta MFCC - Variance of Sum of Absolute Delta MFCC - Duration

3.3 Classifier Evaluation Schemes

The experimentation paradigm used for evaluation of the classifiers is stratified 10-fold cross validation. In N-fold-cross-validation, the labeled corpus S is randomly split into N partitions of about equal size. Next, the classifier is generated using the chosen learning algorithm and conditions, but one of the N subsets is left out of the training set and used as the testing set. The remaining (N-1) subsets are used for training the classifier. This process is repeated N times, every time using a different subset for testing. The overall performance is then taken as the average of the performance achieved in the N runs. In stratified cross-validation, the generated folds contain approximately the same proportions of classes as the original dataset. Ten is the most common value for N used in the literature and was also adopted here.

4 Mono Corpus Experiments

Table 5 summarizes the results of the classifiers trained and tested on each of the individual databases for both of the feature sets and all three machine learning algorithms. Overall, these results compare favourably with the state of the art results reported in literature for the same databases (see Table 6).

Table 5. Percentage classification accuracy in single corpus experiments

	Kismet		BabyEars		Berlin		Danish	
MLA	**AIBO**	**SBA**	**AIBO**	**SBA**	**AIBO**	**SBA**	**AIBO**	**SBA**
SVM	83.7	83.2	**65.8**	67.9	**75.5**	65.5	63.5	56.8
KNN	82.2	**86.6**	61.5	**68.7**	67.7	59.0	49.7	55.6
ADA C4.5	**84.63**	81	61.5	63.4	74.6	46.0	**64.1**	**59.7**

As one can see from Table 5, there is no single best machine learning algorithm or feature vector for all the databases as they all achieve globally comparable results. Comparing the best results that can be obtained with the AIBO and SBA feature sets, we can see that AIBO outperforms SBA for the Berlin and Danish databases, while SBA performs slightly better than AIBO on the Kismet and BabyEars databases. Since the latter two have fewer voiced segments on average per utterance (4 and 2 respectively) than the former (8 and 14), it would seem that segment based approaches, such as SBA, could be advantageous for short utterances, while the global approaches that are more common in classifiers for expressive speech perform better on longer utterances. We should note, however, that the difference in expression styles between the databases also correlates with the type of algorithm that performs best. It is also interesting to note that the best learning algorithm to use with SBA on Kismet and BabyEars seems to be KNN while, on Berlin and Danish, KNN performs poorly. This could be related to the compactness of the SBA feature set and the relevance of each of its member features.

Table 6. Percentage classification accuracies reported in the literature

	Kismet	**BabyEars**	**Berlin**	**Danish**
Machines	82 Breazeal & Aryananda 2002[2]	67 Slaney & McRoberts 2003	void	54 Hammal et al. 2005
Humans	void	7/7 raters correct: 79 Slaney & McRoberts 2003	85 Paeschke & Sendlmeier 2000	67 Ververidis & Kotropolos 2004
Baseline[3]	32	42	34	51

---

[2] This result relates to a subset of the Kismet database that was used here.

[3] The baseline classification results in table 6 were obtained by classifying all test utterances as the most common emotional class in the database.

5 Multi-corpus Experiments

In multi-corpus experiments and for each of the emotional database pairs (Kismet-BabyEars and Berlin-Danish) the following is performed. First, only the emotional classes that are common to both databases in consideration are kept in each of the databases. The remaining classes are removed. Three kinds of experiments are performed on the paired databases. First, within-corpus classification on each of the two databases is performed for comparative purposes. Then off-corpus and integrated-corpus experiments are performed. In off-corpus classification, a classifier is machine learned using one corpus and tested on emotional samples from the other corpus. Integrated-corpus testing involves merging the two corpora into one speech corpus and then performing within-corpus testing on the resulting corpus.

5.1 The BabyEars-Kismet Database Pair

The common classes that remain in the paired BabyEars-Kismet databases are approval, attention and prohibition (see Table 1). Based on the results of the single database experiments, the SBA approach has been found to be more suitable for the Kismet and BabyEars databases. Therefore, it was used for all the multi-database experiments with these databases. The 3-way classification accuracy obtained for the BabyEars corpus was 65.4%; for the Kismet corpus it was 88.3%. The classification results in off-corpus experiments are shown in Table 7.

Table 7. Off-corpus classification results

Training Set	Testing Set	Classification Accuracy	Baseline Accuracy
BabyEars	Kismet	54.40%	34.90%
Kismet	BabyEars	45.00%	41.70%

When testing on the Kismet corpus and training on the BabyEars corpus, the resulting accuracy is higher than the baseline accuracy. This suggests that the learned classifier has captured enough information about the emotional class found in the testing set by learning from the training set even though the two sets come from two different domains and are recorded under different conditions. Training using the more varied BabyEars database and testing on Kismet database is found to be more successful than the other way around. This is probably due to the fact that it is unlikely for a classifier trained on a narrow corpus (Kismet corpus) to generalize for a wider corpus (BabyEars corpus) than the other way around.

In integrated-corpus experiments, the overall classification accuracy obtained when the corresponding classes in the two corpora are merged is 74.7 % correct. For comparison purposes, the resulting accuracy is plotted in Fig. 2 next to the results that were obtained in the within-corpus tests.

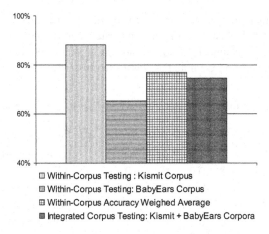

Within-Corpus Testing : Kismit Corpus
Within-Corpus Testing: BabyEars Corpus
Within-Corpus Accuracy Weighed Average
Integrated Corpus Testing: Kismit + BabyEars Corpora

Fig. 2. Comparison of the classification accuracies obtained in different settings

In order to examine in more detail the confusion tendencies in integrated corpus mode, the corresponding emotional classes of the two corpora were kept distinct. Therefore, there were six emotional classes in total, namely: {Approval, Attention Prohibition}_Kismet, and {Approval, Attention, Prohibition}_BabyEars.

The confusion matrix resulting from machine learning and classification with these six classes is shown in Table 8. Shaded cells denote test utterances that have been correctly classified. One can see that Kismet samples are almost never confused with BabyEars samples, but a significant number of BabyEars samples are confused with Kismet samples. The majority of those confused samples are actually classified correctly in terms of the conveyed emotion, which is very encouraging (32 samples with correct emotional label out of 50 corpus-confused samples, or 64 % correct)

Table 8. Confusion matrix in the case of integrated-corpus classification across six classes

a	b	c	d	e	f		class
150	26	8	1	0	0	a	Ap_K
26	140	0	0	0	0	b	At_K
1	0	186	0	1	0	c	Pr_K
20	7	6	105	32	40	d	Ap_B
0	2	3	24	86	32	e	At_B
1	0	9	29	16	93	f	Pr_B

The resulting classification accuracy in integrated corpus mode is somewhere in between the accuracies generated in within-corpus testing. This might suggest that the patterns behind the different emotions in the two corpora do not overlap in the feature space. As an example, the "Approval" class of the Kismet corpus is not getting confused with, let us say, the "Attention" class of the BabyEars corpus and so forth as the confusion matrix in Table 8 shows.

In order to further analyze the proximity of the emotional classes in the above experiment we applied clustering on the integrated corpus of the Kismet/BabyEars databases using the K-means clustering algorithm in Weka. We choose to cluster the instances into 6 clusters and we note the classes to clusters evaluation in Table 9.

Table 9. Classes to clusters evaluation in BabyEars Kismet integrated corpus

A	B	C	D	E	F	←	Assigned to cluster
0	30.9	0	30.3	38.6	0.1		kismet_approval
0	0	99.8	0.2	0	0		kismet_attention
100.0	0	0	0	0	0		kismet_prohibition
0	3.2	0	19.0	31.0	46.7		babyears_approval
4.1	0	46.6	0	0	49.3		babyears_attention
98.3	0	0	0.7	1.0	0		babyears_prohibition

Table 9 shows the classes to clusters evaluations taken as per class percentages (each row adds up to 100%). The Kismet and the BabyEars databases lie on sets of clusters with significant overlap. Specifically, the Kismet database is on clusters {A, B, C, D, E} whereas the BabyEars database is on clusters {A, C, D, E, F}. Furthermore, similar emotions in the two databases frequently lie on the same clusters. For example, *kismet_approval* is on clusters {B, D, E} whereas *BabyEars_approval* falls on {D, E, F}. Similarly, Kismet_prohibition and BabyEars_prohibition lie on cluster {A}. Furthermore, different emotions in the two databases lie on different clusters. In other words, the same cluster does not contain different emotions from different databases. For example, clusters {C, D, E} carry {*attention, approval, approval*} from both databases. Cluster F carries *approval* and *attention* from the BabyEars database only.

5.2   The Berlin-Danish Database Pair

The common classes that remain in the paired Berlin-Danish databases are Angry, Sad, Happy and Neutral (Table 2). Based on the results of Table 5 the AIBO approach has been found to be more suitable for the Danish and the Berlin databases. Therefore, this approach is used here, together with both SVM and KNN classifiers. The results from the performed within-corpus, off-corpus and integrated-corpus experiments are shown in Tables 10, 11 and 12, respectively.

Table 10. Within-Corpus Results Using the AIBO Approach

Database	Classification Accuracy
Danish	64.90%
Berlin	80.7 %

Table 11. Off-Corpus Classification Results using the AIBO approach

Training	Testing	MLA	Classif. Acc.	Baseline Acc.
Berlin	Danish	SVM	20.8%	24.3%
Berlin	Danish	KNN	22.9%	24.3%
Danish	Berlin	SVM	52.6%	46.2%
Danish	Berlin	KNN	38.9%	46.2%

Table 12. Classification Results in Integrated Corpus Tests

MLA	Classif. Acc.
SVM	72.2 %
KNN	66.83 %

As these tables show, the obtained off-corpus classification accuracies are similar to the baseline classification accuracies, implying that generalization was not possible across databases. In integrated corpus experiments, the accuracies obtained are again in-between the accuracies that were obtained in the within-corpus experiments, indicating that learning of the merged databases was possible.

When the same emotional types of the two databases were again assigned a different label, this resulted in eight emotional classes, namely Danish_{neutral, happy, angry, sad} and Berlin_{neutral, happy, angry, sad}. It can be observed in Table 13 that in this 8-way classification experiment in the integrated database, the instances belonging to one database are not at all being confused with instances belonging to the second database, while some confusion occurred within the Kismet/BabyEars database.

Table 13. Confusion Matrix in Integrated Corpus mode using Danish/Berlin Database

a	b	c	d	e	f	g	h	<--	classified as	
74	1	2	1	0	0	0	0	|	a	berlin_neutral
3	36	0	25	0	0	0	0	|	b	berlin_happy
4	0	48	0	0	0	0	0	|	c	berlin_sadness
1	25	0	101	0	0	0	0	|	d	berlin_anger
0	0	0	0	106	2	17	8	|	e	danish_neutral
0	0	0	0	7	29	2	13	|	f	danish_happy
0	0	0	0	21	4	27	0	|	g	danish_sad
0	0	0	0	11	16	2	23	|	h	danish_angry

We applied clustering on the eight-class integrated corpus of the Danish/Berlin databases using the K-means clustering algorithm in order to examine the locations of the emotional classes in the feature space. We choose to cluster the instances into 8 clusters. The results are shown in Table 14. As can be seen, the Danish and Berlin databases lie on sets of clusters with little overlap. Specifically, the Danish is on clusters {A, B, D, E, G} whereas the Berlin database is on clusters {C, F, H}. Consequently, similar emotions in different databases lie on different clusters. For example, *Berlin_happy* is on clusters {F, H} whereas *Danish_happy* falls on {A, B, D, E, G}. Furthermore, the expression of emotional classes in the Berlin database is seen to be less varied and more consistent than the expression of the same emotional classes in the Danish database. For example, the emotions {*happy, neutral, sadness, anger*} are each assigned to a single cluster with a rate of {100%, 71.9%, 78.8%, 80.3%} in the Berlin database as compared to {33.8%, 45.1%, 53.8%, 48.1%} in the Danish database. The higher consistency of emotional expressions in the Berlin database also explains the higher classification accuracy obtained in within corpus testing on the Danish and Berlin databases as reported in Table 10.

It is a well-known fact in machine learning that the more specific and uniform the training corpus is, the more accurate the classifier trained using that corpus. On the other hand, when the classifier is learned using a more heterogeneous corpus, the expected classification accuracy is usually less when the learned classifier is used to classify new instances. In our case, it turned out that using a heterogeneous emotional corpus (Kismet/BabyEars and Berlin/Danish database pairs) for constructing the classifier did not result in a notable deterioration in classification accuracy. In other words, the added robustness of being able to deal with emotions in speech that is recorded in more than one setting is not costly in terms of recognition accuracy.

Table 14. Classes to clusters evaluation in Danish/Berlin integrated corpus

a	b	c	d	e	f	g	h	<= assigned to cluster
0	0	0	0	0	0	0	100	berlin_neutral
0	0	0	0	0	71.9	0	28.1	berlin_happy
0	0	78.8	0	0	0	0	21.2	berlin_sadness
0	0	0	0	0	80.3	0	19.7	berlin_anger
15.0	18.8	0	15.8	33.8	0	16.5	0	danish_neutral
3.9	7.8	0	45.1	7.8	2.0	33.3	0	danish_happy
7.7	7.7	0	26.9	3.8	0	53.8	0	danish_sad
7.7	5.8	0	48.1	9.6	0	28.8	0	danish_angry

6   Conclusions

The difference in performance reported on the same databases between the AIBO and the SBA approach are significant. The AIBO approach seems to be better suited for classification of emotions in emotion databases with long utterances whereas the SBA works better with short utterances.

The choice of the most effective machine learning algorithm (MLA) seems to depend on the approach (AIBO vs SBA). An approach that uses a large feature set of low level statistics such as the AIBO approach seems to work best with an SVM or a

ADA-C4.5 classifier. KNN performs best with the SBA approach, which is based on a more compact feature set than the AIBO approach.

Off-corpus testing on both corpus pairs of parallel emotional classes reveals that there is little generalization happening for the same emotional classes across databases. Fortunately, when the two emotional corpora that share the same emotional classes are merged into one single large corpus, the classification accuracy on the resulting database is only slightly reduced compared to the single database accuracies. Such findings suggest that the existing approaches for the classification of emotions in speech are efficient enough to handle larger amounts of training data without any reduction in classification accuracy. This way, more recordings expressing the same emotions in slightly different domains can continuously be added to the training corpus to produce a more robust classifier for the target emotions

Automatic clustering of the emotional classes in the integrated corpora shows that similar emotions in the databases lie on distinct sets of clusters. If an ideal feature space could be employed, similar emotions belonging to different databases should be assigned to the same clusters. To achieve the desired robustness, an alternative method to the use of more training data perhaps lies in the design of better features and new classification paradigms. For emotion recognition, the integration of knowledge from domains such as psychoacoustics could be one step towards building emotion recognition systems that mimic human emotion perception.

Acknowledgements. The authors thank the owners of the databases for providing their data-bases. Parts of the research reported on in this paper were performed in the context of the IBBT projects Architectures for Mobile Community Content Creation (A4MC3) and Virtual Individual Networks (VIN), as well as in support of the HOA8 project Development of a research methodology and instrument for studying the quality of early parent-child relations at the Vrije Universiteit Brussel and the IWOIB project Anty that enjoys the support of the Brussels region.

References

1. Picard, R.: 1997. Affective Computing. The MIT Press. Wells, J.C.: Accents of English, Cambridge University Press, Cambridge (1982)
2. Nwe, T., Foo, S., De Silva, L.: Speech Emotion Recognition Using Hidden Markov Models. Speech Communication 41-4, 603–623 (2003)
3. Fernandez, R., Picard, R.W.: Classical and Novel Discriminant Features for Affect Recognition from Speech. In: Interspeech 2005, Lisbon, Portugal pp. 473–476 (2005)
4. Cichosz, J., Slot, K.: Low-dimensional feature space derivation for emotion recognition. In: Interspeech 2005, Lisbon, Portugal, pp. 477–480 (2005)
5. Oudeyer, P.: The production and recognition of emotions in speech: features and algorithms. International Journal of Human-Computer Studies 59, 157–183 (2003)
6. Breazeal, C., Aryananda, L.: Recognition of Affective Communicative Intent in Robot-Directed Speech. Autonomous Robots 12, 83–104 (2002)
7. Katz, G., Cohn, J., Moore, C.: A combination of vocal F0 dynamic and summary features discriminates between pragmatic categories of infant-directed speech. Child Development 67, 205–217 (1996)

8. Batliner, A., Fischer, K., Huber, R., Spilker, J., Noth, E.: How to find trouble in communication. Speech Communication 40, 117–143 (2003)
9. Batliner, A., Steidl, S., Hacker, C., Nöth, E., Niemann, H.: Tales of tuning - prototyping for automatic classification of emotional user states. In: Interspeech 2005, pp. 489-492 (2005)
10. Slaney, M., McRoberts, G.: A Recognition System for Affective Vocalization. Speech Communication 39, 367–384 (2003)
11. Shami, M., Kamel, M.: Segment Based Approach to the Recognition of Affective Intents in Speech. Manuscript submitted for publication (2005)
12. Paeschke, A., Sendlmeier, W.: Prosodic characteristics of emotional speech: measurements of fundamental frequency movements. In: Proc. of the ISCA ITRW on Speech and Emotion, Belfast, pp. 75–80 (2000)
13. Engberg, I.S., Hansen, A.V.: Documentation of the Danish Emotional Speech Database (DES). Internal AAU report, Center for Person Kommunikation, Denmark (1996)
14. Witten, I.H., Frank, E.: Data Mining: Practical machine learning tools with Java implementations. Morgan Kaufmann, San Francisco (2000)
15. Frank, E., Xu, X.: Applying Propositional Learning Algorithms to Multi-instance data. Working Paper, Department of Computer Science, University of Waikato (2003), www.cs.waukato.nz/ml/milk
16. Boersma, P., Weenink, D., PRAAT: a system for doing phonetics by computer. Report of the Institute for Phonetic Sciences of the University of Amsterdam 132 (1996), http://www.praat.org
17. Boersma, P.: 1993. Accurate short-term analysis of the fundamental frequency and the harmonics-to-noise ratio of a sampled sound. In: Proceedings of the Institute of Phonetic Sciences of the University of Amsterdam, vol. 17, pp. 97–110 (1993)
18. Shami, M., Kamel, M.: Segment-based Approach to the Recognition of Emotions in Speech. In: IEEE Conference on Multimedia and Expo (ICME05), Amsterdam, The Netherlands (2005)
19. Ververidis, D., Kotropolos, C.: Automatic speech classification to five emotional states based on gender information. In: Proc. Eusipco 2004, Vienna, Austria, pp. 341–344 (2004)
20. Hammal, Z., Bozkurt, B., Couvreur, L., Unay, D., Caplier, A., Dutoit, T.: Passive versus active: vocal classification system. In: Proc. of Eusipco 2005, Antalya, Turkey (2005)

Acoustic Impact on Decoding of Semantic Emotion

Erik J. Eriksson[1], Felix Schaeffler[2], and Kirk P.H. Sullivan[1]

[1] Umeå University, Umeå, Sweden
kirk@ling.umu.se
[2] Queen Margaret University, Edinburgh, Scotland

Abstract. This paper examines the interaction between the emotion indicated by the content of an utterance and the emotion indicated by the acoustic of an utterance, and considers whether a speaker can hide their emotional state by acting an emotion even though being semantically honest. Three female and two male speakers of Swedish were recorded saying the sentences "Jag har vunnit en miljon på lotto" (I have won a million on the lottery), "Det finns böcker i bokhyllan" (There are books on the bookshelf) and "Min mamma har just dött" (my mother just died) as if they were happy, neutral (indifferent), angry or sad. Thirty-nine experimental participants (19 female and 20 male) heard 60 randomly selected stimuli randomly coupled with the question "Do you consider this speaker to be emotionally X?", where X could be angry, happy, neutral or sad. They were asked to respond yes or no; the listeners' responses and reaction times were collected. The results show that semantic cues to emotion play little role in the decoding process. Only when there are few specific acoustic cues to an emotion do semantic cues come into play. However, longer reaction times for the stimuli containing mismatched acoustic and semantic cues indicate that the semantic cues to emotion are processed even if they impact little on the perceived emotion.

Keywords: Emotion identification, acoustic emotion, semantic emotion, perception, Swedish.

1 Introduction

The interaction of acoustics and semantics in spoken language emotion decoding has been widely studied with a range of listener groups, for example, emotionally deficient participants [1], children [2,3], and adults [2,7,6,5,4].

Psychological and neurological dysfunctions have been shown to affect a person's ability to recognize and to produce emotions. For example, [1] studied how a participants with different degrees of alexithymia was evidenced in the participants' automatic emotion decoding by presented the participants with emotionally congruent and non-congruent pairs of visually presented faces and auditorially presented single word verbal targets. They found that, regardless of the participants mood, the greater the degree of alexithymia the less the amount of emotional information that is automatically processed.

C. Müller (Ed.): Speaker Classification II, LNAI 4441, pp. 57–69, 2007.

Visually and auditory congruent and non-congruent stimuli have also been used to study the processing of emotional information by children. Using such stimuli [4] found that children judged verbal negative content presented with a positive visual content more negatively than their parents on a positive to negative scale. The results also indicated an interaction between the acoustic and semantic part of the verbal stimuli modality; positive signals in one part were disregarded in favour of a negative signal in the other channel. This interaction was further found to be impacted upon by the gender of the speaker; females were significantly judged more negatively than males when presenting verbally negative content. More recently, [2] using verbal stimuli that varied in semantic and acoustic emotional connotation found that children primarily use the content (i.e. semantics) of the stimuli to determine the emotional content of the stimuli, whereas adults primarily use other cues, when the task was to judge whether a stimulus was happy or sad. The stimuli used in [2] contained congruent and non-congruent cues for tone and content and were spoken by a single female.

It has been shown [5] that adult participants attend to tone of voice as the primary cue when judging the emotional content in non-congruent stimuli on a one-dimensional scale (positive through negative). Two other effects of note were found. One, that participants were able to disregard the acoustic cues in favour of content ones when instructed to do so and two, a speaker dependent effect; one speaker was more efficient in conveying their emotional disposition. This effect may have been caused by the use of amateur speakers; an effect discussed in detail in [8].

More recently, [6] studied the effect of emotional congruence on response times in a lexical decision task. The participants' task was to decide whether a word was 'real' or not. During complete randomisation trials (i.e. all stimuli were randomised with regards to their presentation) no significant differences were found between response times for congruent and non-congruent stimuli. An effect was found, however, in blocked trials in which each group of participants heard all the stimuli presented with the same tone of voice. Here, congruent stimuli reaction times were significantly shorter than both non-congruent and baseline stimuli (neutral words spoken in neutral tone) reaction times. In their adult trials [2] also found reaction time differences between congruent and non-congruent stimuli. However, this effect diminished over the course of the trials.

Emotional congruence has also been used to study lexical disambiguation [7]. Nygaard and Lunders presented their participants with homophones of one, or two, semantic attitudes (Happy or Sad) spoken in a neutral, positive or negative tone. The participants' task was to transcribe the homophone as they thought they heard it. Nygaard and Lunders found that their participants were more likely to respond with the happy connoted variant of the homophone than the neutral counterpart when spoken with a happy tone of voice and the negative counterpart when spoken with a negative tone of voice.

In this study, the impact of semantic content on acoustic emotion decoding (and vice-versa) of acted emotion was investigated using signal detection theory. The participants were asked to respond YES or NO to a question asking if the stimulus was a particular emotion or not; congruent and non-congruent stimuli were presented. As previous findings indicate that adults decode emotion based solely on acoustic cues, analysis of the participant responses using signal detection theory should show a significantly better signal detection ability for the acoustic cues than the semantic cues in the stimuli set. The signal detection approach also facilitates rapid participant responses, as the decision to be made by participants is binary; reaction time measurement is, therefore, sound and motivated. The task is straightforward and related to the investigated variables (cf. for example, [6]). The stimuli set was constructed in Swedish. It was expected that reaction times for congruent stimuli would be shortest and that signal detection values would show that acoustic cues are used more accurately in emotion detection than semantic cues.

2   Method

2.1   Participants

These were 39 (20 male and 19 female) undergraduate students (mean age= 24.7, sd= 4.1) at Umeå University, Sweden. All were native speakers of Swedish and reported no known hearing problem. Their participation was voluntary and without reimbursement.

2.2   Materials

The stimuli set was created from the recordings of five Swedish speakers (2 male and 3 female; mean age = 23.6, range: 20 – 30 years), who were selected on the basis of a pre-test [9]. In the pre-test 30 listeners specified the acoustic emotional content 30 speakers tried to convey (happy, sad, neutral and angry). The stimulus sentence for the pre-test was the semantically emotional neutral sentence ("Min vän har köpt en ny bil", My friend has bought a new car). The five speakers with highest identification of acoustic emotions were selected.

Each speaker was recorded uttering three sentences: "Jag har vunnit en miljon på lotto" (I have won a million on the lottery; happy sentence content), "Det finns böcker i bokhyllan" (There are books on the bookshelf; neutral sentence content) and "Min mamma har just dött" (My mother just died; sad sentence content). These sentences were also submitted to a written test in which subjects were to specify whether the sentences signalled happiness, sadness, anger or neutrality (i.e. did not convey a specific emotion) [9]. The three sentences selected signalled their intended semantic emotional content clearly: the sad sentence was identified as 'sad' 100% of the time, the happy sentence as 'happy' 96.7% of the time and the neutral sentence as 'neutral' 80% of the time. These three sentences were recorded using four emotional tones of voice: happy, sad, neutral and angry (unspecified hot or cold).

The stimuli were recorded in a sound-attenuated room using a microphone connected to a PC computer through a microphone pre-amp. The analogue signal was digitized at 44.1 kHz in Adobe Audition and subsequently re-sampled at 16 kHz and high-pass filtered at 60 Hz. Each stimulus was circa 3 seconds long.

Table 1. The categorisation of the stimuli. '+' indicates a match between the question emotion the emotion cue and '−' indicates no match between the question emotion the emotion cue.

Stimulus Set Name	Acoustic Cues	Semantic Cues
Baseline	+	+
Noise	−	−
Test	+	−
	−	+

2.3 Procedure

The listeners were positioned in front of a computer and were equipped with headphones. The test was presented using a tcl/tk script and started with demographic questions about gender and age of the listener. The perception test consisted of 60 randomly presented stimuli, each accompanied by a question (presented visually) of the form "Do you consider this speaker to be emotionally X?", where X was randomly chosen as either happy, sad, angry or neutral. Two buttons were displayed below the question, one NO-button and one YES-button. These were disabled until after the stimulus has been played. The listeners then responded to the question by clicking the appropriate button using the computer mouse. The listeners were requested to answer the question as quickly and accurately as possible. Reaction time was measured from the end-point of the file until one of the buttons has been selected. When the participant had not responded after 10 seconds, a new stimulus was automatically presented. The answers (YES or NO) were recorded together with response times, the file played and the question emotion.

2.4 Experimental Design and Analysis

The data was categorised into three sets as shown in Table 1: (i) The baseline set: the semantic emotion matched that of the acoustic one which, in turn, matched that of the question (for example, semantic happy, acoustic happy and question happy); (ii) The test set: the semantic or the acoustic emotion (but not both) matched the question emotion and (iii) noise: the question emotion did not match either of the signalled emotions.

In a yes-no experimental design responses can be grouped into four different categories: HIT, MISS, FALSE ALARM and CORRECT REJECTION. A HIT is when the participant responds YES when the emotion in the question agrees with the emotion of the stimulus, a MISS is when the participant incorrectly responds NO when the question emotion and the emotion of the stimulus are the same; a FALSE ALARM is when the participant responds YES when the question emotion is not the same as the emotion of the stimulus, and a CORRECT REJECTION is when the participant correctly rejects a stimulus as not being the same as the question emotion. This division of responses into these categories makes it possible to calculate the signal detection ability or the discrimination sensitivity of the listener. The signal detection measure, d-prime, is the difference between the HIT RATE (H), the number of HITS relative to the total number of possible HITS, and the FALSE ALARM RATE (F), the number of FALSE ALARMS relative to the total number of possible FALSE ALARMS, after first being transformed into z-values, $d - prime = z(H) - z(F)$ [10], and is based on both the listener's ability to answer YES correctly and inability to say NO when they should. The value of the d-prime is symmetrically distributed around zero, where zero means no difference between hits and false alarms indicating a lack of signal detection ability.

The data categorisation scheme permits testing for above chance emotion identification for the baseline stimuli: d-prime was calculated using baseline and noise stimuli. D-prime can also be calculated for accuracy of acoustic emotion identification using the test and noise stimuli. Here a HIT is scored by the participant when the acoustic emotion matches the question emotion and the participant answers YES, and a FALSE ALARM when the acoustics and question emotion do not match and the participant answers YES. D-prime can be similarly calculated using the test and noise stimuli for accuracy of semantic emotion identification. Here a HIT is scored by the participant when the semantic emotion matches the question emotion and the participant answers YES, and a FALSE ALARM when the semantic and question emotions do not match and the participant answers YES. As angry only occurs as an acoustic emotion, d-prime angry could only be calculated for acoustic emotion detection, and no baseline d-prime could be calculated. Differences in d-prime values were tested using the Student's t-test.

Correct response reaction times for the various test types were compared using the Kolmogorov-Smirnov test as the reaction times were shown to be skewed by a D'Agostino test. Non-responses, when the listener did not respond, were discarded; these accounted for 0.6% of the total stimuli.

3   Results

The signal detection theory d-prime results are presented followed by the reaction time results and response confusion.

3.1   D-Prime Analysis

In order to ensure accurate measures of parametric tests, the normality of the d-prime values was investigated. Shapiro-Wilk's test confirmed a normally distributed set of data (W= 0.99, p= 0.99).

The participants' d-prime scores for the identification of the baseline stimuli were significantly above chance (t(37) = 34.79, p= 0.0001). This indicated that the speakers successfully conveyed the emotional content in a manner the listeners expected using voice alone, semantic cues alone, or both in tandem.

The participants' ability to detect emotion based on the semantics of the test stimuli, as indicated by the d-prime values for semantic emotion identification,

Table 2. Student's t-tests between the acoustic emotions' detectability measured in d-prime

	Angry	Happy	Neutral
Happy	$t(75.5) = -2.9$		
	$p = 0.02$		
Neutral	$t(75.9) = 0.77$	$t(75.5) = 3.08$	
	$p = 0.44$	$p = 0.002$	
Sad	$t(75.9) = 2.93$	$t(75.8) = -0.72$	$t(75.9) = -3.70$
	$p = 0.004$	$p = 0.475$	$p = 0.001$

was significantly lower than for detection of emotion based on the acoustics of the test stimuli, as indicated by the d-prime values for acoustic emotion identification ($t(71.58) = 3.95$, p= 0.001).

When compared to the d-prime values for the baseline stimuli, the d-prime values for acoustic emotion identification were significantly lower ($t(69.1) = 17.6$, p= 0.001). A significant difference was also found for semantic emotion identification ($t(74.8) = 25,1$, p= 0.001). The participants' acoustic emotion detection performance was better than chance ($t(38) = 3.61$, p= 0.001), but the participant's semantic emotion detection was not better than chance ($t(38) = -1.79$, p= 0.081). No gender difference was found in the signal detection (d-prime) rate for either acoustic cues ($t(36) = 1.278$, p= 0.21) or the semantic cues ($t(36) = -0.0001$, p= 0.9998).

The mean d-prime values for each of the test sets (acoustic and emotion) and the baseline set show that the baseline stimuli are most readily identified (mean d-prime $= 2.39$), followed by the acoustic cue test second (mean d-prime $= 0.34$) and the semantic emotion cue test set the least readily recognised with decoding at chance level (mean d-prime $= -0.13$).

The signal detection performance of the listeners in terms of d-prime values was investigated for the question emotions. For each question emotions that was cued by both acoustics and semantics (Happy, Neutral and Sad) a Student's t-test was computed between the semantic and acoustic emotion detection ability. These tests showed significant differences in performance for Happy ($t(75.78) = 2.53$, p= 0.013) and Neutral ($t(73.8) = -2.07$, p= 0.042) but not for Sad ($t(73.7) = 1.55$, p= 0.126).

Using the same d-prime values it is possible to compare emotion detection within the test due to acoustic and semantic cues. The results presented in Table 2 show that for emotion detection based on the acoustic cues there were significant differences in emotion detection depending on the emotion of the question, and those presented in Table 3 show that for emotion detection based on the semantic cues there were no significant differences in emotion detection that were dependent upon the emotion of the question.

D-prime allows comparison of emotion detection; it is possible to use d-prime to determine which of the emotions was most easy to identify (high d-prime) and which was the hardest (low d-prime). The means of d-primes for each acoustic

Table 3. Student's t-tests between the semantic emotions' detectability measured in d-prime

	Happy	Neutral
Neutral	$t(75.9) = 1.43$, $p = 0.156$	
Sad	$t(75.6) = -1.87$, $p = 0.065$	$t(75.9) = -0.42$, $p = 0.674$

Table 4. Descriptives (means and standard deviation (std. dev)) for acoustic emotion detection values (d-prime)

Emotion	$\bar{x}$	std. dev
Angry	−0.098	0.67
Happy	0.237	0.62
Neutral	−0.215	0.67
Sad	0.34	0.65

emotion are shown in Table 4. From this table it can be seen that Sad was the most readily detected emotion, followed by Happy, Angry, and finally Neutral was the least detectable. The values for d-primes for each semantic emotion are not ranked as no significant deference was found.

3.2 Reaction Time Analysis

The reaction times for the correct responses to test set acoustic cues were analysed in relation to the reaction times for correct responses to test set semantic cues and to the correct responses to baseline stimuli set. Due to the skewed distribution of response times (D'Agostino test: skewness= 1.82, p= 0.001 and skewness= 1.81, p=0.001 reaction times to semantic and acoustic cues, respectively. See also Figures 1 and 2) the Kolmogorov-Smirnov non-parametric test was used. (Logaritmization of the variables did not improve skewness.) A difference between the test stimuli set and the baseline set was found. The baseline had significantly lower reaction times (D= 0.145, p= 0.003). No significant difference in response times between correct responses to semantic cues and correct responses to acoustic cues within the test set were found (D=0.039, p= 0.194).

The reaction times for HITS only were also analysed for the reaction times to semantic and acoustic cues. The Kolmogorov-Smirnov test showed a significant difference between the response times (D= 0.328, p= 0.0001). The reaction times to acoustic cues were shorter than the reaction times to semantic cues (median response time to the acoustic cues: 1.51s; median response time to the semantic cues: 2.46s).

Reaction times were also analysed for gender differences. A Kolmogorov-Smirnov test showed a significant difference between male and female response times for HITS for both responses to acoustic cues (D= 0.0848, p= 0.002) and responses to semantic cues (D= 0.1012, p= 0.002). The female group was faster in

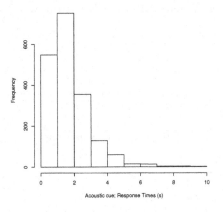

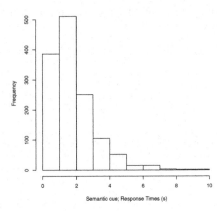

Fig. 1. Histogram showing the distribution of response times for test set acoustic cue HITS and CORRECT REJECTIONS

Fig. 2. Histogram showing the distribution of response times for test set semantic cue HITS and CORRECT REJECTIONS

Table 5. Reaction time descriptives (in seconds) for correct responses by gender and cue

	Female		Male	
	Mean	Median	Mean	Median
Acoustic cue	1.66	1.41	1.80	1.47
Semantic cue	1.74	1.49	1.93	1.57

both cases (see Table 5). No significant difference was found due to the question asked. Figures 3 – 6 show box-plots of the participants' reaction times by gender. Each figure presents a set of triplets that are defined by the question asked. The reaction times in each panel have been subjected to a Kolmogorov-Smirnov test and the significant gender differences within a triplet are marked with a '*'. These figures, thus, provide more detailed information about the participants' reaction times.

3.3   Confusion Analysis

Participant responses were analysed in terms of confusion between the acoustic and the semantic cues. For each question emotion, the responses (either YES and NO) were investigated in terms of which emotion listeners were responding to. This analysis includes yes-responses to noise data. Tables 6 – 9 detail the responses to each semantic emotion cue per acoustic emotion cue. Each table details a question emotion.

The confusion tables reveal that the semantic cues to emotion decoding play little role in the decision process whether the aural stimuli is the same as

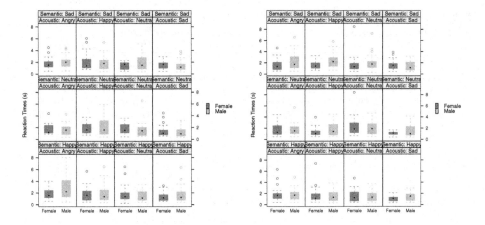

Fig. 3. Box plots of the reaction times for all Question:Angry; Acoustic Cue; Semantic Cue tripets by gender. Significant differences found using a Kolmogorov-Smirnov test are indicated with a '*'.

Fig. 4. Box plots of the reaction times for all Question:Happy; Acoustic Cue; Semantic Cue tripets by gender. Significant differences found using a Kolmogorov-Smirnov test are indicated with a '*'.

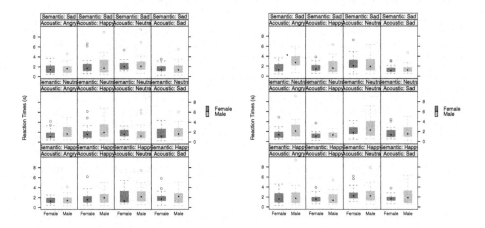

Fig. 5. Box plots of the reaction times for all Question:Neutral; Acoustic Cue; Semantic Cue tripets by gender. Significant differences found using a Kolmogorov-Smirnov test are indicated with a '*'.

Fig. 6. Box plots of the reaction times for all Question:Sad; Acoustic Cue; Semantic Cue tripets by gender. Significant differences found using a Kolmogorov-Smirnov test are indicated with a '*'.

question emotion or not. In Tables 6 – 9, most YES responses are found where the question and acoustic cue agree. The degree of YES confusion for the question Angry is small with only the occasional YES when the acoustic cue is not that of Angry. For the Happy question the semantic cue Happy achieved a few YES

Table 6. Responses to all stimuli for the question emotion Angry

Acoustic Emotion	Semantic Emotion	Yes	No
	Happy	24	16
Angry	Neutral	34	1
	Sad	34	14
	Happy	1	50
Happy	Neutral	9	43
	Sad	3	52
	Happy	2	43
Neutral	Neutral	7	39
	Sad	2	45
	Happy	1	49
Sad	Neutral	1	46
	Sad	0	46

Table 7. Responses to all stimuli for the question emotion Happy

Acoustic Emotion	Semantic Emotion	Yes	No
	Happy	16	42
Angry	Neutral	9	36
	Sad	7	45
	Happy	51	4
Happy	Neutral	42	5
	Sad	40	3
	Happy	3	46
Neutral	Neutral	9	37
	Sad	8	44
	Happy	2	40
Sad	Neutral	1	45
	Sad	0	42

Table 8. Responses to all stimuli for the question emotion Neutral

Acoustic Emotion	Semantic Emotion	Yes	No
	Happy	0	62
Angry	Neutral	3	62
	Sad	2	49
	Happy	13	39
Happy	Neutral	14	33
	Sad	6	42
	Happy	39	13
Neutral	Neutral	48	3
	Sad	30	17
	Happy	11	45
Sad	Neutral	7	40
	Sad	3	48

Table 9. Responses to all stimuli for the question emotion Sad

Acoustic Emotion	Semantic Emotion	Yes	No
	Happy	3	30
Angry	Neutral	3	46
	Sad	9	34
	Happy	0	37
Happy	Neutral	1	46
	Sad	2	47
	Happy	15	27
Neutral	Neutral	10	42
	Sad	25	22
	Happy	38	8
Sad	Neutral	49	4
	Sad	53	2

responses even when the acoustic cue is that of Angry. For the Neutral question, the Happy and Sad acoustic cues manage to attract a few YES responses together with the Happy and Neutral semantic cues. For the Sad question, the sad semantic cue coupled with the neutral acoustic cue managed to attract a relatively large number of YES responses; this was the only combination in the study where it can be claimed that the semantic cues, at times, overrode the acoustic cues.

4 Discussion

Previous research that has investigated emotion decoding has measured this by either asking participants to write down, or select from a given set of emotions, the emotion they perceive from the stimuli or asking participants to rate the emotion they perceive on a given dimension. A detailed discussion of the advantages and disadvantages of these two approaches is given in [8] and [11]. For example, naming emotions (picking them from a set) is flawed as it forces participants into selecting emotions which may not be as clear to the participant as the researcher desires [8].

The experimental design used in this paper did not force participants to pick one emotion from a given set of emotions or to map the perceived stimuli onto arbitrary dimensions. I asked the participants to respond to whether a particular stimuli was a specific emotion or not. This experimental design gave the participants the opportunity to respond YES or NO to the presented emotion stimulus triplets depending upon whether they perceived the oral stimulus as representing the question emotion or not. This use of a binary participant response affords the use of signal detection theory and the calculation of listener discrimination ability as measured by d-prime. Further the set of test stimuli facilitated the calculation of d-prime for acoustic and semantic cues of emotion transmission separately; high d-prime values indicate good signal detection and values close to zero indicate near random signal detection.

Another advantage of the experimental design used is that by asked the participants to evaluate the input and relate it to their emotion prototypes, it is possible to measure emotion matching, and not simple the participant's emotion naming or mapping ability. The approach, however, does not provide any information about how the stimuli are processed, which features the listener attends to or what information is relevant for a specific decoding.

Accuracy for emotion decoding, and thus detection, was high for the baseline stimuli set. This showed that the listeners accurately responded YES when the semantic and acoustic cues in the aurally presented stimulus matched the emotion in question asked and NO when the semantic and acoustic emotion cues did not match the emotion in the question. The detection rate significantly fell when the acoustic and semantic emotion cues contradicted each other. It was, further, found that listeners are more likely to rely on the information given by the acoustic channel.

Thus, when two different channels (in this case acoustic and semantic) signal different emotions, the chance of deducing the speaker's intended emotion is lower, but due to the greater impact of the acoustic cues, individuals acting acoustic cues for an emotion are more likely to succeed in hiding their emotional state than individuals that attend to the semantic content of their speech but ignores the need to act the acoustic cues to emotion. As no gender difference in detection rate was found (cf. [12], was has posited that such a differences exists), males and females appear to attend to the same cues and to the same degree.

The listeners' reaction times were longer for the test set stimuli, where the acoustic and semantic cues did not match, than for baseline stimuli, where the acoustic and semantic cues matched. This result parallels the findings of [6] and [13]. A significant gender difference was found for the reaction time in the test set. The female participants responded faster than the males when responding correctly; this finding is similar to the one reported in [14]. Taken together with the d-prime results, it can be posited that although there is no difference between the genders in their emotion identification and discrimination abilities, the reaction time difference indicates a difference in how the genders neurologically process emotional content (see, for example, [12]) and that females (possibly) handle emotion cue discrepancies earlier than males.

The examination of confusion between the question and the emotion cues revealed that the semantic cues played little, and often no, role in the decision process. However, for Neutral and Sad there is more confusion. This is notable as this differs from the results of [14] who specified sad, anger and fear as the most easily recognized emotions in speech. Scherer's findings indicate that sad semantic cues should impact upon the perceived acoustic cues as suggested by [4]; this is not supported by the findings of this study.

5   Conclusion

The experiment presented here has used a novel, to the emotion in speech research sector, method to measure listeners' ability to decode emotional content in spoken material. The impact of semantic cues to emotion upon acoustically encoded cues to emotion detection has been investigated. The findings show that semantic cues to emotion play little role in the decoding process; the signal detection ability for acoustic cues to emotion is superior to that of semantic cues to emotion. Only when there are few specific acoustic cues (for instance, when the target acoustics cue the emotion Neutral) do semantic cues come into play, and even then only marginally. Thus, from this acted speech material it can be concluded that a speaker can hide their emotional state by acting an emotion even though being semantically honest. However, the longer reaction times for the test data indicate that the semantic cues to emotion are processed. The gender differences in reaction time may reflect a gender-based difference in the processing of emotion information in speech; this warrants further investigation.

Acknowledgements. This research was partly funded by a grant from the Bank of Swedish Tercentenary Foundation Dnr: K2002-1121:1–4 to Umeå University.

References

1. Vermeulen, N., Luminet, O., Corneille, O.: Alexithymia and the automatic processing of affective information: Evidence from the affective priming paradigm. Cognition and Emotion 20(1), 64–91 (2006)
2. Morton, J.B., Trehub, S.E.: Children's understanding of emotion in speech. Child Development 72(3), 834–843 (2001)

3. Friend, M., Bryant, J.B.: A developmental lexical bias in the interpretation of discrepant messages. Merill-Palmer Quaterly 46, 342–369 (2000)
4. Bugental, D., Kaswan, J., Love, L.: Perception of contradictory meanings conveyed by verbal and nonverbal channels. Journal of Personality and Social Psychology 16(4), 647–655 (1970)
5. Mehrabian, A., Wiener, M.: Decoding of inconsistent communications. Journal of Personality and Social Psychology 6(1), 109–114 (1967)
6. Wurm, L.H., Vakoch, D.A., Strasser, M.R., Calin-Jageman, R., Ross, S.E.: Speech perception and vocal expression of emotion. Cognition and Emotion 15(6), 831–852 (2001)
7. Nygaard, L., Lunders, E.: Resolution of lexical ambiguity by emotional tone of voice. Memory & Cognition 30(4), 583–593 (2002)
8. Scherer, K.R.: Vocal communcations of emotion: A review of research paradigms. Speech Communication 40, 227–256 (2003)
9. Lahti, A., Nilsson, C., Nordbert, J., Karlsson, L.R.L.: Har semantiken betydelse i ett emotionellt uttryck? Cognitive Science term paper, Umeå University, Sweden (2002)
10. Green, D.M., Swets, J.A.: Signal detection theory and psychophysics. John Wiley and Sons, Inc. New York (1966)
11. Cowie, R., Cornelius, R.R.: Describing the emotional states that are expressed in speech. Speech Communication 40, 5–32 (2003)
12. Schirmer, A., Kotz, S.A., Friederici, A.D.: Sex differentiates the role of emotional prosody during word processing. Cognitive Brain Research 14, 228–233 (2002)
13. Mullennix, J.W., Bihon, T., Bricklemyer, J., Gaston, J., Keener, J.M.: Effects of variation in emotional tone of voice on speech perception. Language and Speech 45(3), 255–283 (2002)
14. Schirmer, A., Kotz, S.A.: ERP evidence for a sex-specific stroop effect in emotional speech. Journal of Cognitive Neuroscience 15(8), 1135–1148 (2003)

Emotion from Speakers to Listeners: Perception and Prosodic Characterization of Affective Speech

Catherine Mathon and Sophie de Abreu

EA333, ARP, Université Paris Diderot,
UFRL case 7003, 2 place Jussieu, 75251 Paris Cedex
{mathon,deabreu}@linguist.jussieu.fr

Abstract. This article describes a project which aimes at reviewing perceptive works on emotion and prosodic description of affective speech. A study with a spontaneous French corpus, for which a corresponding acted version has been built, shows that native listeners perceive the difference between acted and spontaneous emotions. The results of cross-linguistic perceptual studies indicate that emotions are perceived by listeners partly on the basis of prosody only, proposing the universality of emotions like anger, and partly on the basis of the variability. The latter assumption is supported by the fact that the characterization of anger in degrees is different depending on the mother tongue of the listeners. Finally, a prosodic analysis of the emotional speech is presented, measuring F0 cues, duration parameters and intensity.

Keywords: Emotions, perception, prosody.

1 Introduction

This paper reports on a project we conducted which aimes to contribute to the description of the prosody of affective speech. The choice of a corpus in the study of the vocal characterization of emotion is not trivial: Research is often based on non-natural corpora, played by actors. This allows for a tighter control of the quality of the recordings and also to select the emotion to be acted. Recently, however, more and more studies have insisted on the necessity of using natural and authentic emotional speech. As we wanted to work on spontaneous emotion, we used a corpus extracted from a radio programme which plays hoaxes on unsuspecting members of the public. The emotion we chose to focus on is anger.

The first part of this paper will describe how we validated the emotive charge of the corpus. We show that native listeners can sense the difference between acted and spontaneous emotion, thus confirming us in our decision to work with a spontaneous corpus. Then, we focus on the cultural or universal aspect of the affective behaviour (which differentiates attitudes from emotions). To verify the universality of anger as an emotion, we made a set of cross-linguistic perceptual tests. These tests show that listeners may perceive emotions on the

C. Müller (Ed.): Speaker Classification II, LNAI 4441, pp. 70–82, 2007.

basis of prosody only[1]. The results confirm both universality and variability in the perception of anger: It is identified when participants hear only prosody; moreover, the characterization of anger in various degrees depends on the mother tongue of the listeners. It appears that the global concept of anger may be graduated, and the categories so defined are closed to attitudes: Their perception is cultural. Finally, results from a prosodic analysis of the emotive speech are presented measuring F0 cues, duration parameters, and intensity.

2   The Spontaneous Speech Corpus

For the reasons stated above, we decided to work on spontaneous rather than acted emotions and thus endeavoured to select a suitable spontaneous French corpus. We chose a corpus based on radiophonic hoaxes (online Fun Radio http://www.funradio.fr). The radio presenter calls on institutions (high schools, hospitals) or professionals (bakers, taxidermists, bankers) and, playing the role of a client, asks something which does not fit the situation. For example, the animator calls a taxidermist and asks him for a taxi cab, creating thus a situation where the professional tries to explain the mistake while the animator acts as if he does not understand the problem. Eventually, this miscommunication[2] leads the victim of the hoax to express anger.

The corpus consists of one hour and four minutes of speech (twenty-four hoaxes). Fifteen out of the twenty-four initial dialogues were transcribed using *Transcriber 4.0* which supports reading, selecting and transcribing a sound file (.wav) by speaker turns. We chose not to analyze the production of the animator for two main reasons: First, it is not possible to interpret his speech on the same level as the speech of the victim as he is in fact addressing a "third" party – the public – and his interventions with the victims are shaped by this (see the discussion on communicational tropes in Kerbrat-Orecchioni, 1990 [1]). Secondly, we assume that the type of anger the animator expresses is different to that expressed by the victim: Only the victim, unaware of the hoax, expresses spontaneous emotion, whereas the animator is acting his emotion. Since our aim is to test the perception of authentic spontaneous emotions, we concentrated on the "victim's speech".

In total, the collect corpus comprises twenty-seven speakers: thirteen men and fourteen women. Some of the dialogues were rejected because the speech contained a dialectal or foreign accent or the victims had realised the call was a hoax. Each sentence of each turn was annotated using three types of labels: ANGER , associated with a number from one to five to mark a degree (A1, A2, ..., A5), NEUTRAL ATTITUDE (N), and OTHER EMOTION (OE). These three labels constituted the main categories. Table 1 depicts the distribution of speaker turns based on the annotations. Among the total speaker turns, 305 were annotated as ANGER. The corpus contains 40 % ANGER and 52 % NEUTRAL speakers turns.

---

[1] However, other linguistic information completes this perception and comprehension.
[2] There are two types of miscommunication, one based on the ambiguity of the context, and the other on the meaning of the word.

After this first annotation, the speaker turns were classified and extracted from their context. The first annotation, based on our intuition and our competence as French native speakers, allowed harmonizing the emotive charge of the sentences proposed to the appreciation of native speakers in the pretest presented below.

Table 1. Distribution of speakers turns based on annotations. N= NEUTRAL ATTITUDE; A1,...,A5 = ANGER; OE = OTHER EMOTION (OE).

Annotation	N	A1	A2	A3	A4	A5	OE	Total
Number of speaker turns	399	89	117	70	24	5	61	765
% of speaker turns	52	12	15	9	3	1	8	100

3   Perception Tests

Based on this corpus, we wanted to test how anger is actually perceived by the listeners. In order to do so, a set of perceptual tests was devised. First, the corpus had to be validated by showing that the chosen sentences are really perceived as conveying anger. Then, it was shown that listeners can differentiate spontaneous and acted anger. Finally, the role of prosody in the recognition of anger in a cross-linguistic study is pointed out.

3.1   Validation of the Corpus

To validate the corpus, a pretest was conducted. Five French native speakers were asked to decide if the sentences they heard were said with anger, a neutral attitude, or another emotion.

The first difficulty we had to face was the question of attributing a degree to the emotion conveyed by the sentences of the corpus. If the sentence was pronounced with anger, the subjects had to evaluate the degree of this emotion by graduating from one to five. A total of eighty-one sentences were chosen, five of which being training sentences. The subjects had to listen to the randomized eighty-one sentences. The beginning of each sentence was indicated by a specific sound and they had two seconds to accomplish their task. This pretest revealed that the choice between ANGER and OTHER EMOTION was difficult for the subjects. Particularly, it was difficult for them to conceptualize emotion (for instance, understanding what anger or emotion means), especially in a test situation. We therefore decided to eliminate the category of OTHER EMOTION for the perceptual test.

On the basis of the pretest, a set of sentences was selected: Only those sentences for which the judgments were clearly defined by a majority were kept. For instance, for ANGER, the sentences which scored at least 80 % of answers were kept. From the pretest, thirteen sentences judged as expressing ANGER and thirteen judged as expressing NO ANGER, as well as five training sentences were selected.

3.2   Spontaneous Versus Acted Speech

Almost all the studies about emotions are based on non-natural corpora, usually played by actors (see e.g. Scherer [2], Bänziger [3], Léon [4], Fónagy [5]). The sentences are chosen in order to be semantically neutral, built with written syntax. This type of methodology makes it possible to control the experimental parameters and the sound quality of the corpus. Nonetheless, more and more researchers are interested in studying spontaneous emotion and use corpora collated under natural conditions (see e.g. Campbell [6], [7], Douglas-Cowie [8]). They assume that working with natural speech is the best way to capture the reality of emotion. At this stage, however, it is not known whether there is a significant difference between both types of corpora. Would the difference between spontaneous and acted emotion be perceptible to listeners? And how would it be actually realised? Through prosodic or syntactic parameters? The stuy described in this section attempts to provide an answer to the first question, that is to say: Can we perceive the difference between natural and played emotion?

Generation of an Acted Corpus. In order to verify our hypothesis, we had to get natural and played corpora of emotion. We decided to work from the spontaneous French corpus described above. The acted corpus was built from the spontaneous one. Natural sentences from the corpus of spontaneous speech containing anger were extracted. In order to verify the influence of oral syntax and especially the influence of disfluencies on the perception of emotion (here anger) as natural, we modified the extracted sentences, removing all the disfluencies (such as pauses, repetitions, hesitations, and false starts) with the software *Soundforge 7.0*. Finally, the natural corpus contained the original sentences extracted from hoaxes as well as the same sentences but without the disfluencies.

Three French speakers were asked to read the sentences, acting anger: (1) With exactly the same segmental content as in the original ones. For that purpose, the actors were given exactly the same orthographical transcription of each sentence. The aim was to focus on the prosodic realizations. (2) With sentences having the same content but no disfluencies. The disfluencies were removed to check their influence on the perception of speech and emotion as natural or not. (3) With sentences reworked with written syntax. This was to see if written syntax can be an important cue of non natural emotion.

The French speakers were recorded on a Sony MD recorder in a quiet room in order to avoid any noise. Unfortunately, the recording conditions were not identical with those of the spontaneous corpus. It was considered that this could influence the results of our test. In order to neutralize these effects, the acted stimuli were reworked with Soundforge 7.0: The intensity was lowered until a good homogeneity of intensity between all the stimuli was obtained.

Subjects and Task. The subjects (ten French listeners) were asked decide if, in each of the sentences they listened to, the anger was real or acted. They listened to each sentence only once. They had five seconds to decide and answer. The test set consisted of: ten original sentences extracted from the corpus of hoaxes described above; ten original sentences without disfluencies; ten sentences acted from the exact transcription of the original sentences; ten sentences acted from the transcription of the original sentences without disfluencies and Ten sentences reworked in written syntax. Four training sentences were added. The subjects heard five sentences twice to verify the coherence of the answers.

Results. Table 2 shows the percentage of answers for spontaneous anger for each of the five types of stimuli. Original sentences (with and without disfluencies) have been well recognized at 87 % and 83 % as spontaneous anger. Acted sentences were not recognized as spontaneous anger. Note that acted sentences without disfluencies were perceived as more natural than the other acted sentences. Surprisingly, acted sentences from the original orthographical transcription have been perceived as the least natural. Disfluencies are characteristic of oral speech and it was expected that played sentences with disfluencies would seem more natural. The results do not confirm this hypothesis which might be explained by the fact that our speakers found it difficult to act the sentences with all the disfluencies, to adhere strictly to the marked text.

Table 2. Percentage of times subjects anwered with "spontaneous anger" for each type of stimulus

Original sentences	87 %
Original sentences without disfluencies	83 %
Acted sentences	23 %
Acted sentences without disfluencies	35 %
Acted sentences with written syntax	28 %

3.3   Crosslinguistic Tests on Prosody

With this third test, we wanted to show whether prosody is sufficient to recognize anger or not. Particularly, we were interested to find out about the universality and variability of the perception of anger. The universal character of the emotion is demonstrated if, whatever their mother tongue, the subjects can identify anger with prosodic information only. If universality is effectively demonstrated, it is then interesting to know to what extent the specific mother tongue of the listeners has an effect on the perception. This we refer to as variability.

Stimuli. In order to mask the lexical meaning, from which information about the presence and the intensity of an emotion could be deduced, the prosodic information had to be isolated from other linguistics parameters. One way to proceed is through re-synthesis. However, as we did not want to modify the spontaneous prosody of the sentences, re-synthesis did not represent a satisfying solution: The authentic value of the document would not have been preserved.

Another method to hide linguistic content is low-pass filtering. The problem with this method is that it eliminates the energy in high frequencies. Since energy in high frequencies is an important parameter of anger [9], we assumed that it is a decisive perceptive parameter that could not be dismissed from our analysis. The method of adding white noise to hide the linguistic content of sentences in order to keep only the prosodic information was thus adopted.

A white noise was added to each of the sentences selected after the pretest with the software *Soundforge 7.0*. The white noise created had the same length as the original sentence. The intensity of the white noise was defined according to the intensity of the speakers voice for each sentence. Sentences and white noise were mixed to create our stimuli.

It is worth mentioning that the constructed stimuli were perceived by the listeners as sounds of bad quality. This impressionistic perception can be compared to that obtained with a low-pass filtering which gives the impression of a damaged sound. We believe that adding white noise is a good way to control the effects of segmental information and to evaluate the part played by prosodic parameters in the perception of emotion.

The 26 chosen sentences of the stimuli were doubled in order to verify the coherence of the answers. Then these 52 sentences were randomized and preceded by 5 training sentences. The test which contains 57 stimuli is about 8 minutes long.

Subjects and Task. Three groups of listeners were invited to do the perceptual test: The first one, composed of 10 native speakers of French (6 women and 4 men), represented our control group. The Portuguese speakers, 7 women and 3 men, our test group, were students in the Faculdade de Letras, Universidade de Lisboa, from the same class. Their level of proficiency in French corresponds to B2 level, according to the European portfolio of languages [10].

The Czech speakers, 8 women and 2 men, are students of the University of West Bohemia of Pilsen. They also have a B2 level. It was important to get listeners of an intermediate level of French because we considered that beginners would not have enough knowledge to interpret the various degrees of anger, while advanced students would have too much knowledge about prosody to show significant results.

The task the subjects had to accomplish was double: The listeners had to decide if the stimulus conveyed anger or not, and if so to evaluate the degree of anger. They were previously advised that the quality of the sound was bad, in order to avoid the need for adaptation.

Interface. It is acknowledged that the way a test is presented to the subjects can have an important impact on the results of the test. Careful attention was paid to the directives given to the subjects in this test.

The crosslinguistic perception test was presented on a computer. The interface was written in HTML and was created in EasyPhp. Results were then extracted to a text format.

One of the advantages of the interface is its practical aspect since results can be automatically extracted. Moreover, the main advantage of this kind of interface is that the subjects are in a friendly, albeit restrictive, environment. As opposed to a paper format which would allow hesitations or modifications of choice with deletions, the electronic format allows just one choice restricted in time. We were also interested in testing the robustness of the presentation of such a test on the Internet. The interface was designed in order to control what speakers really do while taking the test: The number of "clicks" they perform, and the time allocated at each step. It was not possible, for instance, for the subjects to go back and redo a step[3]. It was not possible either to give an answer to the subsequent step without having given one for the previous one, and the duration of the step.

Fig. 1. Emonicons used in the study for (from left to right): Neutral, Anger, and "hear the sample" (Neutral state refers to effective neutral state and attitudes different from Anger)

Because of the cross linguistic aspect of this test, extra attention was given to the aspects of learning a foreign language while creating the interface. One main problem was to decide which language to use in the interface. Since the subjects were learners of French, they did understand French perfectly. Instructions were therefore given in their mother tongue so as to avoid any confusion about the task. As for the questions within the test itself, it was important to avoid cognitive overload due to constant alternation between languages. We finally proposed to use visually explicit instructions based on "emoticons" which are easily recognisable icons from their common usage on the internet. We used three emoticons that are depicted in Figure 1. This interface was created so that it could be used to test other languages and other types of emotion.

Results. The coherence of the answers given the first and the second time were verified using a Spearman correlation test on the Software *Statview 5.0.* Table 3 shows that results are significant ($p < 0.0001$). The answers of the three groups tested (white noise condition) were compared with the judgments of the control group (French without white noise). The three groups tend to the same results. The main judgment for all the stimuli and the three language groups is NO ANGER. However, the sum of the other answers is higher. STRONG ANGER (4 and 5) is rare, while MILD ANGER (1 and 2) appear twice as often. As expected, listeners answered NO ANGER more often than other choices, since half of the sentences do not express anger (according to the results of the pretest). Portuguese and Czech chose NO ANGER more often than French (50 % for Portuguese and 44 % for Czech vs. 37 % for French). The 13 ANGER sentences (based on the pretest) have been recognized as such by 84 % of French and Czech

---

[3] An identification number was attributed to each step so that we could check the correct progression of the test.

Table 3. Results of the Spearman correlation test

French (without white noise)	$p_9 = 0.763$	$p < 0.0001$
French (with white noise)	$p_9 = 0.848$	$p < 0.0001$
Portuguese FFL	$p_9 = 0.833$	$p < 0.0001$
Czech FFL	$p_9 = 0.691$	$p < 0.0001$

subjects and by 77 % of the Portuguese. In the light of these first results, we can conclude that prosody is sufficient to recognize anger.

Next we looked at the categorization of anger into degrees by the listeners. French subjects spread their responses more widely into the categories of anger 1, 2 and 3: They detect MILD ANGER (A1 and A2) more than Portuguese and Czech subjects. For A1 and A2, there is a 10 % difference between the French and Portuguese and a 9 % difference between the French and Czech.

These results suggest that French grade their answers more homogeneously in the mild anger categories (A1 and A2), whereas non native listeners (Czech and Portuguese) often perceive mild anger as no anger at all. Czech listeners judged stimuli as 'Strong Anger' (A4-A5) more often than French and Portuguese listeners (8 % and 6 % for A4 and A5 vs. 7 % and 4 %). It is thus fair to conclude that it is easier to judge STRONG ANGER than MILD ANGER for Portuguese and Czech learners of French as a Foreign Language (FFL).

Some t-tests showed there are some differences of perception within the three groups. For example, the difference between the Czechs answers and the French ones (with noise) for ANGER 1 is significant, p=0.0207. Czech listeners have a more approximate perception of the degrees of anger than the French native speakers. The difference between French listeners without noise and Portuguese for ANGER 3 and ANGER 4 is significant too (p=0.0338 and p=0.0301). This result indicates that semantic information probably had an impact on the answers of the French listeners during the test without noise for these categories. Some sentences judged as ANGER 3 and 4 were recategorized as ANGER 1 and 2 when listeners could not reach the semantic meaning any more. This result might indicate that when emotional information is semantically high, the role of prosody is less important to avoid redundancy.

4   Prosodic Analysis

The perceptual test made it possible to categorize anger sentences in degrees. We wanted to measure the acoustic parameters which characterize each degree of anger. To that purpose, all measures were taken from the original sentences without any modifications.

Among the abundant literature about the acoustic cues of emotion, we decided to follow the suggestions for measurements proposed by Bänzinger (2004) [3]. We focused on F0 parameters and rhythm phenomena. Intensity is a very interesting feature to study, but our corpus does not allow reliable measures of intensity:

The recordings circumstances were not controlled at all. Furthermore, the sound was filtered through two channels: telephone and radio channel.

4.1   Automatic Measures of F0

We proceeded to measure F0 automatically with the help of the software Win-PitchPro. This software takes into consideration the transcriptions and signal segmentations first made with Transcriber. WinPitchPro recognises all speakers created with Transcriber and treats them separately in specific layers. It analyzes fundamental frequency (F0 > 0 Hz) automatically. For each speaker and each category of annotations (see above Table 1), we took measures of F0 (sampling rate 20 ms). The measurements are then transferred to an Excel table. Some statistics were then performed, calculating the minimum, maximum, mean and range of F0 for each speaker and each category.

Mean F0. Figure 2 shows the results obtained by comparing the mean F0 for each category, based on the gender of speakers. The F0 mean values in Hertz have been transformed in semitones. As the graph shows, the F0 mean increases depending on the degree of anger. The stronger the anger is, the more the F0 increases. From one degree to another, there is a difference of one or two semitones. The increasing curve is interrupted for the female speakers, at the degree 3 of anger. A fall of both F0 minimum and F0 maximum for ANGER 3 is apparent which implies a fall of the mean. This difference may be due to the few utterances of ANGER 3 compared to ANGER 2.

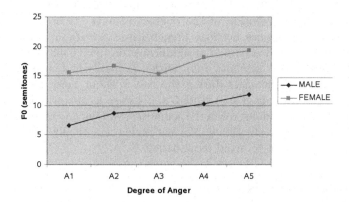

Fig. 2. F0 mean (semitones) shared out by degrees of anger and by gender

Intra-Speaker Measures of Mean F0. We wanted to see the differences of F0 mean between the neutral state and the various degrees of anger, and measure the increasing rate of F0. For each sentence recognized as ANGER by the French subjects of the perceptive test without white noise (see 3.3), we compared the F0 mean of the sentence to the F0 mean of the NEUTRAL speaker turns of the same speaker. Next, the ratio between the two values was calculated and converted into a percentage. The values obtained were assembled depending on the degrees of emotion. For some sentences, the subjects hesitated between two degrees, which explains the repartition of the categories in the graph below. It is worth noting that the listeners always hesitated between two consecutive degrees (for example ANGER 1 and 2, but not ANGER 1 and 3).

As shown in Table 4, in all the categories of anger an increase of mean F0 between ANGER and NEUTRAL can be observed in sentences uttered by the same speaker. Globally, the ratio increases with the degree of anger, which means that there is an increase of mean F0 between ANGER and NEUTRAL and that the increase is higher for STRONG ANGER than for MILD ANGER. The ratio goes from 18 % (A2 A3) to more than 60 % (A4 A5).

Table 4. Ratio (percentage) of mean F0 between NEUTRAL and various degrees of ANGER (grouped in pairs)

Anger Group	A1 A2	A2 A3	A3 A4	A4 A6
Ratio	29.6 %	17.6 %	45 %	65.5 %

4.2 Temporal and Rhythm Measures

Duration Measures. Duration of speaker's turns has been measured. Looking at Table 5 we noticed that the longer the speaker turn was, the less likely it was to be perceived as STRONG ANGER. The longest speaker turns were recognized as MILD ANGER (ANGER 1- ANGER 2), and the shortest ones as STRONG ANGER (ANGER 3-ANGER 4 and ANGER 4-ANGER 5). The presence of disfluencies in the longest speaker turns may also influence the perception of anger as milder. The sentences recognized as ANGER 4 and 5 (from the results of the evaluating test without white noise) contain high semantic information: Even if the duration of these speaker turns is more important than those of ANGER 3 and 4, the semantic information influences the judgment of the subjects.

Table 5. Duration of speaker turns

Anger Group	A1 A2	A2 A3	A3 A4	A4 A6
Duration	6.1	4.7	3.1	4.3

Speech and Pronunciation Rates. The speech and pronunciation rates are presented in Figure 3. Speech rate is defined as the number of syllables uttered in one second of speech, containing pauses, disfluencies etc. Pronunciation rate is measured as the number of syllables in a speech segment without pauses and interruptions.

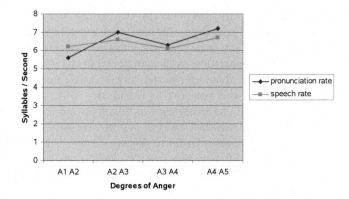

Fig. 3. Pronunciation and speech rates shared out by categories of anger

Pronunciation and speech rates tend to increase with the degrees of anger. The stronger the anger is, the more the speech and pronunciation rates increase, that is to say the faster the speakers speak. But there is a break point on the increasing curves of speech and pronunciation rate for the categories A3 A4.

4.3   Example Sentences

We made a detailed analysis of the sentences. There are some interesting intonation curves, corresponding to the ANGER patterns as defined by Lon [4]. Figure 4 shows an intonation curve obtained with WinPitchPro for a sentence perceived as STRONG ANGER with a score of 84 %. It has been pronounced by a male speaker. There are steep rises and falls. The F0 range for the whole sentence is 237 Hz (about 4 semi-tones). On some syllables, the F0 range is very important, about 100 Hz. Here, on the syllable "con", the average F0 of the syllable is 306 Hz, whereas the average of the sentence is 212 Hz, a difference of 6 semi-tones (94 Hz). The speakers register is very high, from 88Hz to 318 Hz (22 semi-tones) for the same sentence.

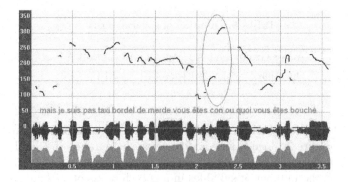

Fig. 4. Intonation curve for a strong ANGER-sentence (WinPitchPro)

Some relevant acoustic features for the analysis of our corpus were checked. The increase of F0 and rhythm rates, depending of the degree of anger perceived, has been verified. A global analysis of our corpus was performed with the help of automatic measurements. Other parameters such as intensity require further investigation in order to analyze fully the way anger is manifested in speech.

5 Conclusions

In this paper, we proposed and described an experimental protocol for the treatment and analysis of emotion in a natural speech corpus. We first evaluated how much emotion was conveyed in our corpus. Then, we made a set of perceptual tests which yielded some interesting results. There is a perceptive difference between played-acted and natural emotional speech. Listeners are clearly able to distinguish between them, notwithstanding the type and the content of the sentences.

Prosody is sufficient to allow listeners to give an appropriate evaluation of the emotion expressed throughout the corpus. In a cross linguistic perspective, the distinction between ANGER and NO ANGER is common to the three language groups tested (Czech, Portuguese and French). Differences appear between the three groups in the evaluation of the degree of Anger (how angry is the person?).

Finally, we tried to describe some prosodic parameters. We showed that F0 average is a good parameter to distinguish anger from a neutral state, and also to distinguish the degrees of anger themselves. We showed that duration of speaker turns and rhythm (i.e. speech and pronunciation rates) had an effect on the perception of the degrees of anger. We then proposed some examples of sentences recognized as Anger with a brief analysis. Further work is required on the parameter of intensity. Work is in progress to finalize the F0 analysis of the syllables. As to the perceptual part of our project, it would be interesting to compare different ways of masking the signal and correlate the results.

Acknowledgments. The authors would like to thank all the subjects who accepted to be submitted to the tests. Special thanks to J.Y. Dommergues for his good advises, to Nicolas Créplet for his help for the interface and to Candide Simard for the time she spent to correct this article.

References

1. Kerbrat-Orecchioni, C.: Les interactions verbales. Armand-Colin, Paris, France (1990)
2. Scherer, K.R.: How emotion is expressed in speech and singing. In: Proceedings of ICPhS 95. Stockholm, vol. 3, pp. 90–96 (1995)
3. Bänziger, T.: In: Communication vocale des émotions. Thèse, Geneva (2004)
4. Léon, P.: Précis de phonostylistique: Parole et expressivité. Nathan, Paris, France (1993)
5. Fnagy, I.: La vive voix. Essais de psychophonétique. Payot, Paris (1983)

6. Campbell, N.: The recording of emotional speech (JST/CREST database research) In: Proceedings from COCOSDA Workshop, Taejon, Korea (2001)
7. Campbell, N.: Databases of expressive speech (JST/CREST database research) In: Proceedings from COCOSDA Workshop, Singapore (2003)
8. Douglas-Cowie, E., Cowie, R., Schröder, M.: A new emotion database: considerations, sources and scope. In: Proceedings from ISCA Workshop 2000, Newcastle, North Ireland (2000)
9. Miller, L.: The intelligibility of interrupted speech. Journal of the Acoustic Society of America 22, 167–173 (1950)
10. de l'Europe, C.: Portfolio Européen des Langues (2007), http://www.enpc.fr/fr/international/eleves_etrangers/portfolio.pdf
11. Bänziger, T., Scherer, K.R.: Relations entre caractéristiques vocales perues et émotions attribuées. In: Actes des Journées Prosodie, Grenoble, France, pp. 119–124 (2003)
12. Bänziger, T., Grandjean, D., Bernard, P.J., Klasmeyer, G., Scherer, K.: Prosodie de l'émotion: étude de l'encodage et du décodage. Cahiers de Linguistique franaise 23, 11–37 (2001)

Effects of the Phonological Contents on Perceptual Speaker Identification*

Kanae Amino[1], Takayuki Arai[1], and Tsutomu Sugawara[2]

[1] Department of Electrical and Electronics Engineering
[2] Graduate Division of Foreign Studies, Sophia University,
7-1 Kioi-cho, Chiyoda-ku, 102-8554, Tokyo, Japan
{amino-k, arai, sugawara}@sophia.ac.jp

Abstract. It is known that the accuracy of perceptual speaker identification is dependent on the stimulus contents presented to the subjects. Two experiments were conducted in order to find out the effective sounds and to investigate the effects of the syllable structures on familiar speaker identification. The results showed that the nasal sounds were effective for identifying the speakers both in onset and coda positions, and coronal sounds were more effective than labial counterparts. The onset consonants were found to be important, and the identification accuracy was degraded in onsetless structures.

Keywords: Perceptual speaker identification, Familiar speaker identification, Nasal sounds, Coronal sounds.

1 Introduction

It is manifest that human beings have the innate ability to recognise speakers by speech sounds alone. This means that speech sounds convey information about the speakers as well as the linguistic contents.

Speech individualities, or speaker-specific characteristics contained in speech sounds, derive either from physiological properties of the speaker or from his/her learned habits. The former includes the length or the thickness of the vocal folds, the length or the volume of the vocal tract, and all other physical properties of a given speaker, and the latter is exemplified by speaking style, speaking rate, and social and regional dialects. The modality of an utterance and the articulatory disorders may also be included among the learned habits.

The term "speaker individuality," called "voice quality" in a broad sense in some studies, can be used to refer to "a quasi-permanent quality running through all the sound that issues from a speaker's mouth [1]," and the "characteristic auditory

---

* This work was originally presented at the 9th European Conference on Speech Communication and Technology Interspeech 2005 (Experiment 1) and at the International Workshop on Frontiers in Speech and Hearing Research 2006 (Experiment 2). For details see references [23] and [24].

C. Müller (Ed.): Speaker Classification II, LNAI 4441, pp. 83–92, 2007.
© Springer-Verlag Berlin Heidelberg 2007

colouring of a given speaker's voice [2]." Ball and Rahilly [3] regard this quality as being responsible for human identification of a speaker or a group of speakers.

In forensic speech sciences, it is important to know about the relationship between speech individualities and how people perceive them. The use of speech materials in court cases has a relatively long history since 1660 [4], though there are still some controversial issues remaining, such as the correspondences between acoustic properties of the sounds and the human percepts, the limits of the human memory, and the effects of the recording conditions and transmission.

If we could find acoustic correlates of speaker individuality, they can be exploited for various fields in speech technology [5]. For example, speaker individuality is extracted and used in automatic speaker recognition and in voice conversion [6, 7]. In automatic speech recognition, on the other hand, speech individuality should be excluded. One way to find the acoustic parameters that indicate speech individuality is to conduct a speaker identification experiment by listening, and to investigate the property of human perception [8]. The factors that affect the perception would be important in defining speech individuality.

When identifying a speaker, listeners abstract speaker-specific characteristics of the utterance, and collate them with the information stored in the brain. It is reported that the processing of speech contents and that of speaker identity occur separately, though they interact with each other [9, 10]. Also, it is also pointed out that listeners use linguistic information in order to identify the speakers, and vice versa [9, 11].

One example of the interaction between the linguistic information and the speaker information is that different speech sounds are more or less effective for perceptual speaker identification [12]. This means that the accurateness of the identification depends on what sounds are presented to the listeners. In previous studies, sonorants, such as vowels and voiced consonants, were reported to be effective for perceptual speaker identification [13-19]. Especially, vowels and nasals are found to be effective [13, 14]. The same results were obtained in automatic speaker recognition tests [20-22].

Investigating the differential effects of the sounds on speaker identification enables us to know about the effects of the physical and physiological speaker variations in speech production, and at the same time, leads to a better understanding of human cognition. In this present study, we carried out two experiments in order to see the differences among the sounds in perceptual speaker identification tests. In the first experiment [23], the differences among the onset consonants in monosyllables which were excerpted from sentences were inspected, and in the second experiment [24], the stimuli with various syllabic structures were compared. The results showed that nasals and coronals in the onset position were effective for the identification.

2   Experiment 1

2.1 Methodology

Speakers and subjects. In human speaker identification tests, the selection of speakers and subjects is one of the most important and difficult tasks. The size of the speaker ensemble is concerned with the difficulty of the test task, and a homogeneous subject group is also necessary for reliable data. Also the speakers' ages, genders and accents must be consistent [12].

Taking these things into account, we selected ten male speakers and five male subjects. All of them were undergraduate students at Sophia University, and they had lived in the same dormitory for more than four years. They were all native speakers of Japanese and none of them had hearing impairments.

Speech materials. The speech materials used in this study were CV Japanese monosyllables excerpted from the carrier sentences. The onset consonants were six oral consonants articulated at coronal area, and three nasal consonants /m/, /n/, and /ɲ/. The vowel was controlled to be /a/, because this vowel was the most effective vowel for speaker identification in previous studies [12-14, 16, 18], and also for making the experiment simple.

The recording sessions were held in a soundproof room. The speakers uttered the sentences shown in Table 1. All the materials were recorded onto a digital audiotape at the sampling frequency of 48 kHz with 16 bit resolution, using the electret-condenser microphone (SONY, ECM-MS957) and DAT recorder (SONY, TCD-D8).

The uttered sentences were /'aCaCaCa' to: o ʃiʒi ʃimasɯ̥/, as shown in Table 1. The carrier sentence means "I support 'aCaCaCa' (political) party," and the first four syllables, /'aCaCaCa'/, are the names of the fictional political parties. The symbol "a" is the Japanese /a/, and "C" stands for one of the following consonants: /t/, /d/, /s/, /z/, /r/, /j/, /m/, /n/ and /ɲ/. The reason for using the names of parties is that the suffix "/-to:/ (party)" forms compound words that do not have accentual nucleus (pitch fall) [25], thus /'aCaCaCa'/ is uttered with relatively stable accent pattern in the last two syllables.

The fourth syllables of the uttered sentences were manually excerpted for making the stimuli presented to the listeners. The excerption was conducted based on the waveform, using the computer software Cool Edit Ver.96 (Syntrillium Software Corporation).

Five tokens for each consonant were selected to be used in the test, and the total number of the stimuli was 450, i.e. corresponding to five tokens, nine consonants and ten speakers. The speech samples were randomly presented to the subjects, and a 500 ms portion of white noise was inserted before each stimulus in order to degrade the auditory memory of the preceding stimulus [26].

Procedures. The experiment was also conducted in the soundproof room. The subjects listened to the stimuli through binaural headphones (SONY, MDR-Z400) at a comfortable loudness level.

The subjects were first informed of the names of the ten speakers, listened to several sample files as to each speaker, and practised the task by use of these samples. These files were different from the samples used in the actual test, and the subjects listened to them and practised only once. During the test, they were told to write the name of the speaker on the answer sheets for each stimulus. They took breaks after every 150 trials, and the total test time was about 40 minutes.

Table 1. List of the recorded sentences in Experiment 1. Combinations of various consonants and the vowel /a/ were read in the carrier sentence.

/aCaCaCa/	+ carrier sentence
/a.ta.ta.ta/	
/a.da.da.da/	
/a.sa.sa.sa/	
/a.za.za.za/	/toː o ʃiʒi ʃimasɯ/
/a.ɾa.ɾa.ɾa/	
/a.ja.ja.ja/	
/a.ma.ma.ma/	
/a.na.na.na/	
/a.ɲa.ɲa.ɲa/	

2.2 Results

The results of the identification test are shown in Table 2. Just as with the results in the previous experiments [13, 14], the nasals are the most effective sounds for the identification of the speakers, followed by the fricatives and the oral stops. Moreover, in the voiceless-voiced pairs of the same places and manners of articulation, /ta/-/da/ and /sa/-/za/, the tendency was seen that the voiced sounds obtained higher scores than the voiceless counterparts. This tendency was also reported in the previous studies [13, 14, 18, 21].

In the statistical analyses, the differences among the consonants were not significant in ANOVA. In t-test, the difference between the nasal and the oral sounds was significant ($p = 0.0044$). There were no other pairs that differed significantly in t-test: for example, the pairs like oral stops-fricatives ($p = 0.25$), obstruents-sonorants ($p = 0.15$), and voiced-voiceless ($p = 0.36$).

Table 2. Identification results for each stimulus. The number of the correct answers (centre column) and the percent correct (right column) are shown. The number of samples for each stimulus (the denominator) is 250.

Stimulus	Percent Correct (%)
/na/	86.0 (215/250)
/ɲa/	85.6 (214/250)
/ma/ /za/	80.8 (202/250)
/sa/	78.8 (197/250)
/ja/	78.4 (196/250)
/da/	78.0 (195/250)
/ɾa/	74.4 (186/250)
/ta/	73.6 (184/250)

3 Experiment 2

3.1 Methodology

In Experiment 1, the effectiveness of the nasals in monosyllabic stimuli was found out. However, only the nasals in the onset position were examined, and the nasals in the coda position were not dealt with. Moreover, the stimuli had an onset consonant followed by a nucleus vowel and therefore the effects of the vowel part or the transition to the following vowel were not inspected.

In Experiment 2, we carried out another speaker identification test in order to investigate the effects of the syllable structures and the contributions of the onset consonant and the transition to the vowel to the speaker identification.

Speakers and speech materials. Eight male students in the age range 22-25 years old (average 23.1 years old) served as the speakers in this experiment. All of them spoke Tokyo Japanese as their native language and had normal hearing.

The recording procedure was exactly the same as in Experiment 1. As shown in Table 3, the speech materials used in this experiment were Japanese non-sense monosyllables of various structures. In order to see how the syllable structures and coda nasals work in the identification of the speakers, the materials covered the following structure types: V, VV, VN, CV, CVV and CVN. This variety of structures enables us to know the influence of the onset consonants, syllable weight and the coda nasals. The speakers read out each kind of material seven times and five of them were selected and used as the stimuli.

In order to examine the contribution of the consonant-to-vowel transitions, we prepared two more structures, –V and –VN, which were cut out from recorded CN and CVN. These two types were edited manually on the computer, using the software Praat [27]. The onset consonants were cut off just before the visible transitions of the second formant of the following vowel began on spectrograms. Thus, the stimuli –V and –VN contained the transition parts to the nucleus vowel. We will indicate it by the notation '–.'

Subjects. Eight students, two males and six females, who belonged to the same research group as the speakers participated in the experiment. They had spent at least one year with the speakers and knew all of the speakers very well. The mean age was 23.1 years old and they were all native speakers of Japanese. None of them had any known hearing impairment.

Procedures. The procedure of the second perception test was almost the same as in the test in Experiment 1 except that the test sessions were performed on a computer. The subjects listened to the test sample, identified the speaker, and then answered by clicking on a rectangle with the name of the speaker to whom s/he thought the speech belonged.

The total number of the test stimuli was 920, i.e. corresponding to 8 speakers, 23 stimuli and 5 different tokens for each stimulus. The total test time was about an hour, and the subjects took breaks after every 230 trials.

Table 3. List of the stimuli used in Experiment 2. Consonants in the parentheses were cut off manually from corresponding samples in CV and CVN.

Syllable structure	Stimuli
V	/a/
VV	/aa/
VN	/aɴ/
CV	/ba/ /da/ /ma/ /na/
CVV	/baa/ /daa/ /maa/ /naa/
CVN	/baɴ/ /daɴ/ /maɴ/ /naɴ/
–V	/(b)-a/ /(d)-a/ /(m)-a/ /(n)-a/*
–VN	/(b)-aɴ/ /(d)-aɴ/ /(m)-aɴ/ /(n)-aɴ/*

3.2 Results

The results of the perception test are summarised according to the syllable structures in Figure 1 and to the onset consonants in Figure 2.

Figure 1 shows that the structures with an onset consonant (shown by black bars) gained higher scores than onsetless structures (grey and striped bars). It also tells us that there is a tendency that the heavier syllables obtained better scores except in /VN/. Coda nasals also seem to be effective for the identification in /CVN/ and /–VN/. As to the influence of the transition, we cannot tell many things only from the results of this study, but the scores of the edited syllables, /–V/ and /–VN/, did not reach those of the structures with an onset.

It is affirmed again in Figure 2 that onset consonants are important. The data here do not include the results of the edited structures. The letter φ indicates the onsetless syllables, /V/, /VV/, and /VN/. The scores of these onsetless syllables were the worst of all structures, though it still gained more than 90 % correct identification.

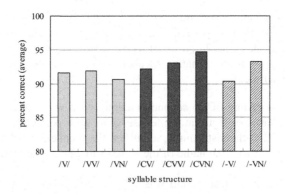

Fig. 1. Percentages of correct speaker identification (as to the syllable structure)

One can also see in Figure 2 that the alveolar consonants in the onset position were more effective than the bilabial consonants in the test. Nasal consonants, /n/ and /m/, were better than their oral counterparts, /d/ and /b/, respectively.

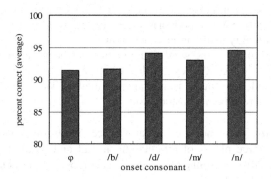

Fig. 2. Percentages of correct speaker identification (as to the onset consonant)

4   General Discussion

In this study, two perception experiments were conducted in order to investigate the differential effects among the stimulus contents on familiar speaker identification by listening.

In Experiment 1, ten male speakers were identified by five listeners who knew the speakers very well, and the identification rate was the highest when the stimuli containing the nasal sounds were presented. The difference between the nasal stimuli and the oral stimuli was significant. In Experiment 2, eight male speakers were identified by eight familiar subjects, and the effects of the syllable structures were examined. The results showed that the structures with onset consonant obtained higher identification scores than onsetless structures, and coda nasals were found to be effective for speaker identification. Furthermore, the identification was more accurate when the onset consonant was one of the nasals, and also the coronal consonants were better than the labial consonants.

In summary, the results in this study yield the following conclusions:

- Onset consonants are important for perceptual speaker identification.
- Nasals are effective for speaker identification both in onset and coda positions.
- Coronal consonants convey more individuality than labial consonants.

Onset consonants. The structures with transition or the onsetless structures in this study gained no higher identification rates. This suggests that the differential effects in the onset consonants come from the consonant parts of the stimuli.

Nasals. The properties of the nasal sounds are speaker-dependent, because the shapes of the resonators involved in the articulations of these sounds are considerably different for individuals [28]. In addition, the shapes of these resonators cannot be changed voluntarily. This means that the resonance properties of nasals rarely change.

The nucleus vowel in the structure that has a nasal sound in onset or coda, or both, position(s) is necessarily nasalised to some extent. This nasalisation process occurs especially in the structure with a coda nasal, and the nasalised vowels are predicted to contain more individuality than non-nasalised vowels.

Japanese coda nasal /N/ in the word-final position has been said to be articulated at the uvula, but recent work [29] reports that the place of articulation of /N/ differs among speakers, and this sound is not always uvular. This variation among the speakers explains the results in this study, too.

Coronal consonants. This tendency is what was seen in our previous experiment, too [14]. Japanese has three places of articulation in oral and nasal stops, i.e. bilabial, alveolar and velar. Alveolar sounds have the largest range of possible articulation of these three, as the phonology of Japanese does not require any contrasts in place feature in the coronal area among the stop sounds. This may lead to inter-speaker differences in articulation of alveolar sounds.

Phonological unmarkedness and speaker individuality. It is interesting that the effective structures or the sounds shown above are all linguistically unmarked ones. The structures with onset consonants are universally unmarked [30]; vowels and nasals are the sounds that children learn in the early stage of language acquisition. There is also a typologically universal tendency that coronal is the dominant place of articulation compared to labial and guttural [31].

The relationship between the linguistically unmarked structures/sounds and speaker individuality has not been clarified yet, but the obtained results here imply that unmarked structures/sounds are more effective for perceptual speaker identification than other structures/sounds. One reason for this may be that unmarked sounds occur more frequently in natural language.

The final goal of our study is to delimit the speaker individuality conveyed by speech sounds and to understand the interaction between human perception of the speaker individuality and the linguistic information.

Our future task will be to look into the acoustic characteristics of the stimuli used in this study, and to show quantitative data for explaining the effectiveness of the phonologically unmarked sounds. We must also test on different kinds of vowels, in order to examine the effects of co-articulation. Speaker identification experiments with reversed speech may also be useful for revealing the properties of human perception. As to the nasal sounds, the acoustic properties are inevitably degraded by flu or other diseases in the supra-laryngeal part, and the study of the influences of these factors will be one of our future tasks.

Acknowledgments. This work was supported by a Grant-in-Aid for Scientific Research (A) 16203041, and by a Grant-in-Aid for JSPS Fellows (17·6901).

References

1. Abercrombie, D.: Elements of General Phonetics. Edinburgh Univ. Press, Edinburgh (1967)
2. Laver, J.: The Phonetic Description of Voice Quality. Cambridge Univ. Press, Cambridge (1980)
3. Ball, M., Rahilly, J.: Phonetics. Arnold, London (1999)
4. Hollien, H.: The Acoustics of Crime. Plenum, New York (1990)
5. Furui, S.: Acoustic and Speech Engineering. Kindai Kagaku-sha Publishing Company, Tokyo (1992)
6. Hashimoto, M., Kitagawa, S., Higuchi, N.: Quantitative Analysis of Acoustic Features Affecting Speaker Identification. J. Acoust. Soc. Jpn. 54(3), 169–178 (1998)
7. Kuwabara, H., Sagisaka, Y.: Acoustic Characteristics of Speaker Individuality: Control and Conversion. Speech Com. 16, 165–173 (1995)
8. O'Shaughnessy, D.: Speech Communications –Human and Machine–. 2nd edn. Addison-Wesley Publishing Company, New York (2000)
9. Nygaard, L.: Perceptual Integration of Linguistic and Nonlinguistic Properties of Speech. In: Pisoni, D., Remez, R. (eds.) The Handbook of Speech Perception, pp. 390–413. Blackwell Publishing, Oxford (2005)
10. Kreiman, J., van Lacker, D., Gerratt, B.: Perception of Voice Quality. In: Pisoni, D., Remez, R. (eds.) The Handbook of Speech Perception, pp. 338–362. Blackwell Publishing, Oxford (2005)
11. Goggin, J., Thompson, C., Strube, G., Simental, L.: The Role of Language Familiarity in Voice Identification. Memory and Cognition 19, 448–458 (1991)
12. Bricker, P., Pruzansky, S.: Speaker Recognition. In: Lass, N. (ed.) Experimental Phonetics, pp. 295–326. Academic Press, London (1976)
13. Amino, K.: The Characteristics of the Japanese Phonemes in Speaker Identification. Proc. Sophia Univ. Ling. Soc. 18, 32–43 (2003)
14. Amino, K.: Properties of the Japanese Phonemes in Aural Speaker Identification. Tech. Rep. IEICE 104(149), 49–54 (2004)
15. Bricker, P., Pruzansky, S.: Effects of Stimulus Content and Duration on Talker Identification. J. Acoust. Soc. Am. 40(6), 1441–1449 (1966)
16. Kitamura, T., Akagi, M.: Speaker Individualities in Speech Spectral Envelopes. J. Acoust. Soc. Jpn (E) 16(5), 283–289 (1995)
17. Matsui, T., Pollack, I., Furui, S.: Perception of Voice Individuality Using Syllables in Continuous Speech. In: Proc. Autumn Meet. Acoust. Soc. Jpn, pp. 379–380 (1993)
18. Nishio, T.: Can We Recognise People by Their Voices? Gengo-Seikatsu 158, 36–42 (1964)
19. Stevens, K., Williams, C., Carbonell, J., Woods, B.: Speaker Authentication and Identification: A Comparison of Spectrographic and Auditory Presentations of Speech Material. J. Acoust. Soc. Am. 44(6), 1596–1607 (1968)
20. Nakagawa, S., Sakai, T.: Feature Analyses of Japanese Phonetic Spectra and Consideration on Speech Recognition and Speaker Identification. J. Acoust. Soc. Jpn. 35(3), 111–117 (1979)
21. Ramishvili, G.: Automatic Voice Recognition. Engineering Cybernetics 5, 84–90 (1966)
22. Sambur, M.: Selection of Acoustic Features for Speaker Identification. Proc. IEEE Trans. ASSP 23(2), 176–182 (1975)

23. Amino, K., Sugawara, T., Arai, T.: Correspondences between the Perception of the Speaker Individualities Contained in Speech Sounds and Their Acoustic Properties. In: Proc. Interspeech, pp. 2025–2028 (2005)
24. Amino, K., Sugawara, T., Arai, T.: Effects of the Syllable Structure on Perceptual Speaker Identification. IEICE Tech. Rep. 105(685), 109–114 (2006)
25. Kindaichi, H., Akinaga, K. (eds.): Meikai Japanese Accent Dictionary, 2nd edn. Sanseido, Tokyo (1981)
26. Repp, B., Healy, A., Crowder, R.: Categories and Context in the Perception of Isolated Steady-State Vowels. J. of Exp. Psychol.: Human Perc. Perf. 5(1), 129–145 (1979)
27. Boersma, P., Weenik, D.: Praat Doing Phonetics by Computer. Ver.4.3.14 (Computer Program) (2005), retrieved from http://www.praat.org/
28. Dang, J., Honda, K.: Acoustic Characteristics of the Human Paranasal Sinuses Derived from Transmission Characteristic Measurement and Morphological Observation. J. Acoust. Soc. Am. 100(5), 3374–3383 (1996)
29. Hashi, M., Sugawara, A., Miura, T., Daimon, S., Takakura, Y., Hayashi, R.: Articulatory Variability of Japanese Moraic-Nasal. In: Proc. Autumn Meet. Acoust. Soc. Jpn, pp. 411–412 (2005)
30. Spencer, A.: Phonology. Blackwell, Oxford (1996)
31. Whaley, L.: Introduction to Typology. Sage Publications, London (1997)

Durations of Context-Dependent Phonemes: A New Feature in Speaker Verification

Charl Johannes van Heerden[1,2] and Etienne Barnard[1,2]

[1] University of Pretoria, Pretoria Gauteng, South Africa
cvheerden@csir.co.za
http://www.ee.up.ac.za
[2] Human Language Technology Group, Meraka Institute, CSIR, Meiring Naude Rd,
Brumeria, Pretoria Gauteng, South Africa
ebarnard@csir.co.za
http://www.meraka.org.za

Abstract. We present a text-dependent speaker verification system based on Hidden Markov Models. A set of features, based on the temporal duration of context-dependent phonemes, is used in order to distinguish amongst speakers. Our approach was tested using the YOHO corpus; it was found that the HMM-based system achieved an equal error rate (EER) of 0.68% using conventional (acoustic) features and an EER of 0.32% when the time features were combined with the acoustic features. This compares well with state-of-the-art results on the same test, and shows the value of the temporal features for speaker verification. These features may also be useful for other purposes, such as the detection of replay attacks, or for improving the robustness of speaker-verification systems to channel or speaker variations. Our results confirm earlier findings obtained on text-independent speaker recognition [1] and text-dependent speaker verification [2] tasks, and contain a number of suggestions on further possible improvements.

Keywords: Speaker verification, triphones, time durations, Hidden Markov Models.

1 Introduction

Speaker verification (SV) is a widely-used biometric, and is useful in several circumstances – e.g. for multilevel access control to prevent unauthorized individuals from gaining access to high security systems [3]. SV is not considered entirely secure and there are several problems which limit the accuracy of such systems, such as noise on telephone channels [4,5], good mimics, recordings of valid speakers' voices etc. To address some of these issues, a new class of features based on temporal information in spoken utterances was proposed in [1] (for text-independent speaker recognition) and [2] (for text-dependent speaker verification). In [2] preliminary tests demonstrated the value of these features in addressing problems due to noise and recordings. The database of speakers in [2] was very small and the claims of temporal information improving the equal

C. Müller (Ed.): Speaker Classification II, LNAI 4441, pp. 93–103, 2007.
© Springer-Verlag Berlin Heidelberg 2007

error rate (EER) of SV systems had to be verified on a larger corpus of data. Here, we report on a set of experiments using the YOHO corpus, which has been used widely to evaluate SV systems [6] and has a structured methodology for performing comparative tests [7]. We then discuss a number of ways in which more sophisticated models may be used to further enhance the accuracy of the duration model, and show preliminary results indicating that such a model may indeed provide a more accurate model of phoneme durations.

2   The YOHO Corpus

The YOHO corpus is a large supervised speaker verification database [6]. It consists of 138 speakers (106 males and 32 females) who spoke prompted utterances from a restricted grammar set of 56 two-digit numbers ranging from 21-97 [8]. The utterances comprised combination-lock phrases (e.g. $21 - 38 - 44$) as proposed by [8]. Four such phrases were prompted during a verification session and 10 such phrases for a training/enrollment session. The YOHO corpus has 4 enrollment sessions per speaker and 10 verification sessions. The data was recorded with a 3.8 kHz bandwidth in an office environment with normal background noise.

3   Testing Procedure

3.1   Background

In order to compare the performance of our proposed speaker verification system to that of other speaker verification systems, a standard testing procedure was employed, similar to that used by others on the same corpus (see [9], [10], [11]). The exact test procedure is most clearly described by Reynolds [9].

Table 1 summarizes the results of several different tests that were performed by Reynolds[9] on the YOHO corpus. (In this table, *msc* denotes "maximally-spread close" and *msf* "maximally-spread far"; these are two different approaches to selecting cohort speakers – see below.) The test *M+F(10 msc)* was used as basis for our comparison, the only difference being that all four enrollment sessions were used for enrolling the speakers. (Reynolds used the fourth session for cohort selection).

Table 1. Equal error rates reported in [9] for different experimental conditions

Test	YOHO(eer)
M(10 msc)	0.20
M(5 msc, 5 msf)	0.28
F(10 msc)	1.88
F(5 msc, 5 msf)	1.57
M+F(10 msc)	0.58
M+F(5 msc, 5msf)	0.51

In order to perform comparable tests using the temporal features, we had to adapt the use of cohorts for score normalization. A cohort set is a small selection of speakers other than the true speaker, which are used to normalize the speaker's score. That is, to determine whether the true speaker $(Pr(\lambda_c|X))$ or an impostor $(Pr(\lambda_{\bar{c}}|X))$ is speaking, we compute the likelihood ratio:

$$likelihoodratio = \frac{Pr(\lambda_c|X)}{Pr(\lambda_{\bar{c}}|X)} \qquad (1)$$

In (1) X denotes the spoken utterance, λ_c the claimed speaker model and $\lambda_{\bar{c}}$ the cohort (also known as background or impostor) model. By applying Bayes' rule and working in the log domain, (1) can be rewritten as

$$\Lambda(X) = \log p(X|\lambda_c) - \log p(X|\lambda_{\bar{c}}) \qquad (2)$$

The speaker is accepted as the claimed speaker if $\Lambda(X) > \theta$ and rejected as an impostor if $\Lambda(X) < \theta$ where θ is an appropriate threshold [9]. θ can be speaker specific (which is computationally more expensive, but also more accurate) or global. The determination of the EER in our test used a global threshold approach, as in [9].

This standard approach to normalization works well if only one type of feature is employed. However, the choice of cohort speakers dictates a group of speakers that cannot be tested as possible impostors, which complicates the procedure when a second feature set is to be used. (If the cohort speakers are based on acoustic features only, they will not necessarily be a good model when using the time feature.) We therefore chose to normalize the temporal features using a universal background model (UBM) rather than a cohort set.

3.2   Detailed Test Description

The HTK 3.2.1 toolkit [12] was used to construct the speaker verification system. MFCCs were used as input features together with delta and acceleration coefficients. HMMs with one Gaussian mixture per state were created for all context-dependent triphones occurring in the restricted grammar set.

A cohort set of 10 speakers were selected for every speaker in the database in accordance with the procedure in[9]. Choices that arise with background speakers are the choice of specific speakers and the number of speakers to employ. The selection can be viewed from two different points of view. Firstly, the background set can be chosen in order to represent impostors that sound similar to the speaker, referred to as dedicated impostors [9]. Another approach is to select a random set of speakers as the background set, thus expecting casual impostors who will try to represent a speaker without consideration of sex or acoustic similarity. By selecting the dedicated impostor background set, in contrast, the system may be vulnerable to speakers who sound very different from the claimed speaker [8].

The selection of the background set was done on a per speaker basis and it was decided to use the dedicated impostor approach. First the $N = 20$ closest speakers to a speaker were determined using pair-wise distances between the speaker and all others. The pair-wise distance between speakers i and j with corresponding models λ_i and λ_j is

$$d(\lambda_i, \lambda_j) = \log \frac{p(X_i|\lambda_i)}{p(X_i|\lambda_j)} + \log \frac{p(X_j|\lambda_j)}{p(X_j|\lambda_i)}, \tag{3}$$

where $\frac{p(X_i|\lambda_i)}{p(X_i|\lambda_j)}$ is a measure of how well speaker i scores with his/her own model relative to how well speaker j scores with speaker i's model. The ratio becomes smaller as the match improves.

These N speakers are known as the close cohort set, which is denoted by $C(i)$ for speaker i. The final background set consists of the $B = 10$ maximally spread speakers from $C(i)$, denoted $B(i)$. To determine $B(i)$, the closest speaker to i is moved to $B(i)$ and B' is set to 1 (1 speaker in the background set). The next speaker c from those left in $C(i)$ to be moved to $B(i)$ is then selected as

$$c = \arg \max_{c \in C(i)} \left\{ \frac{1}{B'} \sum_{b \in B(i)} \frac{d(\lambda_b, \lambda_c)}{d(\lambda_i, \lambda_c)} \right\} \tag{4}$$

This procedure is repeated until $B' = B$. According to [9], the maximal spread constraint is to prevent "duplicate" speakers from being in the cohort set.

For speaker i, all other speakers (excluding i's cohort set of 10 speakers) were then used as impostors and tested using (2). Speaker i's verification data was also tested using (2), resulting in 1270 impostor attacks and 10 true attempts to gain access to the system (since every speaker has 10 verification sessions). This process was repeated for all speakers in the corpus, resulting in 175260 impostor attacks and 1380 true attempts.

In particular, (2) was evaluated as follows using the cohort set and the claimed speaker model: First, $\log p(X|\lambda_c)$ was evaluated as

$$\log p(X|\lambda_c) = \frac{1}{T} \sum_{t=1}^{T} \log p(x_t|\lambda_c), \tag{5}$$

where T is the number of frames in the utterance and $\frac{1}{T}$ is used to normalize the score in order to compensate for different utterance durations.

$\log p(X|\lambda_c)$, the probability that the utterance was from an impostor was calculated using the claimed speaker's cohort set as

$$\log p(X|\lambda_{\overline{c}}) = \log \left\{ \frac{1}{B} \sum_{b=1}^{B} p(X|\lambda_b) \right\}, \tag{6}$$

where $p(X|\lambda_b)$ was calculated as in (5).

The EER was then calculated by creating a list of all the likelihood ratios, sorting it and finding the threshold point where the percentage of true speakers below the threshold is equal to the percentage of false speakers above the threshold.

4 Time Information

The use of context-dependent triphones as a feature is new in speaker verification. It was decided to use separate Gaussian distributions with only one mixture each to model the durations of each context-dependent triphone of a speaker. This is a crude duration model, and we return to other possibilities in the conclusion below. The grammar for the YOHO corpus contains 36 different triphones. An example of a triphone is r-iy+ey, which denotes the phoneme iy preceded by an r and followed by an ey. Other contexts containing this same phoneme were ay-iy+n, p-iy+sh and t-iy+d. The context dependence is an important consideration, since significant contextual variability in phoneme duration was observed.

The models were constructed for each triphone k by calculating the sample mean

$$\bar{x} = \frac{1}{M} \sum_{n=1}^{M} x_n, \tag{7}$$

where M is the number of observations of the triphone and x_n is the duration of the n'th observation. An unbiased estimate of the sample variance σ^2 was also calculated as

$$s_{M-1}^2 = \frac{1}{M-1} \sum_{n=1}^{M} (x_n - \bar{x})^2. \tag{8}$$

Every speaker thus has 36 time models of the form $(\bar{x}, s_{M-1}^2)$. The time models were constructed by using all the extracted time durations from the 4 enrollment sessions. Testing was then performed by first extracting the time durations of the triphones in the test session and then calculating a score

$$P(x|\bar{x}, s_{M-1}^2) = \frac{1}{\sqrt{2\pi s_{M-1}^2}} e^{-\frac{(x-\bar{x})^2}{2s_{M-1}^2}}, \tag{9}$$

where x is the observed duration of a specific triphone. The evaluation of (9) yields a value that occurs on the normal distribution with parameters $(\bar{x}, s_{M-1}^2)$. This value is normalized by evaluating the normal distribution with the same value, but using the UBM parameters (which are the means and variances of the appropriate context-dependent triphone, calculated across all training sessions by all speakers). A score is then generated for a speaker i as

$$Score_i = \frac{1}{L} \sum_{l=1}^{L} log(P(x_l|\lambda_c)) - \frac{1}{L} \sum_{l=1}^{L} log(P(x_l|\lambda_{UBM})) \tag{10}$$

where L is the number of observed triphones in the test session. Tests were again performed on a rotating scheme as before, where one speaker is the claimed "client" and all speakers excluding the (acoustic) cohort set are tested using the claimed speaker's models. Once all scores have been obtained, they were again put in an ordered list and the EER was determined.

Figures 1 and 2 illustrate typical distributions of durations observed in our tests. Figure 1 shows an example of a triphone that provides good discrimination

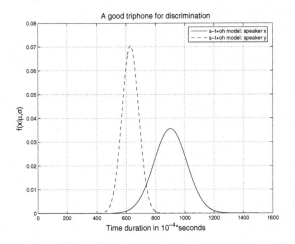

Fig. 1. Probability distributions of a triphone that provides good discrimination between a pair of speakers

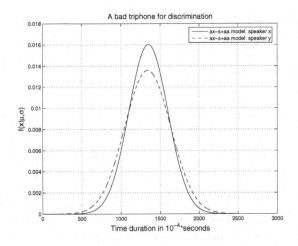

Fig. 2. Probability distributions of a triphone that does not provide good discrimination between a pair of speakers

between two speakers; in other words, time durations of speaker x matched to the model of y would produce a poor score and the general UBM would be chosen, resulting in a correct reject decision of the impostor. Figure 2 illustrates a bad example of a triphone to use, since speaker x's time durations would match speaker y's durations well, resulting in a good score for an impostor.

Since both cases are observed in our data, it is an empirical task to determine whether durations are useful for the task of speaker verification. We therefore now report on tests that were performed to address this question.

5   Results

The results obtained using conventional acoustic scores, temporal features, and a combination of the two types of features, are summarized in table 2, and the corresponding DET curves can be seen in figure 3. The combined EER was obtained by taking a linear combination of the likelihood ratios obtained using time and MFCC features, with the weighting constant determined empirically.

Table 2. Equal error rates obtained on the YOHO database (M+F, msc)

Feature set	eer
MFCCs	0.68%
Time	9.2%
MFCCs and time	0.31%

Several tests have been performed on the YOHO database [6]. ITT's Continuous Speech Rrecognition and Neural Network systems achieved EERs of 1.7% and 0.5% respectively. MIT/LL's Gaussian Mixture Model (GMM) system achieved an EER of 0.51%, Rutgers' Neural Tree Network achieved 0.65% and Reynolds's GMM based system achieved 0.58%. Only the last test can be directly compared to the system described here, since the other tests were performed under different conditions.

Our results with the acoustic (MFCC) features are seen to be comparable to those achieved by other researchers. The temporal features by themselves are significantly less reliable than the acoustic features, but reduce the error rate by a factor of approximately two when combined with those features. This suggests that the temporal features are reasonably uncorrelated with the acoustic features, and the scatter plot in figure 4 confirms this impression. (For clarity, only 400 randomly-selected pairs of acoustic and temporal scores are shown in the figure). The correlation coefficient between the scores using the two types of features was found to be 0.201.

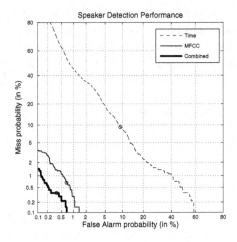

Fig. 3. DET curves for time, MFCC and combined features

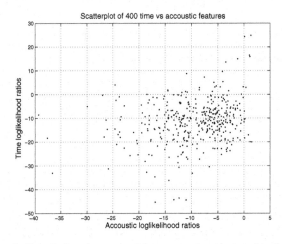

Fig. 4. Correlation between the temporal and acoustic features

6   Models for Predicting Phoneme Duration

In the preceding sections we assumed that the duration of a particular phoneme spoken by a given speaker is described by a normal distribution, independently of the durations of other phonemes in the utterance. This is clearly not realistic - for example, the speaking rate will tend to influence all the phonemes in an utterance [13] in a correlated manner. It is therefore interesting to ask whether a more detailed duration model can be developed, to account for such influences on phoneme durations. Such a model could also include factors such as the position of the phoneme in the word or utterance, but for now we will concentrate on the influence of speaking rate.

To do this, we developed a model for predicting the duration of a phoneme of the form

$$t(ms) = \left[t_{f,global} \cdot \alpha_{f,s} \; \chi_{w,s} \right] \cdot \lambda_{f,s}^{T}, \tag{11}$$

where $t_{f,global}$ is the global mean estimate of the phoneme duration for phoneme f, $\alpha_{f,s}$ is a speaker-specific parameter that was determined to adapt the global phone duration estimate to speaker s, and $\chi_{w,s}$ is the "stretch factor" for a specific word w spoken by s. This is determined as

$$\chi_{w,s} = \frac{\tau - \hat{\tau}}{\sum \sigma_n}. \tag{12}$$

Here τ is the true word length, $\hat{\tau}$ is the estimated word length that was determined by summing the means of the phonemes constituting the word and $\sum \sigma_n$ is the sum of the standard deviations of these phonemes. Finally, $\lambda_{f,s}$ is the vector of parameters obtained from a General Linear Model (GLM) in order to model the effect of the speech rate on the specific phoneme.

In order to evaluate the performance of this simple model of speaking rates, we computed the speaker-specific models on the training data. We then calculated the word-specific factor $\chi_{w,s}$ for each word in the test set, and compared the predicted phoneme accuracy according to our model with two baseline models: (a) the duration of each triphone is assumed to be constant (as in Section 4 above) and (b) the duration of each triphone is assumed to scale linearly with the stretch factor. In Table 3 we show that the GLM indeed is significantly more accurate than the constant stretch factor, which in turn outperforms the constant-duration model used previously.

Table 3. Comparison of three approaches to the modelling of speech rate. The second column contains the mean-squared difference between the actual and predicted phoneme durations (averaged over all test utterances), and third column contains the standard error of this estimate (that is, the standard deviation of all differences divided by the square root of the total number of phonemes in these utterances).

Model	MS error (msec)	Standard error of MS error estimate
Constant speaker-specific duration per phoneme	777.25	65.94
Linear scaling of phoneme durations	522.33	25.46
General linear model of phoneme durations	430.59	16.86

7   Discussion

It has been shown that durations of context-dependent triphones constitute a feature set that can improve the accuracy of speaker verification systems to a significant degree. Although our results were obtained with an HMM in a text-dependent application, it seems likely that an equally low correlation between acoustic and temporal scores will be found with other classes of SV systems. This was indeed true for the text-independent speaker recognition system in [1]. We are therefore confident that similar improvements will be obtained in other SV systems.

Our current system uses the temporal features in a fairly crude fashion: all triphones are modelled with independent Gaussian distributions, and all triphone scores are combined with equal weight. It will be interesting to see how much improvement can be obtained with more sophisticated models (which, for example, assign greater weight to more discriminative triphones or those which have been observed more frequently, or consider correlations between the different triphone durations).

Our initial experiments with more sophisticated duration models (Section 6) suggest that accounting for effects such as speech rate should further improve the discriminative power of duration models, and we are currently investigating how this can be incorporated into our verification system. Modelling effects such as the position of the phoneme in the utterance should produce additional improvements.

Another promising area for further research is related to the relative robustness of temporal and acoustic features to factors such as channel variation and speaker condition [13]. In [2] temporal information was found to be more robust against channel interference than MFCCs, but that result needs to be tested on a more substantial corpus. (Unfortunately, YOHO is not suitable for this purpose, since variable recording conditions were not part of the YOHO protocol.)

Since triphone durations are a very compact descriptor of an utterance, this feature set may also be useful in detecting and deflecting replay attacks. A database of durations during previous verification sessions may be maintained conveniently. One can then calculate the probability of a specific triphone or a sequence of triphones having the same (within some small threshold) time duration, setting a threshold for an acceptable probability and rejecting the speaker as an impostor launching a replay attack if the probability is lower than the threshold.

Overall, it seems as if triphone durations are likely to be a useful addition to almost any toolbox for SV system development.

Acknowledgments. Dr. Marelie Davel assisted in several aspects of this research.

References

1. Ferrer, L., Bratt, H., Gadde, V.R.R., Kajarekar, S., Shriberg, E., Sönmez, K., Stolcke, A., Venkataraman, A.: Modeling duration patterns for speaker recognition. In: Proceedings of Eurospeech, pp. 784–787 (September 2003)
2. van Heerden, C.J., Barnard, E.: Using timing information in speaker verification. In: Proceedings of the Symposium of the Pattern Recognition Association of South Africa, pp. 53–57 (December 2005)

3. Campbell, J.P.: Speaker recognition: A tutorial. Proceedings of the IEEE 85, 1437–1462 (1997)
4. Martin, A.: Evaluations of Automatic Speaker Classification Systems. In: Müller, C. (ed.) Speaker Classification. Lecture Notes in Computer Science / Artificial Intelligence, vol. 4343, Springer, Heidelberg (2007) (this issue)
5. Koreman, J., Wu, D., Morris, A.C.: Enhancing Speaker Discrimination at the Feature Level. In: Müller, C. (ed.) Speaker Classification. Lecture Notes in Computer Science / Artificial Intelligence, vol. 4343, Springer, Heidelberg (2007) (this issue)
6. Campbell, J.: Testing with the YOHO CD-ROM voice verification corpus. In: Proceedings of the International Conference on Acoustics, Speech and Signal Processing (ICASSP), vol. 1, pp. 341–344 (May 1995)
7. Campbell, J.P., Reynolds, D.A.: Corpora for the evaluation of speaker recognition systems. In: Proceedings of the International Conference on Acoustics, Speech and Signal Processing (ICASSP), vol. 2, pp. 829–832 (March 1999)
8. Higgens, A., Bahler, L., Porter, J.: Speaker verification using randomized phrase prompting. Digital Signal Processing 1(2), 89–106 (1991)
9. Reynolds, D.A.: Speaker identification and verification using gaussian mixture speaker models. Speech Communication 17, 91–108 (1995)
10. Liou, H.-S., Mammone, R.J.: A subword neural tree network approach to text-dependent speaker verification. In: Proceedings of the International Conference on Acoustics, Speech and Signal Processing (ICASSP), vol. 1, pp. 357–360 (May 1995)
11. Rosenberg, A.E., DeLong, J., Lee, C.-H., Juang, B.-H., Soong, F.K.: The use of cohort normalized scores for speaker verification. In: Proceedings of the International Conference on Spoken Language Processing (ICSLP), vol. 1, pp. 599–602 (October 1992)
12. Young, S., Evermann, G., Gales, M., Hain, T., Kershaw, D., Moore, G., Odell, J., Ollason, D., Povey, D., Veltchev, V., Woodland, P.: The HTK Book, Cambridge University Engineering Department (2005), http://htk.eng.cam.ac.uk/
13. Huckvale, M.: How is individuality expressed in voice? An introduction to speech production & description for speaker classification. In: Müller, C. (ed.) Speaker Classification. LNCS(LNAI), vol. 4343, Springer, Heidelberg (2007) (this issue)

Language–Independent Speaker Classification over a Far–Field Microphone

Jerome R. Bellegarda

Apple Inc., Two Infinite Loop, Cupertino, California 95014, USA
jerome@apple.com

Abstract. The speaker classification approach described in this contribution leverages the analysis of both speaker and verbal content information, so as to use two light-weight components for classification: a spectral matching component based on a global representation of the entire utterance, and a temporal alignment component based on more conventional frame-level evidence. The paradigm behind the spectral matching component is related to latent semantic mapping, which postulates that the underlying structure in the data is partially obscured by the randomness of local phenomena with respect to information extraction. Uncovering this latent structure results in a parsimonious continuous parameter description of feature frames and spectral bands, which then replaces the original parameterization in clustering and identification. Such global analysis can then be advantageously combined with elementary temporal alignment. This approach has been commercially deployed for the purpose of language-independent desktop voice login over a far-field microphone.

Keywords: Spectral matching, global representation, latent structure, distance metric, desktop voice login.

1 Introduction

1.1 Background

This contribution is an outgrowth of an approach we proposed a few years ago for the dual verification of speaker identity and verbal content in a text-dependent voice authentication system [6]. Voice authentication, the process of accepting or rejecting the identity claim of a speaker on the basis of individual information present in the speech waveform [9], has received increasing attention over the past two decades, as a convenient, user-friendly way of replacing (or supplementing) standard password-type matching [7]. The application considered in [6] was desktop voice login, where access to a personal computer can be granted or denied on the basis of the user's identity. In that context, the authentication system must be kept as unintrusive as possible, which normally entails the use of a far-field microphone (e.g., mounted on the monitor) and a very small amount of enrollment data (5 to 10 seconds of speech). In addition, the scenario of choice is text-dependent verification, which is logistically closest to that of a typed password.

C. Müller (Ed.): Speaker Classification II, LNAI 4441, pp. 104–115, 2007.

In the setup of [6], each speaker was allowed to select a keyphrase of his or her own choosing, and enrollment was limited to four instances of the keyphrase, each 1 to 2 seconds of speech. Assuming the user maintains the confidentiality of the keyphrase, this offers the possibility of verifying the spoken keyphrase in addition to the speaker identity, thus resulting in an additional layer of security (cf. [11]). We thus developed an approach which leveraged the analysis of both speaker characteristics and verbal content information. The resulting technique had an innovative spectral matching component based on a global representation of the entire utterance.

As has since become clear, the paradigm underlying this component is related to latent semantic mapping (LSM), a data-driven framework for modeling meaningful global relationships implicit in large volumes of data [4]. LSM operates under the assumption that there is some latent structure in the data, which is partially obscured by the randomness of local phenomena with respect to information extraction. Algebraic techniques are brought to bear to estimate this structure and get rid of the obscuring "noise." In the present case, this results in a parsimonious continuous parameter description of feature frames and spectral bands, which then replaces the original parameterization in clustering and identification. The outcome of such global analysis can then be advantageously combined with the outcome of elementary temporal alignment.

1.2   Contrast with HMM Solutions

The above approach can be viewed as following a divide and conquer strategy with a slightly unconventional division of labor. While existing systems tend to directly target speaker information on the one hand and verbal content information on the other, the method of [6] deliberately blurs the line between the two, instead targeting more explicitly spectral and temporal information. This has several advantages over standard classification techniques using HMM technology with Gaussian mixture distributions (see, e.g., [14], [15]), for which the typical framework is illustrated in Fig. 1.

For the verification of speaker identity, speaker-dependent (SD) language-independent (LI) sub-word HMMs must be constructed from the training data available. This is liable to suffer from scarce data problems when enrollment is severely limited, as is the case here. Variance estimation is of particular concern, as the underlying Gaussian mixture distributions run the risk of being too sharp and overfitting the training data [12]. And speaker adaptation is not really a viable option, as it tends to reduce the amount of discrimination that can be achieved between different speakers.

In addition, handling verbal content with the HMM paradigm normally entails using large vocabulary speech recognition to recognize the uttered word sequence. This requires speaker-independent (SI) HMMs to characterize the acoustics, and a large vocabulary language model (commonly an n-gram) to characterize the linguistics. As a result, the overall solution is language-dependent (LD). Note that each component must perform a combination of spectral matching and time alignment to compute its own, separate likelihood score associated

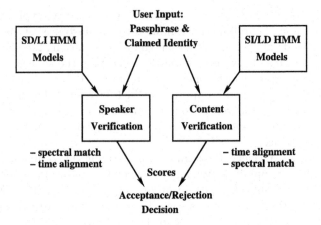

Fig. 1. Standard Approach to Text-Dependent Verification

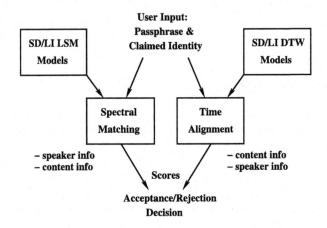

Fig. 2. Alternative Approach to Text-Dependent Verification

with the input speech. The accept/reject decision is then based on the combination of these two scores.

In contrast, the approach advocated in [6] largely decouples spectral matching and time alignment, as illustrated in the framework of Fig. 2. Spectral matching is performed primarily based on an LSM-based utterance-level representation obtained by integrating out frame-level information, while temporal alignment, based only on this frame-level information, no longer involves specific word knowledge. This allows the two components to remain speaker-dependent and language-independent (SD/LI). On the other hand, it is likely that each component now relies on a combination of speaker and content information.

Note that for temporal alignment, we use simple dynamic time-warping (DTW). Although HMMs can more efficiently model statistical variation in spectral features, here DTW is sufficient, because the LSM approach already takes care of spectral matching, and therefore the requirements on DTW are less stringent than usual. As before, each component produces a likelihood score, and the accept/reject decision is based on the combination of the two scores.

1.3 Organization

The purpose of this contribution is to go over the above approach, with special emphasis on its LSM component, in order to illustrate, on this case study, the potential benefits to more general speaker classification tasks. The material is organized as follows. The next section briefly mentions feature extraction, which is common to the two components. Section 3 reviews the LSM framework and its relevance to the problem at hand. In Section 4, we derive a distance measure specifically tailored to speaker classification, and present the ensuing LSM component. Section 5 discusses integration with the DTW component. Finally, in Section 6 we report experimental results underscoring the performance of the integrated LSM+DTW system.

2 Feature Extraction

Feature extraction largely follows established procedure (see, e.g., [14], [15]). We extract spectral feature vectors every 10ms, using short-term Fourier transform followed by filter bank analysis to ensure a smooth spectral envelope. (This is important to provide a stable representation from one repetition to another of a particular speaker's utterance.) To represent the spectral dynamics, we also extract, for every frame, the usual delta and delta-delta parameters [16].

After concatenation, we therefore end up with a sequence of M feature vectors of dimension N. To distinguish them from other types of vectors discussed later, we will refer to these feature vectors as *frames*. For a typical utterance, $M \approx 200$ and $N \approx 40$. The resulting sequence of frames is the input to both the LSM component and the DTW component of the proposed method.

3 LSM Framework

3.1 Single–Utterance Representation

Upon re-arranging this input sequence, each utterance can be represented by a $M \times N$ matrix of frames, say F, where each row represents the spectral information for a given frame and each column represents a particular spectral band over time. We then compute the singular value decomposition (SVD) of the matrix F, as [10]:

$$F = U S V^T, \tag{1}$$

where U is the $(M \times R)$ matrix of left singular vectors, S is the $(R \times R)$ diagonal matrix of singular values, V is the $(N \times R)$ matrix of right singular

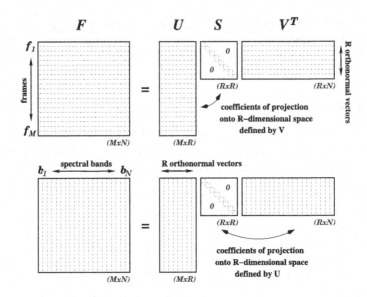

Fig. 3. Two Equivalent Views of SVD Decomposition

vectors, $R < \min(M, N)$ is the order of the decomposition, and ^{T} denotes matrix transposition. As is well known, both U and V are column-orthonormal, i.e., $U^TU = V^TV = I_R$, the identity matrix of order R [8]. For reasons to become clear shortly, we refer to (1) as the decomposition of the utterance into *single-utterance* singular elements U, S, and V.

Such whole utterance representation has been considered before: see, e.g., [1]. The resulting parameterization can be loosely interpreted as conceptually analogous to the Gaussian mixture parameterization in the HMM framework. The main difference is that the Gaussian mixture approach is implicitly based on a sub-word unit (such as a phoneme), whereas the LSM approach operates on the entire utterance, which introduces more smoothing.

3.2   Interpretation

It is intuitively reasonable to postulate that some of the singular elements will reflect more speaker information and some others more verbal content information. But it is not completely clear exactly which reflects what. In [1], a case was made that speaker information is mostly contained in V. Speaker verification was then performed using the Euclidean distance after projection onto the "speaker subspace" defined by V, on the theory that in that subspace utterances from the true speaker have greater measure. This is illustrated in Fig. 3, top figure. In that interpretation, each row of V^T can be thought of as a basis vector spanning the global spectral content of the utterance, and each row of US represents the degree to which each basis vector contributes to the corresponding frame.

But an equally compelling case could be made under the (dual) assumption that verbal content information is mostly contained in U. In this situation speaker verification could conceivably be performed after projection onto the "content subspace" spanned by U. One would simply compute distances between reference and verification utterances in that subspace, on the theory that a large distance between two utterances with the same verbal content would have to be attributed to a speaker mismatch. This is illustrated in Fig. 3, bottom figure. In that interpretation, each column of U is a basis vector spanning the verbal content of the utterance, and each column of SV^T represents the degree to which each basis vector contributes to the corresponding spectral band.

In the standard LSM framework (cf. [3] – [5]), such discussion normally leads to a common representation in terms of the rows vectors of US and VS, which correspond to the coordinates of the M frames and N spectral bands in the space of dimension R spanned by the singular vectors (usually referred to as the LSM space $\mathcal{L}$). As this mapping, by definition, captures the major structural associations in F and ignores higher order effects, the "closeness" of vectors in $\mathcal{L}$ is determined by the overall speech patterns observed in the utterance, as opposed to specific acoustic realizations. Hence, two frames whose representations are "close" would tend to have similar spectral content, and conversely two spectral bands whose representations are "close" would tend to appear in similar frames.

In the present situation, however, this level of detail is not warranted, because we are not trying to compare individual frames (or spectral bands, for that matter). Rather, we are interested in assessing potential changes in global behavior across utterances. The problem, then, is not so much to analyze a given LSM space as it is to relate distinct LSM spaces to each other. For this, we need to specify some distance measure suitable to compare different global representations. The following justifies and adopts a new metric specifically tailored to the LSM framework.

4   LSM–Tailored Metric

4.1   Multiple–Utterance Representation

Assume, without loss of generality, that (1) is associated with a particular training utterance, say the jth utterance, from a given speaker, and consider the set of all training utterances from that speaker. This set will be represented by a $\tilde{M} \times N$ matrix, with $\tilde{M} \approx JM$, where J is the number of training utterances for the speaker. Denoting this $\tilde{M} \times N$ matrix by $\tilde{F}$, it can be decomposed as:

$$\tilde{F} = \tilde{U}\,\tilde{S}\,\tilde{V}^T, \tag{2}$$

with analogous definitions and properties as in (1). In particular, (2) defines a similar, though likely distinct, LSM space from the one obtained via (1), which was only derived from a single utterance.

Obviously, the set of all training utterances contains the jth utterance, so by selecting the appropriate M rows of $\tilde{F}$, we can define:

$$\tilde{F}_{(j)} = F = \tilde{U}_{(j)} \tilde{S} \tilde{V}^T, \tag{3}$$

where the subscript $_{(j)}$ serves as an index to the jth utterance. Presumably, from the increased amount of training data, the matrices $\tilde{S}$ and $\tilde{V}$ are somewhat more robust versions of S and V, while $\tilde{U}_{(j)}$ relates this more reliable representation (including any embedded speaker information) to the original jth utterance. We refer to (3) as the decomposition of the utterance into *multiple-utterance* singular elements $\tilde{U}_{(j)}$, $\tilde{S}$, and $\tilde{V}$, and similarly to the LSM space associated with (2) as the underlying *multiple-utterance* LSM space.

4.2   Mapping Across LSM Spaces

This opens up the possibility of relating the two LSM spaces to each other. Note that the equality:

$$\tilde{U}_{(j)} \tilde{S} \tilde{V}^T = U S V^T \tag{4}$$

follows from (1) and (3). To cast this equation into a more useful form, we now make use of the (easily shown) fact that the matrix $(V^T \tilde{V})$ is (both row- and column-) orthornormal. After some algebraic manipulations, we eventually arrive at the expression:

$$\tilde{S} (\tilde{U}_{(j)}^T \tilde{U}_{(j)}) \tilde{S} = (V^T \tilde{V})^T S^2 (V^T \tilde{V}). \tag{5}$$

Since both sides of (5) are symmetric and positive definite, there exists a $(R \times R)$ matrix $D_{j|\tilde{S}}$ such that:

$$D_{j|\tilde{S}}^2 = \tilde{S} (\tilde{U}_{(j)}^T \tilde{U}_{(j)}) \tilde{S}. \tag{6}$$

Note that, while $\tilde{U}^T \tilde{U} = I_R$, in general $\tilde{U}_{(j)}^T \tilde{U}_{(j)} \neq I_R$. Thus $D_{j|\tilde{S}}^2$ is closely related, but not equal, to $\tilde{S}^2$. Only as the single-utterance decomposition becomes more and more consistent with the multiple-utterance decomposition does $D_{j|\tilde{S}}^2$ converge to $\tilde{S}^2$.

Taking (6) into account and again invoking the orthonormality of $(V^T \tilde{V})$, the equation (5) is seen to admit the solution:

$$D_{j|\tilde{S}} = (V^T \tilde{V})^T S (V^T \tilde{V}). \tag{7}$$

Thus, the orthonormal matrix $(V^T \tilde{V})$ can be interpreted as the rotation necessary to map the single-utterance singular value matrix obtained in (1) onto (an appropriately transformed version of) the multiple-utterance singular value matrix obtained in (2). Clearly, as V tends to $\tilde{V}$ (meaning U also tends to $\tilde{U}_{(j)}$) the two sides of (7) become closer and closer to a diagonal matrix, ultimately converging to $S = \tilde{S}$.

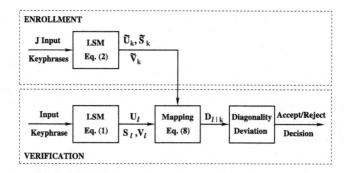

Fig. 4. Overview of LSM Component

4.3 LSM Component

The above observation suggests a natural metric to evaluate how well a particular utterance j is consistent with the (multiple-utterance) speaker model: compute the quantity $D_{j|\tilde{S}} = (V^T \tilde{V})^T S (V^T \tilde{V})$, per (7), and measure how much it deviates from a diagonal matrix. For example, one way to measure the deviation from diagonality is to calculate the Frobenius norm of the off-diagonal elements of the matrix $D_{j|\tilde{S}}$.

This in turn suggests an alternative metric to evaluate how well a verification utterance, uttered by a speaker ℓ, is consistent with the (multiple-utterance) model for speaker k. Indexing the single-utterance elements by ℓ, and the multiple-utterance elements by k, we define:

$$D_{\ell|k} = (V_\ell^T \tilde{V}_k)^T S_\ell (V_\ell^T \tilde{V}_k), \qquad (8)$$

and again measure the deviation from diagonality of $D_{\ell|k}$ by calculating the Frobenius norm of its off-diagonal elements. By the same reasoning as before, in this expression the matrix $(V_\ell^T \tilde{V}_k)$ underscores the rotation necessary to map S_ℓ onto (an appropriately transformed version of) $\tilde{S}_k$. When V_ℓ tends to $\tilde{V}_k$, $D_{\ell|k}$ tends to $\tilde{S}_k$, and the Frobenius norm tends to zero. Thus, the deviation from diagonality can be expected to be less when the verification utterance comes from speaker $\ell = k$ then when it comes from a speaker $\ell \neq k$. Clearly, this distance measure is better tailored to the LSM framework than the usual Euclidean (or Gaussian) distance. It can be verified experimentally that it also achieves better performance.

The LSM component thus operates as illustrated in Fig. 4. During enrollment, each speaker $1 \leq k \leq K$ to be registered provides a small number J of training sentences. For each speaker, the enrollment data is processed as in (2), to obtain the appropriate right singular matrix $\tilde{V}_k$. During verification, the input utterance is processed as in (1), producing the entities S_ℓ and V_ℓ. Then $D_{\ell|k}$ is computed as in (8), and the deviation from diagonality is calculated. If this measure falls within a given threshold, then the speaker is accepted as claimed. Otherwise, it is rejected.

5 Integration with DTW

5.1 DTW Component

The LSM approach deliberately discards a substantial amount of available temporal information, since it integrates out frame-level information. Taking into account the linear mapping inherent in the decomposition, it is likely that the singular elements only encapsulate coarse time variations, and smooth out finer behavior. Unfortunately, detecting subtle differences in delivery is often crucial to thwarting non-casual impostors, who might use their knowledge of the true user's speech characteristics to deliberately mimic his or her spectral content. Thus, a more explicit temporal verification should be added to the LSM component to increase the level of security against such determined impersonators.

We adopt a simple DTW approach for this purpose. Although HMM techniques have generally proven superior for time alignment, in the present case the LSM approach already contributes to spectral matching, so the requirements on any supplementary technique are less severe. As mentioned earlier, here DTW suffices, in conjunction with the LSM component, to carry out verbal content verification.

The DTW component implements the classical dynamic time warping algorithm (cf., e.g., [2]). During training, the J training utterances provided by each speaker are "averaged" to define a representative reference utterance s_{avg}. This is done by setting the length of s_{avg} to the average length of all J training utterances, and warping each frame appropriately to come up with the reference frame at that time. During verification, the input utterance, say s_{ver}, is acquired and compared to the reference model s_{avg}. This is done by aligning the time axes of s_{ver} and s_{avg}, and computing the degree of similarity between them, accumulated from the beginning to the end of the utterance on a frame by frame basis. Various distance measures are adequate to perform this step, including the usual Gaussian distance. As before, the speaker is accepted as claimed only if the degree of similarity is high enough.

5.2 System Integration

For each verification utterance, two scores are produced: the deviation from diagonality from the LSM component, and the degree of similarity from the DTW component. There are therefore several possibilities to combine the two components. For example, it is possible to combine the two scores into a single one and base the accept/reject decision on that single score. Alternatively, one can reach a separate accept/reject decision for each component and use a voting scheme to form the final decision.

For simplicity, we opted for the latter. Thus, no attempt is made to introduce conditional behavior in one component which depends on the direction taken by the other. The speaker is simply accepted as claimed only if both likelihood scores are high enough.

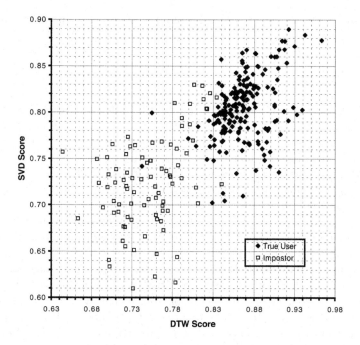

Fig. 5. Performance Space of LSM+DTW Approach

6   Performance Validation

Experiments were conducted using a set of 93 speakers, $K = 48$ true users and $K' = 45$ impostors. True users enrolled by speaking their keyphrase $J = 4$ times. They also provided four instances of a voice login attempt, collected on different days. This resulted in a total of 191 true test utterances, across which the minimum, average, and maximum sentence length were 1.2, 1.8, and 3 seconds, respectively.

To increase the severity of the test, each impostor was dedicated to a particular speaker, and was selected on the basis of his/her apparent "closeness" to that user, as reflected in his/her speech characteristics. For example, to impersonate a male speaker who grew up in Australia, we chose another male speaker with an Australian accent. Further, each impostor was given access to the original enrollment keyphrases from the true speaker, and was encouraged to mimic delivery as best as s/he could. This was to reflect the high likelihood of deliberate imposture in desktop voice login, where the true user is typically known to the impostor. (On the other hand, given this application and in view of Apple's target market, we deemed unnecessary to consider more sophisticated attempts like technical imposture [13].) Each impostor provided two distinct attempts, for a total of 90 impostor test utterances.

The results are plotted in Fig. 5. For the appropriate combination of thresholds, the above system leads to 0 false acceptances and 20 false rejections (10.4%).

After tuning to obtain an equal number of false acceptances and false rejections, we observed approximately a 4% equal error rate.

7 Conclusion

Desktop voice login is a challenging application due to several inherent constraints: (i) the acoustic signal is normally acquired via a low-quality far-field microphone, (ii) enrollment is limited to an average of about, in our case, 7 seconds of speech, and (iii) ideally the solution has to work across multiple languages. Although in typical text-dependent mode it also allows for the dual verification of speaker identity and verbal content, this is not an ideal environment for HMM-based methods using Gaussian mixtures.

We have discussed an alternative strategy which leverages the analysis of both speaker and keyphrase information, and uses a light-weight component to tackle each: an LSM component for global spectral matching, and a DTW component for local temporal alignment. Because these two components complement each other well, their integration leads to a satisfactory performance for the task considered. An equal error rate figure of approximately 4% was obtained in experiments including deliberate imposture attempts. The resulting LSM+DTW system was commercially released several years ago as part of the "VoicePrint Password" feature of Mac OS^{TM}.

This case study illustrates the potential viability of the LSM approach, and associated deviation from diagonality metric, for more general speaker classification tasks. By integrating out temporal aspects across an entire utterance, the LSM representation uncovers information that is largely orthogonal to more conventional techniques. Incorporating this information with other sources of knowledge can therefore prove useful to increase the robustness of the overall classification system.

References

1. Ariki, Y., Doi, K.: Speaker Recognition Based on Subspace Methods. In: Proc. Int. Conf. Spoken Lang. Proc. Yokohama, Japan, pp. 1859–1862 (September 1994)
2. Assaleh, K.T., Farrell, K.R., Zilovic, M.S., Sharma, M., Naik, D.K., Mammone, R.J.: Text–Dependent Speaker Verification Using Data Fusion and Channel Detection. In: Proc. SPIE, San Diego, CA, vol. 2277, pp. 72–82 (August 1994)
3. Bellegarda, J.R.: Exploiting Latent Semantic Information in Statistical Language Modeling. In: Juang, B.H., Furui, S. (eds.) Proc. IEEE, Spec. Issue Speech Recog. Understanding, vol. 88(8), pp. 1279–1296 (August 2000)
4. Bellegarda, J.R.: Latent Semantic Mapping. In: Deng, L., Wang, K., Chou, W. (eds.) Signal Proc. Magazine, Special Issue Speech Technol. Syst. Human–Machine Communication, vol. 22(5), pp. 70–80 (September 2005)
5. Bellegarda, J.R.: A Global Boundary–Centric Framework for Unit Selection Text–to–Speech Synthesis, IEEE Trans. Speech Audio Proc. vol. SAP–14(4) (July 2006)

6. Bellegarda, J.R., Naik, D., Neeracher, M., Silverman, K.E.A.: Language–Independent, Short–Enrollment Voice Verification over a Far–Field Microphone. In: Proc. 2001 IEEE Int. Conf. Acoust. Speech, Signal Proc. Salt Lake City, Utah (May 2001)
7. Campbell Jr., J.P.: Speaker Recognition: A Tutorial. Proc. IEEE 85(9), 1437–1462 (1997)
8. Cullum, J.K., Willoughby, R.A.: Lanczos Algorithms for Large Symmetric Eigenvalue Computations – vol. 1 Theory, Ch. 5: Real Rectangular Matrices, Boston: Brickhauser (1985)
9. Doddington, G.: Speaker Recognition—Identifying People by their Voices, Proc. IEEE, vol. 73 (November 1985)
10. Golub, G., Van Loan, C.: Matrix Computations, 2nd edn. Johns Hopkins, Baltimore, MD (1989)
11. Higgins, A., Bahler, L., Porter, J.: Digital Signal Processing, 1, 89–106 (1991)
12. Li, Q., Juang, B.-H., Zhou, Q., Lee, C.-H.: Automatic Verbal Information Verification for User Authentification. IEEE Trans. Speech Acoust. Proc. 8(5), 585–596 (2000)
13. Lindberg, J., Bloomberg, M.: Vulnerability in Speaker Verification—A Study of Technical Impostor Techniques. In: Proc. EuroSpeech, Budapest, Hungary, pp. 1211–1214 (September 1999)
14. Matsui, T., Furui, S.: Speaker Adaptation of Tied-Mixture-Based Phoneme Models for Text-Prompted Speaker Recognition. In: Proc. 1994 ICASSP, Adelaide, Australia, pp. 125–128 (April 1994)
15. Parthasaraty, S., Rosenberg, A.E.: General Phrase Speaker Verification Using Sub–Word Background Models and Likelihood–Ratio Scoring. In: Proc. Int. Conf. Spoken Language Proc. Philadelphia, PA (October 1996)
16. Rabiner, L.R., Juang, B.H., Lee, C.-H.: An Overview of Automatic Speech Recognition. In: Lee, C.-H., Soong, F.K., Paliwal, K.K. (eds.) Chapter 1 in Automatic Speech and Speaker Recognition: Advanced Topics, pp. 1–30. Kluwer Academic Publishers, Boston, MA (1996)

A Linear-Scaling Approach to Speaker Variability in Poly-segmental Formant Ensembles

Frantz Clermont

JP French Associates,
Forensic Speech and Acoustics Laboratory,
York, United Kingdom
akustikfonetiks@yahoo.com.au

Abstract. A linear-scaling approach is introduced for handling acoustic-phonetic manifestations of inter-speaker differences. The approach is motivated (i) by the similarity commonly observed amongst formant-frequency patterns resulting from different speakers' productions of the same utterance, and (ii) by the fact that there are linear-scaling properties associated with similarity. In methodological terms, formant patterns are obtained for a set of segments selected from a fixed utterance, which we call *poly-segmental formant ensembles*. Linear transformations of these ensembles amongst different speakers are then sought and interpreted as a set of scaling relations. Using multi-speaker data based on Australian English "hello", it is shown that the transformations afford a significant reduction of inter-speaker dissimilarity by inverse similarity. The proposed approach is thus able to unlock regularity in formant-pattern variability from speaker to speaker, without prior knowledge of the exact causes of the speaker differences manifested in the data at hand.

Keywords: Poly-Segmental Ensembles, Linear Scaling, Speaker Variability, Formant-Frequency Patterns.

1 Introduction

The work presented in this paper draws its motivation from a familiar observation, which holds promise for apprehending acoustic-phonetic manifestations of speaker-to-speaker variations and, eventually, gaining better control of their effects in speaker classification. The motivating observation is that, for the same utterance produced by different speakers, formant-frequency patterns retain a certain similarity despite the variations expected from vocal-tract structures and articulatory habits (Nolan, 1983). According to Ohta and Fuchi's (1984) "constancy" interpretation, the similarity may be thought of as a manifestation of different speakers tending to utilise similarly-shaped vocal-tracts while producing the same utterance. A long-foreshadowed (Chiba and Kajiyama, 1958) implication of the similarity proposition is that, irrespective of inter-speaker differences, there should be some hope for predictable regularity in the way in which formant-patterns for a fixed utterance vary from speaker to speaker. Here we develop an approach for characterising this regularity.

C. Müller (Ed.): Speaker Classification II, LNAI 4441, pp. 116–129, 2007.
© Springer-Verlag Berlin Heidelberg 2007

Our first step is to treat the acoustic-phonetic segments selected from *a fixed utterance* as components of a *speaker-dependent ensemble*. Each component contributes towards some global behaviour, whose similarity amongst a set of different speakers becomes quantifiable by comparing their corresponding ensembles to one another. *Ensemble similarity* is a key aspect of our approach, which is explained through the schematic representation shown below in Fig.1.

Along the ordinate axis, each rectangle contains a set of dots which, for *a given speaker* and *a fixed utterance*, schematise relative positions of a formant's frequencies for a set of phonetic segments selected from that utterance. The unequal spacing between dots simulates the variation expected from segment to segment for a given formant. Such a data set forms a *"Poly-Segmental formant Ensemble"* (an ensemble or a PSE in short). The abscissa is a "speaker axis", along which each rectangle represents a PSE for each speaker with a constant, relative spacing of the dots.

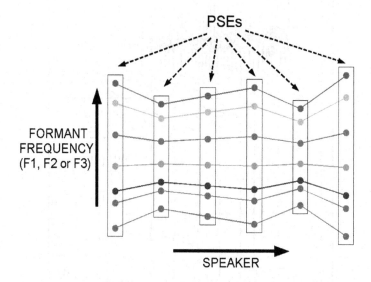

Fig. 1. Idealisation of inter-speaker similarity: Formants (F1, F2 or F3) for an ensemble of (colour-coded) segments plotted for a set of speakers. Rectangles enclose each speaker's PSE.

Figure 1 essentially portrays a *systemic* organisation of multi-speaker, segmental data as a set of speaker-dependent PSEs which, formant by formant, are geometrically interpretable in terms of *scaling relations*. This is the methodological basis for our approach, which is evaluated in the remainder of the paper.

Section 2 describes the poly-segmental data employed for this work. The scaling methodology is presented in Section 3, where the procedures outlined include a pre-scaling check of the extent of similarity in the data, a motivating glimpse at actual ensembles, and the calculation of ensemble scales. In Section 4 the per-formant scales are examined numerically, and their effectiveness as measures of similarity is evaluated. In Section 5 it is shown that pre-normalisation of the raw ensembles by vocal-tract length brings the per-formant scales in line with the notion of uniform scaling. Finally, the effects of direct and inverse scaling are contrasted and discussed.

2   Poly-segmental, Formant Parameterisation of Spoken "hello"

The poly-segmental data used for illustrating and evaluating the ensemble-scaling approach outlined above, originate from a previous study of the spoken word "hello" (Rose, 1999). The data embody interesting features that are described below.

In addition to being a frequent lexical item in spoken English, the word "hello" embodies a situational sensitivity that facilitates elicitation with spontaneous variability. Several situational tokens were thus produced (at one sitting) by 6 adult-male speakers: DM (17 tokens), EM (3 tokens), JM (6 tokens), MD (12 tokens), PS (4 tokens) and RS (7 tokens). They all are native speakers of Australian English with accents ranging from general to slightly broad.

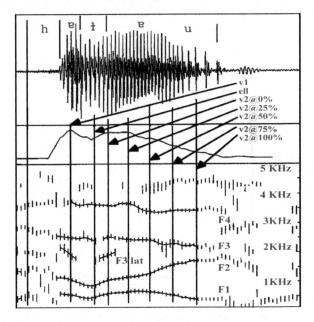

Fig. 2. Acoustic-phonetic analysis of "hello" [Rose's (1999: 9) Fig. 1]. ***Top panel***: Waveform. ***Middle panel***: Energy contour. ***Bottom panel***: Linear-prediction "pole-gram" & F-patterns at 7 segments (*Phonetic* labels on top of waveform; O*perational* labels at right of arrows).

The acoustic-phonetic structure for the word "hello" is adopted from Rose (1999). It consists of 7 segmental landmarks (v1, ell, v2@0%, v2@25%, v2@50%, v2@75%, v2@100%), which span a small subset of the phonetic space but whose realisations involve a range of vocal-tract configurations – one at the initial monophthongal target (v1), one in the middle of the lateral consonant (ell), and five at equidistant instants of the final diphthong (v2). For each segment and every token, the 4 lowest formant-frequencies (F1, F2, F3, and F4) were extracted using linear-prediction analysis.

Per speaker and per formant-frequency, a poly-segmental ensemble is here defined numerically as the set of token-averaged values obtained for each of the 7 segments.

Thus, there are 6 speaker-dependent ensembles for each formant. In the next sections, the scaling methodology is developed and illustrated using F1-, F2- and F3-ensembles.

3 Ensemble-Scaling Technique

The scaling technique is based on Broad and Clermont's (2002) analogous development for characterising the frame-to-frame similarity of co-articulation effects of consonantal context on vowel-formant ensembles. Under the first-order assumption of linearity, the same technique is applicable to poly-segmental formant ensembles defined for different speakers, provided the ensemble data at hand exhibit a certain consistency from speaker to speaker. In Section 3.1 it is shown that the scaling technique affords a preliminary diagnostic for lack of consistency in ensemble similarity. In Section 3.2 the technique itself is described.

3.1 Pre-scaling Diagnostic

A basic aspect of the scaling technique is the use of the speaker-averaged PSE (the mean PSE), as a reference ensemble with respect to which individual PSEs are to be scaled. It stands to reason that the mean ensemble should be desirable for its representative behaviour and its statistical robustness. However, it is its objective role that is paramount in seeking a relative measure of ensemble-to-ensemble similarity. This quest can be pursued more confidently if, indeed, there is evidence of consistent similarity in the data at hand.

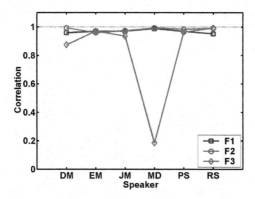

Fig. 3. Profile of correlations between each of the 6 speaker-dependent PSEs and the mean PSE, providing a diagnostic for consistency in ensemble similarity. The very weak correlation of **0.18** for speaker **MD**'s **F3**-ensemble indicates his departure from similarity in F3.

One approach to detecting departure from similarity is to look at the strength of correlation between individual PSEs and the mean PSE. The numerical profile of such correlations is given in Fig. 3 for the 6 speakers and the 3 lowest formants.

Whilst there is strong indication of similarity amongst all speakers' F1- and F2-ensembles, there is also strong evidence against the inclusion of the F3-ensemble for speaker MD. Rather than include him only for F1 and F2, we retain the 5 speakers for whom all formant ensembles are consistently similar, thus avoiding a confounding factor in the evaluation of the scaling technique as a tool for expressing similarity.

Table 1. Correlations between each of the remaining 5 speaker-dependent PSEs, and 2 mean PSEs: one excluding speaker **MD** (values outside parentheses), and the other including speaker **MD** (values within parentheses)

SPEAKERS	F1	F2	F3
DM	0.96 (0.96)	0.99 (0.99)	0.89 (0.87)
EM	0.96 (0.96)	0.99 (0.96)	0.97 (0.97)
JM	0.98 (0.97)	0.97 (0.97)	0.93 (0.93)
PS	0.97 (0.97)	0.98 (0.98)	0.96 (0.96)
RS	0.95 (0.95)	0.99 (0.99)	0.99 (0.99)

The correlations re-calculated (see Table 1) for the non-problematic, 5-speaker set remain quite strong with even a slight improvement for speaker EM's F2-ensemble and speaker DM's F3-ensemble. It is with this 5-speaker set of PSEs that the scaling technique is developed in the next section.

3.2 Ensemble Scaling Via Linear Regression

The strong correlations reported above have confirmed the existence of a consistent similarity amongst 5 of the 6 speakers' ensembles examined, thereby paving the way for the scaling implementation. However, the procedure is more directly motivated by first taking a glimpse at the actual ensemble data shown in Fig. 4.

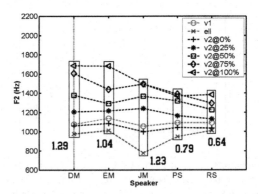

Fig. 4. Actual Poly-Segmental Ensembles obtained for F2. Ensemble scales are shown at bottom of rectangles. Fig. 5 illustrates how the ensemble scale for speaker DM was obtained.

Glimpse at Poly-Segmental Ensembles for F2. On the "speaker axis" of Fig. 4 are juxtaposed the F2-ensembles obtained from the 5-speakers' data. Perhaps the first observation worth noting is that the ensembles are translated with respect to one another. While this may be a useful factor of discrimination amongst speakers, it is inconsequential to scaling. Instead, the crucial factor of similarity is the ensemble-to-ensemble regularity in relative position and spacing of the segments' formants. Although the ensembles shown in Fig. 4 do not appear to be exactly linearly-scaled copies of each other, there is a sufficiently noticeable trend to warrant the next step leading to scaling relations.

Linear-Regression Procedure. Ensemble scaling is achieved using linear-regression fits of each speaker's PSE translated by its mean against the mean of all translated PSEs. This is illustrated in Fig. 5 for speaker DM, where the slope of the fitted line is an estimate of the scaling factor, justly referred to as an *ensemble scale* describing a proportion with respect to the mean ensemble. The scales obtained for DM and the other 4 speakers are shown in Fig. 4 at the bottom of the corresponding rectangles.

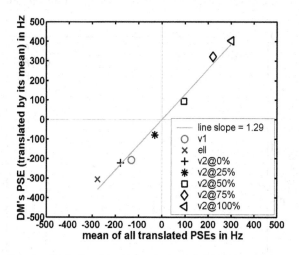

Fig. 5. Linear-regression fit through DM's PSE against the mean PSE

The regression procedure also yields a measure of goodness-of-fit expressed as the Root-Mean-Squared (RMS) deviations of the fitted lines from the ensemble data. Table 2 gives these measures with ranges for F1 ([17-28]), F2 ([17-57]) and F3 ([23-63]) that lie comfortably within the range of difference limens for human perception.

Table 2. Root-Mean-Squared (RMS) deviations (Hz) of fitted lines

SPEAKERS	F1	F2	F3
DM	25	34	42
EM	28	55	63
JM	17	57	47
PS	22	33	36
RS	24	17	23

4 Ensemble Scales

The ensemble scales obtained for all speakers and all formants are now brought together for a close evaluation of their plausibility and effectiveness as measures of similarity. This is undertaken from the viewpoint of uniformity across formants.

4.1 Preliminary Observations

Figure 6 displays the ensemble scales derived from the original (token-averaged) PSEs. For this reason, they will be referred to as *raw* scales. One striking observation concerns the F3-scales for speakers DM and EM, which seem to be at odds with the patterns for the other speakers. In addition to this apparent aberration, there are only very weak correlations between F1- and F2-scales (0.02) and between F1- and F3-scales (-0.17). The raw scales clearly exhibit a strong non-uniformity that is inconsistent with similarity. A deeper investigation is therefore warranted.

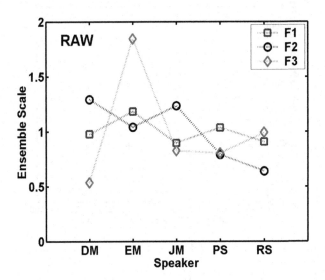

Fig. 6. Per-formant profile of **RAW** scales for the 5-speakers' ensembles. "**RAW**" signifies that the scales shown are based on the original (token-averaged) PSEs.

4.2 Insights from Vocal-Tract Length

Thus far, the scaling technique has yielded insights that might have been obscured if it had simply encompassed all formants in the first place. Nor does it need to as a tool for expressing similarity. The non-uniformity noted above is therefore investigated by independently evaluating the formant ensembles before and after ensemble scaling.

To understand the possible causes of the non-uniformity observed above, we first appeal to a closed-form expression proposed by Paige and Zue (1970: Eq. (13) on p. 169) for estimating vocal-tract length directly from formant frequencies. The expression extends the quarter-wave formula by implicating all available formants, thereby providing length estimates that are more realistic for speech sounds other than the neutral vowel. The length estimates presented here are referred to as L4 in short, as they are based on our measured F1, F2, F3 and F4.

The left panel of Fig. 7 displays, speaker by speaker, the raw L4 as a function of phonetic segment. The per-speaker patterns show encouraging consistency in the way in which variations from segment to segment reflect differing degrees of lip rounding and, conceivably, concomitant adjustments of larynx height. Whilst the overall pattern is indeed "similar" from speaker to speaker, it is visibly different in absolute terms. An intriguing question thus arises – Is the non-uniformity manifest in the raw scales partly attributable to differences in vocal-tract length patterns?

4.3 Post-scaling Effects on Vocal-Tract Length

The question raised above is approached by looking at the inter-speaker variations in vocal-tract length that remain after ensemble scaling. This is readily achieved by using the reciprocals of the raw scales for inverse scaling the ensembles formant by formant. The L4 measure is then re-applied to the inversely-scaled formants, yielding the new pattern shown on the right panel of Fig. 7.

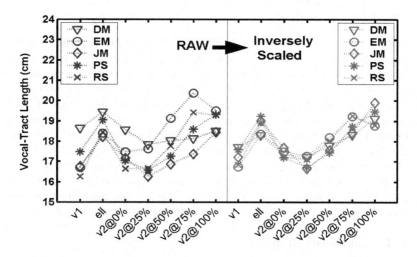

Fig. 7. Vocal-Tract Lengths (L4) based on F1, F2, F3 and F4 (Paige and Zue, 1970). *Left panel*: L4s for *raw* ensembles. *Right panel*: L4s for *inversely-scaled* ensembles.

The new pattern is revealing in several ways. The spread in L4 amongst the 5 speakers is now much smaller, indeed causing a typical behaviour to emerge from segment to segment. This behaviour indicates that, without any knowledge of inter-formant relationships, the *inverted raw scales* already account for a large proportion of the ensemble-to-ensemble variations. It seems reasonable to expect further improvements if the raw ensembles are pre-normalised using the L4 measure, which is inclusive of all available formants. A pre-normalisation stage involving L4 is described in the next section, where the results presented yield a more definite perspective on the similarity proposition.

5   Ensemble Scales and Pre-normalisation Involving L4

The argument put forward in the previous section has brought into focus the fact that the scaling technique operates on a per-formant basis and, therefore, it should not be able to completely handle the speaker-to-speaker differences in vocal-tract length patterns observed in Fig. 7. Our aim here is to secure a fairer outcome of the scaling technique by pre-normalising the raw PSEs using L4.

In Section 5.1 we describe the pre-normalisation method and, in Section 5.2, we compare the resulting ensemble scales with those shown in Fig. 6. For the sake of completeness, we will return to the patterns of vocal-tract length from speaker to speaker, and compare them with those shown in Fig. 7. Finally, the two stages put in place will take us to Section 5.3, where inversely-scaled data (L4-normalised) and raw data are contrasted in the traditional planes spanned by F1 and F2, and F2 and F3.

5.1   Pre-normalisation Method

The method employed for pre-normalising our raw ensembles was inspired by the encouraging results reported in a previous experiment (Wakita, 1977), where vocal-tract length was used as a normalisation parameter for automatic classification of 9 American English vowels uttered by 14 men and 12 women. Wakita's motivation for exploiting vocal-tract length is encapsulated in his statement that it is "not unreasonable as a first step toward inter-speaker normalisation in consideration of the structural similarity of the human vocal organs from individual to individual" (p. 184). Since this echoes some of the motivational arguments presented earlier, the algorithmic procedure developed by Wakita was adapted to our ensemble data.

Here the normalisation factor is defined as the ratio of raw L4s to their average. For each of the 7 phonetic segments, there are 5 such ratios corresponding to our 5 speakers, which are then used to normalise each formant for that segment.

5.2   Scaling Uniformity Revisited

Figure 8 shows the results of applying the scaling technique to L4-normalised PSEs. A more meaningful picture now emerges from the new ensemble scales.

The aberrant behaviour observed earlier for DM's and EM's raw scales has now disappeared and, as a result, the new scales follow a much more consistent pattern across all speakers. DM's and EM's ensembles are relatively larger by comparison with the other 3 speakers' ensembles, and the downward trend from left to right in Fig. 8 is also consistent for the 3 formants. Inter-formant correlations have expectedly grown stronger: from 0.02 to 0.68 between F1- and F2-scales, from 0.76 to 0.89 between F1- and F3-scales, and from -0.17 to 0.93 between F2- and F3-scales.

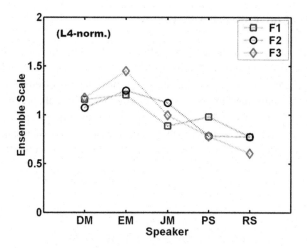

Fig. 8. Per-formant profile of the 5-speakers' ensemble scales resulting from *pre-normalisation* of the raw PSEs by vocal-tract length (L4)

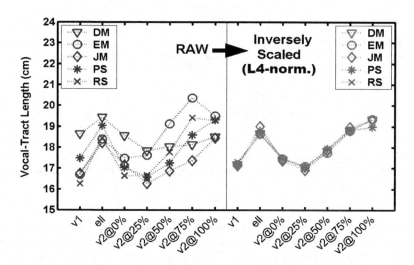

Fig. 9. Vocal-Tract Lengths (L4). *Left panel*: L4s for *raw* ensembles. *Right panel*: L4s for **L4-normalised** and then *inversely-scaled* ensembles.

The emergent perspective is indeed clearer. Some proportion of the variations manifest in the 5-speakers' PSEs appears to be caused by inter-speaker differences in vocal-tract length through the 7 segments representing the word "hello". This is confirmed in Fig. 9, where the residual variation in the new L4s is now negligible. Collectively, the results given in Fig. 8 and Fig. 9 show a believable tendency towards uniformity, thereby lending support to the similarity proposition.

5.3 Reduction of Inter-speaker Differences

The process described above is summarised in Fig. 10. There are two stages leading
to ensemble scales: (i) pre-normalisation of the raw F-patterns by vocal-tract length,
and (ii) linear regression of every ensemble against the mean ensemble. The final
stage (iii) is built into the linear-scaling approach, in that the reciprocals of the
ensemble scales are readily obtained for inverse scaling the ensembles formant by
formant. The following question remains to be elucidated: To what extent the inter-
speaker differences in our "hello" ensembles have indeed been accounted for through
the entire process?

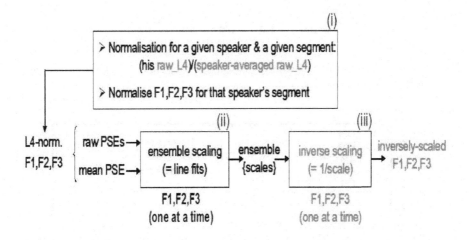

Fig. 10. Implementation stages of the linear-scaling approach: (i) pre-normalisation by vocal-
tract length; (ii) ensemble scaling via linear regression; and (iii) inverse scaling using
reciprocals of ensemble scales

Formant Spaces: Pre- and Post-Scaling. To address the above question in
graphical terms, the raw and the inversely-scaled ensembles are independently re-
grouped as convex hulls in the planar spaces of F1-F2 and F2-F3. There are 7
convex hulls corresponding to the 7 segments representative of "hello", while
every hull encloses 5 data points corresponding to the 5 speakers. The convex hulls
obtained from raw and inversely-scaled ensembles are superimposed in Fig. 11
(F1-F2 plane) and Fig. 12 (F2-F3 plane). The reduction of inter-speaker
differences is significant, as evidenced by the substantial shrinkage of each of the
7 hulls, causing the latter to be completely separated from each other. Whilst this
result provides a clear illustration of the effectiveness of the approach, a more
quantitative assessment is next sought in terms of the residual spread amongst the
5 speakers.

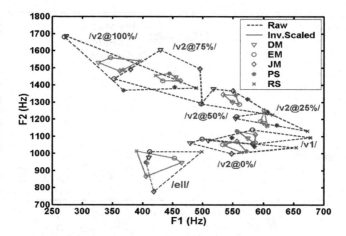

Fig. 11. F1-F2 plane for 7 segments from "hello" spoken by 5 male speakers of Australian English (DM, EM, JM, PS, RS). Raw convex-hulls are shown in blue (dashed lines). Hulls obtained by inverse scaling of L4-normalised ensembles are shown in red (solid lines).

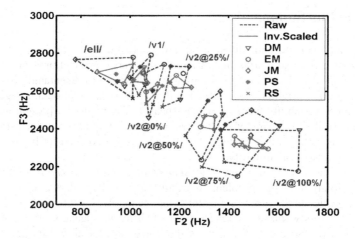

Fig. 12. F2-F3 plane with same labelling conventions as for Fig. 11

Inter-Speaker Spread: Pre- and Post-Scaling. The results plotted in Figs 11 and 12 are re-considered on a per-formant basis to yield the profiles of speaker spreads given in Fig. 13. The spread values represent the root-mean-squared (RMS) deviations from the 5-speaker averages corresponding to individual segments. The RMS profiles indicate that inter-speaker differences have been brought down to the level of inter-token variation expected for F1, F2 and F3, with slightly-differing consistency from segment to segment. On the whole, the post-scaling evidence presented supports our interpretation that the bulk of inter-speaker variations have been accounted for.

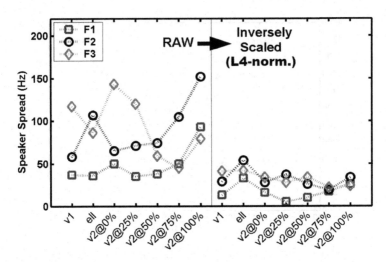

Fig. 13. Speaker spread (RMS) based on the raw (*left panel*), and the L4-normalised and inversely-scaled ensembles (*right panel*)

6 Concluding Summary

This paper has outlined and illustrated a new approach to the problem of handling acoustic-phonetic dissimilarity in different speakers' productions of the same utterance. Instead of focusing on dissimilarity and trying to disentangle its multiple sources, relations of similarity are sought with a view towards unlocking regularity in formant-pattern variability from speaker to speaker. The regularity present in the data at hand is described in terms of scaling factors, which require only linear-regression operations. Dissimilarity amongst speakers is indirectly accounted for by inverting the linear-scaling factors of similarity.

Central to the approach is the *poly-segmental formant ensemble* (PSE), which provides a solution to the problem of capturing global regularity amidst the fine acoustic-phonetic details of its segmental components. As a result, the PSE facilitates the search for (dis)similarity amongst a set of different speakers. These long-term properties of the PSE are reminiscent of those associated with a "setting", which is also poly-segmental by definition (Laver, 1980).

The results achieved thus far have shown the potentiality of approaching the problem of inter-speaker dissimilarity by way of similarity. It will be instructive to evaluate the approach with differing PSEs, with a wider-ranging set of speakers, or with forensically-oriented cases involving, for example, different imitations by the same speaker (Clermont and Zetterholm, 2006). Progress along these paths will help to focus on the pending question of how speakers (or their imitations) are distributed in a space defined by ensemble scales. Ultimately, it is conjectured that PSE scaling should prove useful for handling inter-speaker variations in speaker classification, as there is no requirement for *a priori* knowledge of the exact sources of the variations.

Acknowledgments. I owe the existence of this work[1] to Dr David J. Broad's encouragement and to many inspiring discussions with him. I thank Dr Phil Rose for giving me access to his raw data on Australian English "hello".

References

1. Broad, D.J., Clermont, F.: Linear Scaling of Vowel Formant Ensembles (VFEs) in Consonantal Contexts. Speech Communication 37, 175–195 (2002)
2. Chiba, T., Kajiyama, M.: The Vowel – Its Nature and Structure. Phonetic Society of Japan, Tokyo (1958)
3. Clermont, F., Zetterholm, E.: F-Pattern Analysis of Professional Imitations of "hallå" in three Swedish Dialects, Working Papers, Department of Linguistics and Phonetics, University of Lund 52, 25–28 (2006)
4. Laver, J.: The Phonetic Description of Vowel Quality. Cambridge University Press, Cambridge (1980)
5. Nolan, F.: The Phonetic Bases of Speaker Recognition. Cambridge University Press, Cambridge (1983)
6. Ohta, K., Fuchi, H.: Vowel Constancy on Antimetrical Vocal Tract Shapes between Males and Females. Progress Report on Speech Research, Bulletin of the Electrotechnical Laboratory (Ibaraki, Japan) 48, 17–21 (1984)
7. Paige, A., Zue, V.W.: Calculation of Vocal Tract Length. IEEE Transactions on Audio. ElectroAcoustics 18, 268–270 (1970)
8. Rose, P.: Differences and Distinguishability in the Acoustic Characteristics of "hello" in Voices of Similar-Sounding Speakers: A Forensic Phonetic Investigation. Australian Review of Applied Linguistics 22, 1–42 (1999)
9. Wakita, H.: Normalization of Vowels by Vocal-Tract Length and its Application to Vowel Identification. IEEE Transactions on Acoustics, Speech and Signal Processing 25, 183–192 (1977)

---

[1] Expanded Version of 2004 Paper – Clermont, F.: Inter-Speaker Scaling of Poly-Segmental Formant Ensembles. Proceedings of the 10th Australian International Conference on Speech Science and Technology, Sydney (2004): 522-527.

Sound Change and Speaker Identity:
An Acoustic Study

Gea de Jong, Kirsty McDougall, and Francis Nolan

Department of Linguistics, University of Cambridge,
Sidgwick Avenue, Cambridge CB3 9DA, United Kingdom
{gd288,kem37,fjn1}@cam.ac.uk

Abstract. This study investigates whether the pattern of diachronic sound change within a language variety can predict phonetic variability useful for distinguishing speakers. An analysis of Standard Southern British English (SSBE) monophthongs is undertaken to test whether individuals differ more widely in their realisation of sounds undergoing change than in their realisation of more stable sounds. Read speech of 20 male speakers of SSBE aged 18-25 from the DyViS database is analysed. The vowels /æ, ʊ, uː/, demonstrated by previous research to be changing in SSBE, are compared with the relatively stable /iː, ɑː, ɔː/. Results from Analysis of Variance and Discriminant Analysis based on F1 and F2 frequencies suggest that although 'changing' vowels exhibit greater levels of between-speaker variation than 'stable' vowels, they may also exhibit large within-speaker variation, resulting in poorer classification rates. Implications for speaker identification applications are discussed.

Keywords: Speaker identification, sound change, vowels, formant frequencies, Standard Southern British English.

1   Introduction

The system of sound contrasts in a language is constantly in flux. Linguistic variation leads to change as new realisations of existing contrasts become established, as old contrasts are subject to merger, and as new contrasts are formed. At any point in time, certain sounds are changing while others appear more stable. The variationist approach to linguistic change suggests that, at a given point in time, members of a speech community will realise particular sounds in their language system with two or more realisations (see e.g. Labov 1994: Ch. 14 [1]). These co-existing variants may be categorically different realisations phonetically, e.g. [ʔ] versus [t] for /t/, or, as is typically the case for vowel change, a number of gradiently different realisations along a continuum.

The present study examines such variation as a potential source of speaker-distinguishing information. We hypothesise that, within a given homogeneous speech community, those sounds which are undergoing diachronic change are more likely to exhibit individual variation than sounds which are relatively stable. It is likely that speakers within the group will differ in terms of their realisations of variables which are undergoing change. Certain speakers may exhibit

C. Müller (Ed.): Speaker Classification II, LNAI 4441, pp. 130–141, 2007.

more conservative or more novel realisations than others. Although in the longer term a particular change would be expected to characterise all members of a speech community, in the shorter term patterns of usage may be valuable in distinguishing different speakers (Moosmüller 1997 [2]). This paper explores this possibility through an analysis of speaker-distinguishing properties of monophthongs in SSBE.

Changes in the vowel system of SSBE (Standard Southern British English), Received Pronunciation (RP), and related accents of British English have received attention from a number of phoneticians over the years (e.g. Wells 1962 [3], 1982 [4], 1984 [5]; Gimson 1964 [6], 1980 [7], 1984 [8]; Trudgill 1984 [9], 1990 [10]; Deterding 1990 [11], 1997 [12]; Hughes and Trudgill 1996 [13]; Harrington *et al.* 2000[14]; Cruttenden 2001 [15]; Fabricius 2002 [16]). A recent and comprehensive acoustic study of diachronic change in RP monophthongs is provided by Hawkins and Midgley (2005) [17]. These authors analysed the F1 and F2 frequencies of monophthongs produced in /hVd/ contexts by male speakers of RP in four age groups: 20-25 years, 35-40 years, 50-55 years and 65-73 years. There were five speakers in each age group and directional patterns of differences in formant frequencies across successive age groups were interpreted as evidence of a time-related shift in the acoustic target for the relevant vowel.

The monophthongs Hawkins and Midgley identified as having undergone the largest changes were /ɛ, ʤ1, uː, ʊ/. For /ɛ/, and even more so for /æ/, the frequency of F1 was progressively higher for younger cohorts of speakers. These two vowels also exhibited a slight lowering in their F2 frequencies for successively younger age groups. Phonetic lowering of /æ/, the vowel found in HAD, is consistent with the auditory observations of Cruttenden (2001: 83) [15], Wells (1982: 291-2) [4] and Hughes and Trudgill (1996: 44) [13]. The frequency of F2 for /uː/, as in the word WHO'D, was progressively higher for younger speakers in Hawkins and Midgley's study, consistent with the percept of /uː/ becoming increasingly centralised, or even fronted and less rounded (cf. Hughes and Trudgill 1996: 45 [13]; Wells 1982 [5]; Cruttenden 2001: 83 [15]). Finally, Hawkins and Midgley's data for /ʊ/, as in HOOD, showed a trend for the youngest group of speakers to realise this vowel with a higher F1 frequency and a much higher F2 frequency than the three older groups. Fig. 1 shows a schematic overview of the trends described above and the mean values measured by Deterding in 1990.

Overall these changes appear to contribute to a picture of anti-clockwise movement of the peripheral monophthongs of RP in the vowel quadrilateral. However, this is not the case for all RP monophthongs, with certain vowel qualities showing apparent resistance to change. The vowels /iː, ɪ, ɑː, ɒ, ɔː, ʌ, ɜː/ exhibited similar F1 and F2 frequencies across the four age groups in Hawkins and Midgley's data, with /iː, ɑː, ɔː/ appearing particularly stable.

This chapter examines individual variation in three 'changing' vowels and three 'stable' vowels in SSBE, produced by a group of speakers of the same sex and similar age. The speech analysed is drawn from the DyViS database, developed as part of the research project 'Dynamic Variability in Speech: A Forensic Phonetic Study of British English' at the University of Cambridge.

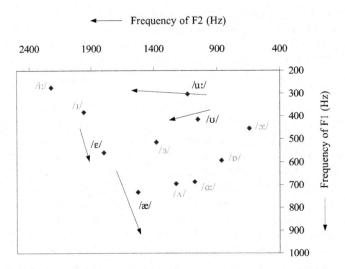

Fig. 1. A schematic representation that shows the SSBE vowels currently undergoing change. The data points are derived from Deterding (1990) [11] and the arrows represent the trends described in Hawkins and Midgley (2005) [17], where the starting-points are the mean values reported for the oldest cohort (65-73 years) and the end-points are the mean values of the youngest cohort (20-25 years) in the study.

Formant frequency measurements of /æ, ʊ, uː/ (changing) and /iː, ɑː, ɔː/ (stable) are compared, to investigate whether patterns of sound change may inform the selection of indices useful for speaker identification.

2   Method

2.1   Database and Subjects

The DyViS database is a large-scale database of speech collected under simu-lated forensic conditions. When completed, it will include recordings of 100 male speakers of SSBE aged 18-25 to exemplify a population of speakers of the same sex, age and accent group. Each speaker is recorded under both studio and tele-phone conditions, and in a number of speaking styles. The following tasks are undertaken by each subject:

1. simulated police interview (studio quality)
2. telephone conversation with 'accomplice' (studio and telephone quality)
3. reading passage (studio quality)
4. reading sentences (studio quality)

A subset of the speakers are participating in a second recording session to enable analysis of non-contemporaneous variation. Further details about the content of the database and elicitation techniques are given in Nolan *et al.* (2006)

[18]. In the present study, data from the fourth task, the read sentences, are analysed for 20 speakers, who were recorded between February and April 2006. The subjects had no history of speech or hearing problems, and their status as speakers of SSBE was judged by a phonetician who is a native speaker of that variety. The subjects are henceforth referred to as S1, S2, S3, etc.

2.2 Materials and Elicitation

The data analysed are six repetitions per speaker of the vowels /iː, æ, ɑː, ɔː, ʊ, uː/ in hVd contexts with nuclear stress. Each hVd word was included in capitals in a sentence, preceded by schwa and followed by *today*, as below:

/iː/	It's a warning we'd better HEED today.
/æ/	It's only one loaf, but it's all Peter HAD today.
/ɑː/	We worked rather HARD today.
/ɔː/	We built up quite a HOARD today.
/ʊ/	He insisted on wearing a HOOD today.
/uː/	He hates contracting words, but he said a WHO'D today.

Six instances of these sentences were arranged randomly among a number of other sentences. The sentences were presented to subjects for reading one at a time using PowerPoint. Subjects were asked to read aloud each sentence at a normal speed, in a normal, relaxed speaking style, emphasising the word in capitals. They practised reading a few sentences at the start before the actual experimental items were recorded. Subjects were encouraged to take their time between sentences and asked to reread any sentences containing errors.

Subjects were recorded in the sound-treated booth in the Phonetics Laboratory in the Department of Linguistics, University of Cambridge. Each subject was seated with a Sennheiser ME64-K6 cardioid condenser microphone positioned about 20 cm from the subject's mouth. The recordings were made with a Marantz PMD670 portable solid state recorder using a sampling rate of 44.1 kHz.

2.3 Measurements

Analysis was carried out using Praat (Boersma and Weenink 2006 [19]). Wideband spectrograms were produced for each utterance. LPC-derived formant tracks were generated by Praat, and formant frequency values written to a log file for the time-slice judged by eye to be the centre of the steady state of each vowel. In cases where no steady state for the vowel was apparent, the time-slice chosen was that considered to be the point at which the target for the vowel was achieved, according to movement of the F2 trajectory (i.e. a maximum or minimum in the F2 frequency). All measurements were compared with visual estimates based on the spectrogram, values from adjacent time-slices, and the peak values of the frequency-amplitude spectrum at the target time-slice. When values generated by Praat were judged to be incorrect, they were replaced by correct values from a time-slice immediately preceding or following the slice being measured.

3 Results

Fig. 2 compares the mean F1 and F2 frequency values of /iː, æ, ɑː, ɔː, ʊ, uː/ from Deterding (1990) [11] with the means across the 20 speakers from the DyViS project. The Figure mainly confirms the changing patterns noted by other authors: for both /uː/, the vowel in WHO'D, and /ʊ/, the vowel in HOOD, the F2 frequency is increasing, indicative of a more fronted pronunciation of those vowels. For /uː/ the change is most marked, going from Deterding's average F2 frequency of around 1100 Hz to a value of almost 1600 Hz. This difference is even larger if compared with the data for Wells (1962) [3] and the 65+ cohort for Hawkins and Midgley (2005) [17]. An increase in the frequency of F1 is observed for the vowel /æ/, giving the vowel a more open articulation. Also here, this increase is larger when compared with the formant values reported by Wells and by Hawkins and Midgley. The DyViS data confirm that indeed the pronunciations of /iː, ɑː, ɔː/ are remaining quite stable: the differences are relatively small when compared with Deterding's data and even smaller when compared with the other authors' data. Overall, the data confirm the research findings described earlier in Fig. 1 for /iː, æ, ɑː, ɔː, ʊ, uː/.

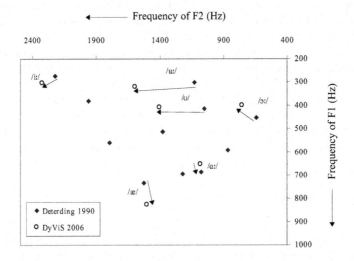

Fig. 2. Mean F1 and F2 frequency values for the recordings used in Deterding (1990) in diamonds (recorded in the late 1980s) and for the DyViS recordings in circles (recorded in 2006)

The mean values of the frequencies of F1 and F2 of /iː, æ, ɑː, ɔː, ʊ, uː/ for each individual speaker are shown in Fig. 3. Each data point represents the average realisation of the relevant vowel for a given speaker across 6 tokens. The Figure shows that these vowels differ considerably from one another in the degree of between-speaker variation they exhibit. For example, /ɔː/, the vowel in HOARD,

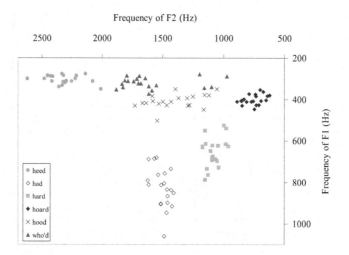

Fig. 3. Mean F1 and F2 frequency values for 20 SSBE speakers for the vowels in HEED(/iː/), HAD(/æ/), HARD(/ɑː/), HOARD(/ɔ/), HOOD(/ʊ/), WHO'D(/uː/). For each speaker, the mean consists of the formant values of 6 tokens per vowel.

is tightly clustered in the vowel space, with F1 mean values only ranging from approximately 350 to 445 Hz and F2 mean values from 625 to around 900 Hz. In other words, this vowel exhibits an F1 range of 95 Hz and an F2 range of 275 Hz. The /æ/ vowel in HAD, on the other hand, has a similar spread for F2 (i.e. 215 Hz), but a much larger one, 380 Hz, for the F1 dimension: means for F1 start around 680 Hz and the highest mean found was 1060 Hz. Overall, /æ, ʊ, uː/ exhibit the widest ranges of realisations when comparing speakers.

Consistent with the hypotheses based on patterns of sound change in SSBE, the vowels in HOOD and WHO'D demonstrate extensive variation in the F2 dimension and /æ/, as in HAD, varies widely in the F1 dimension. A result not predicted by sound change data for SSBE is that of considerable differences among speakers in their average F2 frequency of /iː/. The vowel /ɑː/ is also more variable in the F1 dimension than might be expected. However, formant frequencies are of course not only influenced by vowel quality but also by vocal tract size; we return to this issue briefly in the discussion section.

The observations described above were confirmed statistically by running a Univariate Analysis of Variance with Speaker as a random factor (20 levels) on the F1 and F2 frequencies of each vowel. The resulting F-ratios are displayed in Fig. 4. The vowels demonstrating the greatest ratio of between- to within-speaker variation in F1 are those in HAD and HARD. In the second formant the vowels yielding the highest F-ratios are HEED, HOOD and WHO'D.

To test the degree of speaker-specificity exhibited by the F1 and F2 frequencies in combination for each vowel Discriminant Analysis was carried out. This multivariate technique enables us to determine whether a set of predictors can be combined to predict group membership (Tabachnick and Fidell 1996: Ch. 11

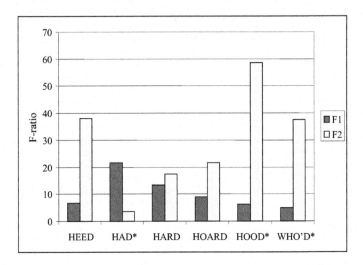

Fig. 4. Univariate ANOVA F-ratios for F1 and F2 frequencies ($df = 19$, 100; $p = 0.000$ in all cases). The vowels with * are the SSBE vowels undergoing change as reported in previous research.

[20]). For the present study a 'group' is a speaker, or rather the set of utterances produced by a speaker. The Discriminant Analysis procedure constructs discriminant functions, each of which is a linear combination of the predictors that maximises differences between speakers relative to differences within speakers. These functions can be used to allocate each token in the data set to one of the speakers and determine a 'classification rate' according to the accuracy of the allocation. In the present study, this is done using the 'leave-one-out' method, where each case is classified by discriminant functions derived from all cases except for the case itself. The higher the classification rate, the more useful the vowel for distinguishing speakers.

Direct discriminant function analyses were performed for each vowel category, using the frequencies of F1 and F2 as predictors of membership of twenty groups, S1, S2, S3, ... etc. ($k = 20$). The data set for each vowel contained the twenty speakers' six tokens, a total of 120 tokens. The resulting classification rates are shown in Fig. 5.

The Discriminant Analyses yielded rates of classification of 25% to 41%, rates much higher than chance (for all vowels $1/20 = 5\%$). However, certain vowel qualities performed better than others. The best classification rate, 41%, was achieved for the HOOD vowel. The vowels in HEED and HAD also performed quite well: 35%. Poorer rates were noted for the vowels of HARD, HOARD and WHO'D with scores ranging from 25% to 28%. These initial findings seem only partially to support our hypothesis that sounds undergoing change (HAD, HOOD, WHO'D) are particularly useful for speaker identification, at least when applying Discriminant Analysis.

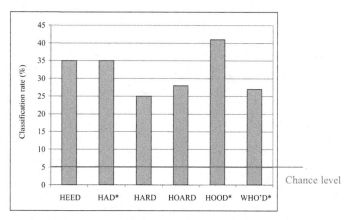

Fig. 5. Discriminant Analysis classification rates using F1 and F2 frequencies as predictors of group (speaker) membership. The vowels with * are the SSBE vowels undergoing change as reported in previous research. Chance level is 5% as there are 20 speakers. The maximum rate is found for HOOD: 41%.

So why does the vowel in WHO'D which has, as seen in Fig. 3, a relatively large between-speaker variation in F2, perform less well in Discriminant Analysis, and why does the supposedly stable vowel in HEED do better? Reasons for the differing degrees of discrimination achieved become clearer when data for individual speakers are examined with respect to within-speaker variation and vocal tract size and the effect that may have on a particular vowel. First, consider the different scenarios for /iː/, /ɔː/ and /uː/ represented by the six tokens of each vowel produced by five speakers shown in the F1-F2 plot in Fig. 6.

For the HOARD vowel /ɔː/, each speaker's tokens are clustered closely together in the vowel space. However, for the WHO'D vowel/uː/, some speakers (S15 and S22) produce very consistent realisations, whilst others (S2, S4 and S9) vary widely especially in the frequency of F2. The situation for /iː/ is different again, with speakers exhibiting large between-speaker variation and small within-speaker variation. Overall, for vowels where a speaker's average (F2, F1) realisation differs widely from one person to the next, but each individual is relatively consistent across his own productions, classification rates are higher. This is the situation for /iː/ (35%) and /ʊ/ (41%), especially due to the contribution of the F2 frequency. However, vowel qualities which exhibit large within-speaker variation for certain speakers perform less well in the classification tests. This is the case for /uː/ (particularly due to F2) and /ɑː/ (particularly due to F1), with classification rates of 27% and 25% respectively. The HAD vowel /æ/ with a rate of 35% also exhibits large within-speaker variation in the F1 direction, but this is compensated for by an extremely large between-speaker variation. The most tightly clustered vowel in Fig. 3, /ɔː/, has low within-speaker variation, but its low between-speaker variation explains its lower classification rate of 28%.

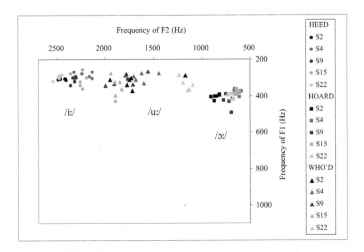

Fig. 6. F1 and F2 frequencies of /iː/, /ɔː/ and /uː/ produced by S2, S4, S9, S15 and S22 (6 tokens each)

As noted above, the vowel /iː/ did not conform to the pattern of a tight cluster of data points expected for a 'stable' vowel; rather, the vowel exhibited a wide spread in the F2 dimension. Combined with low within-speaker variation this leads to a higher level of speaker discrimination (35%) than expected. So, is the vowel in HEED currently undergoing change after all? And is it incorrect to label it as 'stable'? Auditory examination of the data for /iː/ and also for /æ/ indicates that the differences in the frequency of F2 (and F1 for /æ/) observed here do not correspond systematically to auditorily distinct vowel qualities; some vowel tokens with similar coordinates do not sound the same in terms of vowel quality. Also, speakers who tended to give the impression of having either a small or large vocal tract size were usually found at the extreme ends of the formant scale for both /iː/ and /æ/ and their vowel quality in terms of the phonetic vowel quadrilateral usually did not 'correspond' to their location in the F2-F1 graph.

These two vowels may be sensitive to the relation between a speaker's anatomy and the achievement of auditory phonetic goals in rather different ways however. F1 of /æ/ is highly sensitive to pharynx length (Nolan 1983: 171-172 [21]; Stevens 1998: 270-271 [22]) and will vary fairly directly with anatomy. In the case of /iː/, the frequency of F2 may be less crucial perceptually than a weighted average of F2, F3, and F4, which is because they may merge auditorily into one spectral prominence and as a result compensate for each other in the achievement of a specific phonetic quality (Carlson *et al.* 1975 [23]; Nolan 1994: 337-341 [24]). These are matters on which further research in the domain of speaker identity is clearly needed.

The individual differences in the means for the other vowels were more consistent with the predictions based on diachronic change, and this was confirmed by further auditory examination of tokens with formant values at the extreme ends of the clusters for these vowels.

4   Discussion

The F1 and F2 frequencies of the vowels studied achieved differing levels of speaker discrimination. Patterns of sound change are relevant to the degree of speaker-specificity exhibited by a vowel; however, vocal tract differences are also important, as explained above.

The only 'stable' vowel to yield a tight cluster of data points in F1-F2 space was /ɔː/, the vowel in HOARD; both between- and within-speaker variation were small for this vowel and the classification rate yielded was relatively low (28%). Large between-speaker variation was observed in the means for individual speakers for the F2 frequency of /iː/ as in HEED, /uː/ as in WHO'D, /ʊ/ as in HOOD and the F1 frequency of /æ/ (HAD) and to some extent /ɑː/ (HARD) (see Fig. 3). The changing vowels /æ/ and /ʊ/ yielded two of the highest classification rates on the Discriminant Analysis (35% and 41% respectively). However, /uː/ performed less well (27%) due to the large within-speaker variation in this vowel for some speakers (see Fig. 6). Although the data support the observation that /uː/ is changing in SSBE, this vowel did not perform equally well in Discriminant Analysis, since for some individuals the diachronic change is reflected in considerable within-speaker instability. The vowel /ɑː/ was not as stable as /iː/ in terms of the range of F1 values exhibited by different speakers, although the cluster of data points for /ɑː/ was smaller than those of the three 'changing' vowels, and its classification rate was the lowest of the vowels tested (25%).

The vowel /iː/ did not conform to the pattern of a tight cluster of data points expected for a 'stable' vowel; rather, the vowel exhibited a wide spread in the F2 dimension. Nevertheless, stability in /iː/ was evidenced by its low within-speaker variation, which, combined with the vowel's large between-speaker variation, led to a relatively high level of speaker discrimination (35% classification rate). The 'changing' vowel /æ/ with a rate of 35% also exhibits large within-speaker variation in the F1 direction, but this is compensated for by an extremely large between-speaker variation, probably due to a combination of the changing character of the vowel quality and vocal tract size differences.

5   Conclusion

This chapter has provided an analysis of read data from 20 speakers of Standard Southern British English from the DyViS database investigating whether sounds which are undergoing change are those most likely to differ among speakers. The results showed that this is true to an extent, but that some qualifications are needed. The historically stable vowels of SSBE /ɔː/, as in HOARD, and /ɑː/, as in HARD, offered the least reliable discrimination, and the rapidly fronting /ʊ/, as in HOOD, provided the best discrimination. The vowel in WHO'D, /uː/, is fronting, and was able to separate many speakers, but its within-speaker variation was large for some speakers, leading to a lower discrimination. A vowel particularly affected by differences in vocal tract size was /æ/. The HAD vowel showed large between-speaker variation in F1, but this conflated slight audible phonetic quality variation with the acoustic consequences of (inferred) vocal tract size differences. The /iː/ vowel, as in HEED, was auditorily stable but provided good discrimination due to low within-speaker variation (presumably linked to both historical and proprioceptive stability) and large between-speaker vara-tion (perhaps licensed by the possibility of auditory compensation in the higher formants).

The work reported here underlines just how far the speech signal is from being a straightforward biometric. To understand the speech signal fully, and therefore to be able to exploit its potential for the identification of an individual to best effect, we need to appreciate not only its complex relation to the vocal tract which produces it, but also its determination by a linguistic system set in a social and historical context. We hope this paper will stimulate progress towards an integrated theory of vocal identity which will provide principles to underpin practical work in speaker identification.

Acknowledgements

This research is supported by the UK Economic and Social Research Council [RES-000-23-1248]. Thanks are due to Toby Hudson for his assistance in the research and the preparation of this chapter, Mark J. Jones for his initial in-volvement in designing the project and for ongoing feedback, and to Geoffrey Potter for technical assistance.

References

1. Labov, W.: Principles of Linguistic Change. Internal Factors. Blackwell, Oxford (1994)
2. Moosmüller, S.: Phonological variation in speaker identification. Forensic Linguis-tics. The International Journal of Speech, Language and the Law 4(1), 29–47 (1997)
3. Wells, J.C.: A Study of the Formants of the Pure Vowels of British English. Master's thesis, University College, London (1962)
4. Wells, J.C.: Accents of English. Cambridge University Press, Cambridge (1982)
5. Wells, J.C.: English accents in England. In: Trudgill, P. (ed.) Language in the British Isles, pp. 55–69. Cambridge University Press, Cambridge (1984)
6. Gimson, A.C.: Phonetic change and the RP vowel system. In: Abercrombie, D., Fry, D.B., MacCarthy, P.A.D., Scott, N.C., Trim, J.L.M. (eds.) In Honour of Daniel Jones: Papers Contributed on the Occasion of his Eightieth Birthday. Longman, London, pp. 131–136 (1964)
7. Gimson, A.C.: An Introduction to the Pronunciation of English, 3rd edn. Edward Arnold, London (1980)
8. Gimson, A.C.: The rp accent. In: Trudgill, P. (ed.) Language in the British Isles, pp. 45–54. Cambridge University Press, Cambridge (1984)
9. Trudgill, P.: Language in the British Isles. Cambridge University Press, Cambridge (1984)
10. Trudgill, P.: The Dialects of England. Basil Blackwell, Oxford (1990)
11. Deterding, D.: Speaker Normalisation for Automatic Speech Recognition. PhD thesis, University of Cambridge (1990)

12. Deterding, D.: The formants of monophthong vowels in Standard Southern British English pronunciation. Journal of the International Phonetic Association 27, 47–55 (1997)
13. Hughes, A., Trudgill, P.: English Accents and Dialects, 3rd edn. Oxford University Press, New York (1996)
14. Harrington, J., Palethorpe, S., Watson, C.I.: Monophthongal vowel changes in Received Pronunciation: an acoustic analysis of the Queen's Christmas broadcasts. Journal of the International Phonetic Association 30(1), 63–78 (2000)
15. Cruttenden, A.: Gimson's Pronunciation of English, 6th edn. Arnold, London (2001)
16. Fabricius, A.: Weak vowels in modern RP: an acoustic study of happY-tensing and KIT/schwa shift. Language Variation and Change 14(2), 211–238 (2002)
17. Hawkins, S., Midgley, J.: Formant frequencies of RP monophthongs in four age-groups of speakers. Journal of the International Phonetic Association 35(2), 183–199 (2005)
18. Nolan, F., McDougall, K., de Jong, G., Hudson, T.: A forensic phonetic study of 'dynamic' sources of variability in speech: the DyViS project. In: Proceedings of the Eleventh Australasian International Conference on Speech Science and Technology (SST '06), Auckland, Australasian Speech Science and Technology Association, pp. 13–18 (2006), Available via http://www.assta.org
19. Boersma, P., Weenink, D.: Praat 4.4.01 (1992-2006), http://www.praat.org
20. Tabachnick, B.G., Fidell, L.S.: Using Multivariate Statistics. Harper Collins, New York (1996)
21. Nolan, F.: The Phonetic Bases of Speaker Recognition. Cambridge University Press, Cambridge (1983)
22. Stevens, K.N.: Acoustic Phonetics. MIT Press, Cambridge, MA (1998)
23. Carlson, R., Fant, G., Granström, B.: Two-formant models, pitch and vowel perception. In: Fant, G., Tatham, A.A. (eds.) Auditory Analysis and Perception of Speech, pp. 55–82. Academic Press, London (1975)
24. Nolan, F.: Auditory and acoustic analysis in speaker recognition. In: Gibbons, J. (ed.) Language and the Law, pp. 326–345. Longman, London (1994)
25. Fant, G.: Speech Sounds and Features. MIT Press, Cambridge, MA (1973)
26. Peterson, G.E., Barney, H.L.: Control methods used in a study of the vowels. Journal of the Acoustical Society of America 24(2), 175–184 (1952)

Bayes-Optimal Estimation of GMM Parameters for Speaker Recognition

Guillermo Garcia, Sung-Kyo Jung, and Thomas Eriksson

Communication System Group
Department of Signals and Systems
Chalmers University of Technology
412 96 Göteborg Sweden
garciap@s2.chalmers.se

Abstract. In text-independent speaker recognition, Gaussian Mixture Models (GMMs) are widely employed as statistical models of the speakers. It is assumed that the Expectation Maximization (EM) algorithm can estimate the optimal model parameters such as weight, mean and variance of each Gaussian model for each speaker. However, this is not entirely true since there are practical limitations, such as limited size of the training database and uncertainties in the model parameters. As is well known in the literature, limited-size databases is one of the largest challenges in speaker recognition research. In this paper, we investigate methods to overcome the database and parameter uncertainty problem. By reformulating the GMM estimation problem in a Bayesian-optimal way (as opposed to ML-optimal, as with the EM algorithm), we are able to change the GMM parameters to better cope with limited database size and other parameter uncertainties. Experimental results show the effectiveness of the proposed approach.

Keywords: Estimation, Bayes procedures, speaker recognition, Gaussian distributions, modeling.

1 Introduction

Speaker recognition is the process of automatic recognizing who is speaking based on the statistical information provided by speech signals [1]. The main technique is to find a set of features that represents a specific speaker voice. The speaker identity is correlated with the physiological and behavioral characteristics of the speech production system [2, 1], and these characteristics can be captured by short- and long-term spectra. The most common spectral features are Linear Prediction coefficients (LPC), Mel frequency cepstral coefficients (MFCC)[3] and Line Spectral Frequencies (LSF) [4, 5].

State-of-the-art text-independent speaker recognition systems commonly use the Expectation Maximization (EM) algorithm [6, 7] to estimate Gaussian Mixture Models (GMMs) for each speaker. The EM algorithm provides Maximum Likelihood (ML) estimates for the unknown model parameters from a training database. When a GMM for each speaker is computed, Bayesian classification

C. Müller (Ed.): Speaker Classification II, LNAI 4441, pp. 142–156, 2007.
© Springer-Verlag Berlin Heidelberg 2007

is used to determine the most probable speaker for test speech samples with an unknown voice.

A problem in speaker recognition is the mismatch between training and test data. Such mismatches will lead to severe performance loss, and should be avoided if possible. As we show in Section 3, the performance loss due to such mismatches is quite severe in current state-of-the-art systems. One of the reasons for the large performance loss is the two-stage procedure described above, with first ML estimation of the GMM parameters (using the EM algorithm), and then Bayesian classification. In this paper, we propose instead a Bayes-optimal parameter estimation procedure, and we show that our algorithm leads to improved performance, particularly when the amount of training data is small.

The paper is organized as follow. The state-of-the-art of text-independent speaker identification is discussed in Section 2. The Problem of mismatches between training and testing is presented is presented in Section 3. The Bayesian approach for estimation of uncertain parameters is discussed in Section 4. Finally the results and conclusions are discussed in Section 5 and 6, respectively.

2 State-of-the-Art Speaker Identification

We divide the task of speaker identification into two phases; the design phase, where a database of speech samples from known speakers are used to optimize the parameters of the system, and the classification phase, in which speech samples from unknown speakers are classified based on the parameters from the design phase.

2.1 The Design Phase

The first step in the design phase is to determine what features to use. Commonly used features are Mel-frequency cepstrum coefficients (MFCC), linear-prediction cepstrum coefficients (LPCC) or line spectrum frequencies (LSF). The MFCC features are based on the known variation of the human ear's critical bandwidths, with filter-banks that are linear at low frequencies and logarithmic at high frequencies [3]. The LPCC and LSF features are based on an all pole model used to represent a smoothed spectrum [8].

In many text-independent speaker identification systems, Gaussian Mixture Models (GMMs) are used as a statistical model of each speaker. The GMM for a speaker s is defined as

$$f^{(s)}(x) = \sum_{k=1}^{M} w_k^{(s)} g\left(x, \mu_k^{(s)}, C_k^{(s)}\right) \tag{1}$$

$$= \sum_{k=1}^{M} w_k^{(s)} \frac{1}{(2\pi)^{D/2} |C_k^{(s)}|^{1/2}} exp- \left(x - \mu_k^{(s)}\right)^T (C_k^{(s)})^{-1} (x - \mu_k^{(s)}) \tag{2}$$

i.e. it is a weighted sum of Gaussian distributions $g(x, \mu_k^{(s)}, C_k^{(s)})$, where μ_k is the mean and C_k is the covariance matrix of the k-th Gaussian distribution, and D is the dimensionality of the models. Each speaker has a unique model, describing the particular features of his/her voice.

We will use GMM models with diagonal covariance matrices in this report,

$$C_k^{(s)} = \text{diag}(\sigma_{k,1}^{(s)}, \ldots, \sigma_{k,D}^{(s)}). \tag{3}$$

In other reports, it has been shown that for similar complexity diagonal covariances works at least as good, sometimes better, than full covariance matrices, and it will simplify our equations later. The resulting GMM can then be written

$$f^{(s)}(x) = \sum_{k=1}^{M} w_k^{(s)} \prod_{d=1}^{D} \frac{1}{(2\pi)^{D/2}\sigma_{k,d}^{(s)}} \exp\left(-\frac{(x - \mu_{k,d}^{(s)})^2}{2\sigma_{k,d}^{2\,(s)}}\right) \tag{4}$$

To determine the parameters of the GMMs, the EM algorithm can be used. The EM algorithm is an iterative algorithm that uses a training database to find maximum-likelihood (ML) estimates of the weights, means and covariances in the GMM (or at least approximations to the ML estimates).

2.2   The Classification Phase

In the classification phase, the GMMs that are found in the design phase are used to compute the log-likelihood (LL) of a sample from an unknown speaker, $\{x_t\}_{t=1}^{N}$, with respect to the actual speaker GMM parameters θ,

$$
\begin{aligned}
LL^{(s)} &= \sum_{t=1}^{N} \log f^{(s)}(x_t|\theta) \\
&= \sum_{t=1}^{N} \log \sum_{k=1}^{M} w_k^{(s)} \prod_{d=1}^{D} \frac{(2\pi)^{\frac{-D}{2}}}{\sigma_{k,d}^{(s)}} \exp\left(-\frac{(x_{t,d} - \mu_{k,d}^{(s)})^2}{2(\sigma_{k,d}^{2\,(s)})}\right),
\end{aligned} \tag{5}
$$

The speaker with the highest log-likelihood is declared as the winner,

$$\hat{s} = \underset{s\in\{1...S\}}{\arg\max} LL^{(s)} \tag{6}$$

The goal of speaker recognizer is to minimize the probability of error given by $P_e = \Pr[s \neq \hat{s}]$ [9]. We will use this state-of-the-art speaker recognizer as a baseline system, against which the performance of the proposed schemes are compared.

3   The Problem: Mismatches Between Training and Testing

As discussed in the introduction, we (and others) claim that speaker recognition research suffers from performance losses due to mismatches between the training database (the data used in the design phase) and the testing database (the data used in the classification phase, to evaluate the performance of the recognizer). The mismatches are clearly observed by evaluating performance both using an independent test database (which is the standard procedure), and using the training database. We will in this document refer to performance evaluation with the training database as *closed testing*, and using an independent database as *open testing*[1]. The results using closed and open tests are given in Figure 1.

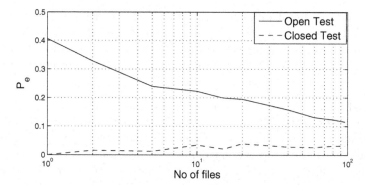

Fig. 1. Comparison of open and closed tests for various database size

The performance gap is quite large, indicating severe mismatches between the training and testing databases. As expected, the performance in an open test improves as the training set increases. In contrast, the closed test gives best result for a small training set, since the GMMs then become over-fitted to the specific details of that set. In the limiting case of an infinitely large training set, the open and closed test would converge to equal performance. In the current experiment in Figure 1, we can extrapolate the curves to at least 1000 files[2] for convergence. This is of course quite unrealistic, and we will have to suffer with small databases and, consequently, a large performance loss for the open test.

There are several possible reasons for the huge difference between closed and open tests, e.g. small training database, over-fitting of GMM parameters etc. Part of the reason for the huge performance penalty for the open test is the separation of the training phase (optimizing GMMs by the EM algorithm) from the evaluation phase (Bayesian classification). In the evaluation phase it is assumed that the pre-determined GMMs are true speaker pdfs, and no consideration is given to possible mismatches. In reality, the mismatches are severe as shown

---

[1] *Closed testing/open testing* should not be confused with *closed-set identification*, which refers to that the unknown voice comes from a fixed set of known speakers, or *open-set identification*, where a previously unknown (not within the training set) speaker is allowed.

[2] One file corresponds here to a short (1 to 2 seconds) phrase by one of the speakers.

above, and there is room for performance improvements if the mismatches are considered.

In this paper, we present a theoretical framework for optimization of the GMM parameters, and we show that Bayesian parameter estimation can partly overcome database mismatches.

4   The Solution: Bayesian Estimation of GMM Parameters

In the baseline system, if we assume that the GMM of each speaker is a perfect estimate of the speaker pdf, the system is the best possible; it is easily verified that the Bayesian classifier achieves the minimum classification error probability when the correct pdfs are given [10]. However, the GMMs are not perfect pdf estimators, due to two reasons:

- The EM algorithm cannot estimate the optimal GMM parameters, due to limited size of the training database, and also due to local minima in the EM optimization. There will always be uncertainties in the estimated parameters, and different parameters will have different uncertainty.
- A Gaussian mixture of finite order cannot describe arbitrary densities perfectly. If the GMM order is increased, the fitting ability of the GMM is increased, but the limited database size sets an upper limit to the number of model parameters that can be accurately determined.

The optimal way to deal with the uncertain parameter values is to use Bayesian estimation [11][12](as opposed to the ML approach of the EM algorithm), where the parameter values Θ are assumed to be random variables with a given pdf $f_\Theta(\theta)$. This is in contrast to the baseline system, where the parameters Θ as determined by the EM algorithm are assumed to be the true values. To compute the Bayes-optimal pdf estimate for a speaker s, the corresponding GMM $f^{(s)}(x|\theta)$ must be modified according to

$$f^{(s)}(x) = \int\limits_{-\infty}^{\infty} f^{(s)}(x|\theta) f_\Theta(\theta) d\theta \qquad (7)$$

In the simplest case, with perfectly estimated parameters Θ, the pdf $f_\Theta(\theta)$ is a Dirac delta function centered at its mean Θ_0 [13]. This simple "perfect" case gives pdf equivalent to the unmodified ML-optimized GMM, as is easily verified by inserting the Dirac function into the integral in (7). Hence, the baseline speaker recognition system is a special case of the Bayes-optimal system, where perfectly estimated parameters are assumed.

To compute our proposed Bayesian estimate, we must either know or assume some characteristics about the pdf of the parameters. The parameters of interest in this case are the weights $w_k^{(s)}$, the means $\mu_{k,d}^{(s)}$ and the variances $\sigma_{k,d}^{(s)}$ for each user s, each mixture component k and each dimension d in the the GMMs as is shown in (4). We will study the Bayesian estimation of each of these parameters one by one in the following.

4.1 GMM Modeling with Uncertain Mean Estimates

First, we assume that the weights and the variances of the GMMs are perfectly estimated by the EM algorithm, while the mean values $\mu_{k,d}^{(s)}$ are treated as uncertain. We model the GMM mean vectors as stochastic variables with Gaussian distributions with known mean μ_μ ("the mean of the mean") and variance σ_μ^2,

$$f_\mu(\mu) = g(\mu, \mu_\mu, \sigma_\mu^2). \tag{8}$$

There are several μ values to be estimated (for each speaker s, for each mixture component k, and for each dimension d) but if we use diagonal covariance matrices in the GMM, we can treat the problem as a set of independent scalar Bayes optimizations. Thus, the problem is simplified to find the Bayes-optimal pdf estimate when the original pdf is Gaussian with mean μ and variance σ^2, and the pdf of μ is also Gaussian, with mean μ_μ and variance σ_μ^2.

Inserting the pdf of μ into (7), we get

$$f(x) = \int_{-\infty}^{\infty} g(x|\mu, \sigma^2) g(\mu|\mu_\mu, \sigma_\mu^2) d\mu \tag{9}$$

$$= \int_{-\infty}^{\infty} g(x - \mu|0, \sigma^2) g(\mu|\mu_\mu, \sigma_\mu^2) d\mu \tag{10}$$

$$= g(x|\mu_\mu, \sigma^2 + \sigma_\mu^2) \tag{11}$$

where we note that we have the convolution of two independent Gaussians, yielding a new Gaussian with added means and variances.

Generalizing the expression to the original multidimensional GMM, the resulting Bayes-optimal pdf is a new GMM where each variance is replaced with the sum of the original variance (as given by the EM algorithm) and the uncertainty variance,

$$f^{(s)}(x) = \sum_{k=1}^{M} w_k^{(s)} g(x|\mu_k^{(s)}, C_k^{(s)} + C_{\mu,k}^{(s)}) \tag{12}$$

where $C_{\mu,k}^{(s)}$ is a diagonal matrix with the uncertainty variances $\sigma_\mu^2(s, k, d)$ where s is the speaker index, k is the mixture component, and d indicates the dimension index of the extracted features.

A main conclusion that can be drawn from the above analysis is that *Bayes-optimal estimates of GMM mean values can be modeled as an added variance compared to the ML estimate*, a conclusion that is intuitively satisfying. In reality, it can be quite difficult to find optimal values of each variance, but we propose a solution in Section 5.1.

4.2 GMM Modeling with Uncertain Weight Estimates

Now we assume that the means and the covariances of the GMM components are perfectly estimated by the EM algorithm, while the weights of the GMMs are treated as stochastic variables. As in the previous case, we will model the weights of the GMMs vectors as stochastic variables with Gaussian distributions with known mean μ_w and variance σ_w^2,

$$f_w(w) = g(w, \mu_w, \sigma_w^2) \tag{13}$$

Several w values to be estimated, one for each mixture component in each speaker model. For simplicity we drop s that corresponds to the speaker since the analysis is valid for all the speakers. Using (1) and the pdf of w into (7) we get

$$f(x) = \sum_{k=1}^{M} \int_{-\infty}^{\infty} w_k g(x|\mu_k, C_k) g(w_k|\mu_{w_k}, \sigma_{w_k}^2) \partial w_k \tag{14}$$

$$= \sum_{k=1}^{M} g(x|\mu_k, C_k) \int_{-\infty}^{\infty} w_k g(w_k|\mu_{w_k}, \sigma_{w_k}^2) \partial w_k \tag{15}$$

$$= \sum_{k=1}^{M} E(w_k) g(x|\mu_k, C_k) = \sum_{k=1}^{M} \mu_{w_k} g(x|\mu_k, C_k) \tag{16}$$

where $E(w_k)$ is the expectation of the weights for the k component of the GMM. In our case, the expectation is equal to the mean value of the weights for the k component of the GMM (μ_{w_k}), which is the same as the output from the EM algorithm. The main conclusion of this analysis is that the Bayesian estimate with noisy weights is the same as the ML estimate, as given by the EM algorithm[7].

4.3 GMM Modeling with Uncertain Variance Estimates

As in the previous sections of Bayes-optimal parameter estimation, we will assume that the weights and the means of the GMM components are perfectly estimated by the EM algorithm, while the covariances of the GMM are treated as uncertain. We will model the variances of the GMM as stochastic variables with Gaussian distributions with known mean μ_{σ^2} and variance $\sigma_{\sigma^2}^2$

$$f_{\sigma^2}(\sigma^2) = g(\sigma^2, \mu_{\sigma^2}, \sigma_{\sigma^2}^2). \tag{17}$$

We use diagonal covariance matrices in the GMMs and treat the problem as a set of independent scalar Bayes optimizations,

$$f(x) = \int_{-\infty}^{\infty} g(x|\mu, \sigma^2) g(\sigma^2|\mu_{\sigma^2}, \sigma_{\sigma^2}^2) \partial \sigma^2. \tag{18}$$

However, this integral does not have a simple closed form. For one extreme case, when $\sigma^2 >> \sigma_{\sigma^2}^2$, the parameter uncertainties in the variance are very low, and we can observe that the Bayesian estimator for this case is equivalent to the unmodified ML-optimized estimator with variance ($\sigma^2 = \mu_{\sigma^2}$). Otherwise, the integral is difficult to compute, and we have not included this case further in the document.

4.4 Optimal Estimation of GMM Parameters

In the previous section, we described the estimation of the parameter uncertainties in the weights $w_k^{(s)}$, the variances $\sigma_k^{2\,(s)}$ and the means $\mu_k^{(s)}$ of the GMMs. The analysis of the estimation shows that the uncertainties in the means can be simply modeled as an additional variance to the GMMs, and including them only requires a small increase in the total complexity of a speaker recognition system. The modified GMM including the model of the uncertainties in the means is expressed as:

$$f^{(s)}(x) = \sum_{k=1}^{M} w_k^{(s)} g(x|\mu_k^{(s)}, C_k^{(s)} + C_{\mu,k}^{(s)}), \qquad (19)$$

where $C_{\mu,k}^{(s)}$ is a diagonal matrix with the variances of the means $\sigma_\mu^2(s,k,d)$ where s is the speaker index, d is the dimension, and k is the component of the GMM. Using (19) in (5), the log-likelihood can be determined as :

$$LL^{(s)} = \sum_{t=1}^{N} \log \sum_{k=1}^{M} w_k^{(s)} g(x_t|\mu_k^{(s)}, C_k^{(s)} + C_{\mu,k}^{(s)}).$$

$$= \sum_{t=1}^{N} \log \sum_{k=1}^{M} w_k^{(s)} \prod_{d=1}^{D} \frac{(2\pi)^{\frac{-D}{2}}}{(\sigma_{k,d}^{2\,(s)} + \lambda_{s,k,d})^{\frac{1}{2}}} \exp\left(-\frac{(x_{t,d} - \mu_{k,d}^{(s)})^2}{2(\sigma_{k,d}^{2\,(s)} + \lambda_{s,k,d})}\right), \qquad (20)$$

where we rename the variances of the means $\lambda_{s,k,d} = \sigma_\mu^2(s,k,d)$. The modeling of the parameter uncertainties can be addressed in different ways. In the followings, we describe the methods and optimization procedures to optimize $\lambda_{s,k,d}$ in function of the speaker (s), the dimension (d) and the component of the GMM (k).

Description of the algorithm for optimization. We perform a full grid search for the optimal set of variances over speaker (s), dimension (d) and mixture component (k) using (20). Note that the optimization over speaker (λ_s), dimension (λ_d) and mixture component (λ_k) is performed independently, one at a time.

To optimize the model as a function of the speaker, we simplify (20) by restricting the variances $\sigma_\mu^2(s,k,d)$ to be equal for all the dimensions and for all the mixtures, but varying between speakers s, such that $\sigma_\mu^2(s,k,d) = \lambda_s$. Introducing λ_s instead of $\lambda_{s,k,d}$ in (20) and applying the afore-mentioned restrictions, we run the optimization algorithm to find the optimal λ_s. The algorithm is initialized with $\lambda_s = 0$ for all the speakers and then sequentially searches for the optimal λ_s for each of the speakers, until convergence of the log-likelihood is achieved.

The algorithm is applied in a similar way for the optimization in function of the dimension (d) and component of the GMM (k), where again we simplify (20), by restricting the variances $\sigma_\mu^2(s, k, d)$ to vary only over the axis we want to optimize (d or k).

5   Results

In this section, we present the results obtained from the optimization of λ. As discussed, the parameter we play with is the extra variance λ as a function of speaker, mixture number or dimension. We separately discuss the three cases, with λ a function of the speaker (λ_s), the mixture (λ_k) and the dimension (λ_d).

5.1   Experimental Setup

For the realization of the experiments, we used the YOHO database. In all tests, 32 speakers were used, whereof 16 were males and 16 females. The database was divided into three parts; one part was used for optimization of the GMMs by the EM algorithm, a second part was used for optimization of the λ values, and a third part was used for evaluation of the performance. Since the YOHO database contains a total of 138 speakers, we were able to create four independent 32-person databases, so that the robustness of the method could be evaluated.

For each speech file in the database, we removed silence at the beginning and end, and apply a 25 ms Hamming window with an overlap of 10 ms. Then, 20th-dimensional mel-frequency cepstrum coefficients (MFCC) was created.

In all of the tests, we use one database for optimization of the GMM's with the EM algorithm, another for optimization of $\lambda_{s,k,d}$, and a third for evaluating the results.

5.2   Experiments with Speaker-Dependent λ

Here we optimize λ_s, with a different value of λ for each speaker. A high value of λ_s for a particular speaker s indicates that the speaker has a high variability and uncertain estimates, and vice versa. When the optimal λ_s values were found, these values were used with an independent database to test their performance.

Table 1. Results for optimal λ_s for each speaker

No of Mixtures	Pe Baseline	Pe with optimal λ_s
4	0.1836	0.1835
8	0.0898	0.0781
16	0.0625	0.0531
32	0.0461	0.0367

Table 1 shows the results of the optimization of λ_s for the 32 speakers in the database. The second column illustrates the probability of error considering a 20 dimensional baseline GMM with 4-32 mixtures, and the third column shows the probability of error using the optimal λ_s for the same database. We see that the performance improvement is not so impressing, but that in the case of 32-order GMM, the relative improvement is 20%-25%.

5.3   Experiments with Dimension-Dependent λ

Here we present the results from the optimization of λ_d, with a different λ for each dimension of the 20-dimensional MFCC vectors. The interpretation of a high λ_d for a particular dimension d (a particular position in the MFCC vectors) is that the dimension under study suffers from a high uncertainty and should be given lower weight in the log-likelihood. Figure 2 shows an example of the estimated λ_d, illustrating the optimal λ_d for each dimension of the 20-dimensional MFCC vectors, using a 16-mixture GMM. The results are a bit noisy, but we can observe that for higher dimensions, the λ_d values are pretty high, indicating that those dimensions are more suspectable to have uncertain estimates in the EM algorithm.

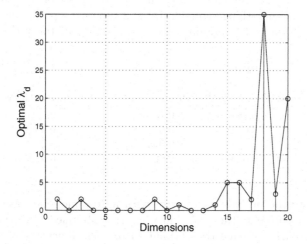

Fig. 2. Values of optimal λ_d for different dimensions used for testing

Due to the noisy behavior of Figure 2, we have also reduced the number of parameters to estimate by approximating the λ_d curves with a parametric sigmoid function, determined as

$$\lambda_d(d) = \frac{A_1}{1 + \exp(-A_2(d + A_3))}, \tag{21}$$

$$d \in \{1 \ldots D\}$$

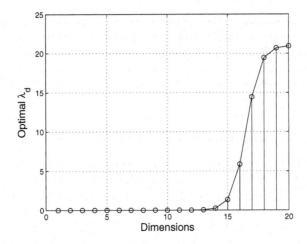

Fig. 3. Smoothing of the λ_d for different dimensions for a 16-mixture GMM

where we optimize the magnitude A_1, the slope A_2, and the offset A_3. The optimization process is based on a full search algorithm on a dense grid. The sigmoid function gives a smooth approximation to the λ_d values, and Figure 3 shows the optimal smoothing sigmoid. Table 2 shows the probabilities of error with different GMM order, obtained by optimization of the λ_d. The second

Table 2. Dimension optimization

No of Mixtures	Pe Baseline	Pe with optimal λ_d	Pe with smoothing of λ_d
4	0.1836	0.1391	0.1328
8	0.0898	0.0633	0.0547
16	0.0625	0.0352	0.0225
32	0.0461	0.0227	0.0195

column illustrate the probability of error with a 20 dimension baseline GMM, the third column presents the probability of error obtained from the Bayesian estimation approach applying the optimal λ_d to each dimension of the GMMs and the forth column shows the probability of error using a smoothing sigmoid curve for the optimal λ_d. The results are convincing, with the results for the smooth sigmoid being the best.

5.4   Testing the Robustness of the Method of Optimization of λ_d

In order to test the robustness of the method, we have used the λ_d values of Figure 3 with independent databases, with different sets of speakers. The results are shown in Table 3. We see that the probability of error in different databases varies considerably, but that the Bayesian approach is always substantially better. Since the λ_d curve was derived using one database, and the results in Table 3

Table 3. Results of the smoothing curve of λ_d for databases of 16 GMMs

Database	Pe Baseline	Pe with smoothing of λ_d
DB2	0.0688	0.0500
DB3	0.1492	0.1094
DB4	0.1734	0.0930

were obtained with three independent databases, we have illustrated the robustness of the proposed method.

5.5   Experiments with Mixture-Dependent λ

Here, we present the results of the optimization of λ_k, for different mixture component of the GMMs. Before applying the method, we ordered the mixtures according to the determinant of the corresponding covariance matrix, placing the component with the lowest determinant first. The interpretation of a high λ_k for a particular mixture component means that this component suffers from a high uncertainty and should be given lower weight in the log-likelihood. Following the same approach used for the optimization in function of the dimension, we observe that the λ_k suffers from noise. Thus, we try to find a smoothing function that can approximate the pattern of the estimated λ_k. The smoothing function used is the sum of two sigmoid functions, determined as:

$$\lambda_k(k) = B_1 \frac{1}{1 + \exp(-B_2(k + B_3))} + R_1 \frac{1}{1 - \exp(-R_2(k + R_3))} \quad (22)$$

$$k \in \{1 \dots M\}$$

where we optimize the amplitudes B_1, R_1, the slopes B_2, R_2 and the offset B_3, R_3 of each of the sigmoid functions, respectively. The optimization is based on a full search algorithm on a dense grid.

Figures 4 and 5 show the optimal λ_k and the smoothing curves for a 16 and 32-mixture GMM. We only apply the smoothing curves to these number of mixture due to fewer mixtures GMM provide patterns difficult to approximate. Table 4 shows the results of the Bayesian approach, the second column illustrates the baseline system, the third column presents the results of using the optimal λ_k and the fourth column shows the results of approximating λ_k with the

Table 4. Comparison of different number of GMMs for optimal λ_ks for each component of the GMM, weighting each distribution by the determinant

No of Mixtures	Pe Baseline	Pe with optimal λ_k	Pe with smoothing of λ_k
4	0.1836	0.1859	
8	0.0898	0.0758	
16	0.0625	0.0570	0.0531
32	0.0461	0.0398	0.0391

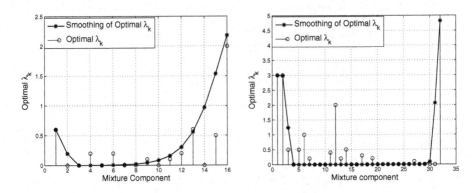

Fig. 4. Comparison of optimal λ_k for a 16 GMMs and the smoothing curve

Fig. 5. Comparison of optimal λ_k for a 32 GMMs and the smoothing curve

smoothing curve. These results shows a maximum improvement of 17% which is good since we are using the same optimal λ_k for all the dimensions of the mixture component.

5.6   Testing the Robustness of the Method for Optimization of λ_k

In order to test the robustness of the method, we have used the λ_k values of Figure 4 with independent databases, with different sets of speakers. Table 5 shows the results of the application of smoothing λ_k to databases of 16 GMMs. We can see that even with this simple implementation of the Bayesian estimation approach, we were able to gain some improvement in the performance.

5.7   Results of the Application of the Bayesian Approach to Different Databases Sizes

As it was shown in Section 3, there is a problem of mismatches between training and testing databases which is reduced as the size of the training set increases. By applying the Bayesian approach, we can make the effect of the mismatch smaller. Figure 6 shows the results of the application of the Bayesian approach to the dimensions for varying training set size and different number of mixtures. We observe that the Bayesian approach leads to an improved performance for all training set sizes.

Table 5. Results showing the Robustness of the method

Database	Pe Baseline	Pe with smoothing of λ_k
DB2	0.0688	0.0672
DB3	0.1492	0.1422
DB4	0.1734	0.1547

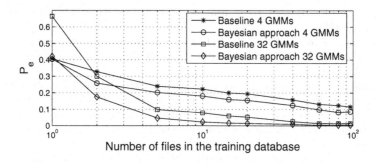

Fig. 6. Comparison of Bayesian approach and conventional method for different size of database using 4 GMMs and 32 GMMs

6   Conclusion

In this work, we illustrate that current state-of-the-art speaker recognition suffers from a mismatch between training and verification data, mainly due to small training sets. We propose a Bayesian approach to reduce the effect of such mismatches, and show that by applying the new method to the parameters of a GMM, we can improve the performance significantly.

As an example of the use of Bayes-optimized GMM for speaker recognition, we focus on estimating accuracy of different speaker models, different dimensions and on different components of GMMs . It is clearly shown that some components are of less importance than other, and by exploiting these differences, large performance gains can be reached.

References

[1] Furui, S.: Recent advances in speaker recognition. Acoustics, Speech, and Signal Processing, ICASSP-89 1, 429–440 (1989)
[2] Atal, B.S.: Automatic recognition of speakers from their voices. Proceedings of the IEEE 64(4), 460–475 (1976)
[3] Davis, P., Mermelstein, S.: Comparison of parametric representations for monosyllabic word recognition in continuously spoken sentences. Acoustics, Speech, and Signal Processing, IEEE Transactions on 28(4), 357–366 (1980)
[4] Premakanthan, P., Mikhael, W.B.: Speaker verification/recognition and the importance of selective feature extraction: review. Circuits and Systems, 2001. MWS-CAS 2001. In: Proceedings of the 44th IEEE 2001 Midwest Symposium, vol. 1, pp. 57–61 (2001)
[5] Mammone, R.P., Xiaoyu Zhang Ramachandran, R.J.: Robust speaker recognition: a feature-based approach. Signal Processing Magazine, IEEE 13(5), 58–71 (1996)
[6] Reynolds, D.A., Rose, R.C.: Robust text-independent speaker identification using Gaussian mixture speaker models. Speech and Audio Processing, IEEE Transactions 3, 72–83 (1995)
[7] Zhang, Y., Alder, M., Togneri, R.: Using Gaussian Mixture Modeling in Speech Recognition. Acoustics, Speech, and Signal Processing, 1994. IEEE International Conference (ICASSP-94) i, 613–616 (1994)

[8] Campbell, J.P.: Speaker recognition: A tutorial. Proceedings of the IEEE 85, 1437–1462 (1997)

[9] Eriksson, T., Kim, S., Kang, H.-G., Lee, C.: An information-theoretic perspective on feature selection in speaker recognition. IEEE Signal Processing Letters 12(7), 500–503 (2005)

[10] Douglas, R.: Experimental evaluation of features for robust speaker identification. IEEE Transactions on Speech and Audio Processing 2(4), 639–643 (1994)

[11] Roberts, S.J., Husmeier, D., Rezek, I., Penny, W.D.: Bayesian approaches to gaussian mixture modeling. IEEE Transactions on Pattern Analysis and Machine Intelligence 20(11), 1133–1142 (1998)

[12] Kay, S.M.: Fundamentals of Statistical Signal Processing, Estimation Theory, Prentice Hall Signal Processing Series, 2nd edn (1993)

[13] Duda, R., Hart, P., Stork, D.: Pattern Classification, 2nd edn. Wiley-Interscience Publishers, Chichester (2001)

Speaker Individualities in Speech Spectral Envelopes and Fundamental Frequency Contours

Tatsuya Kitamura[1] and Masato Akagi[2]

[1] ATR Cognitive Information Science Laboratories,
2-2-2 Hikaridai, Seika-cho, Soraku-gun, Kyoto 619-0288, Japan
kitamura@atr.jp
[2] School of Information Science, Japan Advanced Institute of Science and Technology,
1-1 Asahidai, Nomi-shi, Ishikawa 923-1292, Japan
akagi@jaist.ac.jp

Abstract. Perceptual cues for speaker individualities embedded in spectral envelopes of vowels and fundamental frequency (F0) contours of words were investigated through psychoacoustic experiments. First, the frequency bands having speaker individualities are estimated using stimuli created by systematically varying the spectral shape in specific frequency bands. The results suggest that speaker individualities of vowel spectral envelopes mainly exist in higher frequency regions including and above the peak around 20–23 ERB rate (1,740–2,489 Hz). Second, three experiments are performed to clarify the relationship physical characteristics of F0 contours extracted using Fujisaki and Hirose's F0 model and the perception of speaker identity. The results indicate that some specific parameters related to the dynamics of F0 contours have many speaker individuality features. The results also show that although there are speaker individuality features in the time-averaged F0, they help to improve speaker identification less than the dynamics of the F0 contours.

Keywords: spectral envelopes, fundamental frequency contours, perceptual speaker identification, psychoacoustic experiments.

1 Introduction

How humans perceive speaker individualities in speech waves and how humans identify phonemes overcoming speaker-to-speaker differences in the physical characteristics in speech waves are the most fundamental issues in speech science. Despite earnest studies in this area over the years, such human abilities have not yet been clarified. It is probable that the identification of perceptual cues for speaker identification and the clarification of the relationship between the cues and perceptual distance are required to elucidate the mechanism of the human abilities. In this study, we thus attempt to explore possible perceptual cues in spectral envelopes and fundamental frequency (F0) contours, assuming that the physical characteristics used by humans to identify speakers are significant ones that represent the speaker individuality.

C. Müller (Ed.): Speaker Classification II, LNAI 4441, pp. 157–176, 2007.
© Springer-Verlag Berlin Heidelberg 2007

Speaker characteristics consist of two major elements: static and dynamic physical variations in speech waves. The former is based on physical variations of speech organs and the latter reflects behavioral variations, and the two appear both in speech spectra and F0 contours. Since Takahashi and Yamamoto [1], in their study of Japanese vowels, reported that higher spectral regions contain speaker characteristics, perceptual contributions of speech spectra for speaker identification have been studied by many researchers. Furui and Akagi [2], for example, showed that speaker individualities are mainly in the frequency band from 2.5 to 3.5 kHz, within the range of telephony. Zhu and Kasuya [3] studied the perceptual contribution of static and dynamic features of vocal tract characteristics for speaker identification, and demonstrated that the static feature is significant. In the present study, we aim to investigate the relationship between specific frequency regions of speech spectra and speaker identification, focusing on static features of speech spectral envelopes.

It is well known that time-averaged F0 affects auditory speaker identification and is often used for automatic speaker identification or verification; previous studies, however, have addressed only the static characteristics of F0, but not the dynamic properties. The dynamic features of F0 could be significant for speaker perception in continuous speech, and we thus also investigate potential perceptual cues in dynamic physical characteristics of F0 contours in the present study.

We describe two separate sets of psychoacoustic experiments concerning static characteristics of speech spectral envelopes and static and dynamic features of F0 contours. In Experiments 1, 2, and 3, we attempted to identify the specific frequency regions in which speaker individualities exist. Spectral envelopes of stimuli for several frequency regions were varied using the log magnitude approximation (LMA) analysis-synthesis system [4]. The effectiveness of the specific frequency regions for automatic speaker recognition is also discussed. Experiments 4, 5, and 6 were performed to clarify the relationship between static and dynamic physical characteristics of F0 and speaker perception. In the three experiments, we employed the analysis method proposed by Fujisaki and Hirose [5] (Fujisaki F0 model) to extract and control the physical characteristics of F0, and the LMA analysis-synthesis system was used to synthesize stimuli.

2   Speaker Individualities in Speech Spectral Envelopes

2.1   Analysis of Spectral Envelopes

To identify the frequency bands containing speaker individualities in the spectral envelope, we calculated the variance for the spectral envelopes of the five Japanese vowels, /a/, /e/, /i/, /o/, and /ɯ/, for ten male speakers from the ATR speech database [6]. The sampling frequency of the data is 20 kHz. The spectral envelopes were smoothed with 60th-order fast Fourier transform (FFT) cepstra and the frequency axis was converted to the equivalent rectangular bandwidth (ERB) rate [7].

Let $E_{ijk}(n)$ be the kth-frame log-power spectrum of the jth vowel uttered by the ith speaker at ERB rate n, where $n = 1 \cdots N$, $k = 1 \cdots K$, $j = 1 \cdots J$, and $i = 1 \cdots I$. The variance of $E_{ijk}(n)$ with respect to i is given by

$$\sigma_j^2(n) = \frac{1}{I-1} \sum_{i=1}^{I} \left\{ \frac{1}{K} \sum_{k=1}^{K} E_{ijk}(n) - \frac{1}{IK} \sum_{i=1}^{I} \sum_{k=1}^{K} E_{ijk}(n) \right\}^2, \qquad (1)$$

and the frequency bands having large quantities of $\sigma_j^2(n)$ are regarded to reflect the speaker individualities. The variance of $E_{ijk}(n)$ with respect to j is given by

$$\sigma_i^2(n) = \frac{1}{J-1} \sum_{j=1}^{J} \left\{ \frac{1}{K} \sum_{k=1}^{K} E_{ijk}(n) - \frac{1}{JK} \sum_{j=1}^{J} \sum_{k=1}^{K} E_{ijk}(n) \right\}^2. \qquad (2)$$

The frequency bands having large quantities of $\sigma_i^2(n)$ are regarded to reflect the vowel characteristics.

The variances $\sigma_j^2(n)$ and $\sigma_i^2(n)$ shown in Fig. 1 indicate that potential speaker individualities are mainly above the 22 ERB rate (2,212 Hz) and that vowel characteristics mainly exist from the 12 ERB rate (603 Hz) to the 22 ERB rate. Similar results were reported by Li and Hughes [8] and Mokhtari and Clermont [9].

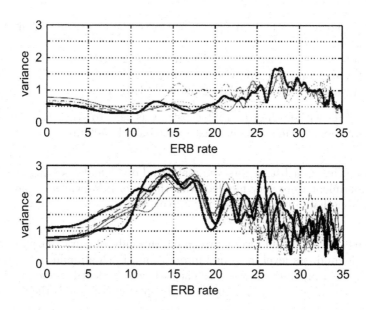

Fig. 1. Variance of spectral envelopes: the upper panel shows interspeaker variance $\sigma_j^2(n)$ and the lower panel shows intervowel variance $\sigma_i^2(n)$

2.2   Experiment 1

On the basis of the analysis results described above, we assumed that the frequency band above the 22 ERB rate contains the speaker individualities and the band from 12 to 22 ERB rate contains the vowel characteristics. Experiment 1 was designed to test this assumption from a psychoacoustic viewpoint [10].

Stimuli. Five male native Japanese speakers recorded the five Japanese vowels at a sampling rate of 20 kHz with 16-bit resolution. When uttering the vowels, the speakers were forced to tune the pitch of their voices to the same height as that of a 120 Hz pure tone in order to avoid the influence of F0 on the speaker identification tests.

The four types of stimuli used in Experiment 1 were LMA analyzed-synthesized speech waves with fixed power. The F0 contour of the stimuli was fixed, as shown in Fig. 2, and frame sequence was randomized to synthesize stimuli in which only the static feature of the spectral envelope depends on the speakers. The spectral envelopes of three of the stimuli were varied in the frequency domain. Two varying methods were used: in Method 1 the spectral envelopes were reversed symmetrically with respect to their autoregressive line, and in Method 2 the spectral envelopes were replaced by their autoregressive line. Figure 3 shows the spectral envelopes varied above 22 ERB rate by these two methods. The following types of stimuli were used:

1a LMA analyzed-synthesized speech waves without varying their spectral envelopes,

1b speech waves varied by Method 1 from 12 to 22 ERB rate,

1c speech waves varied by Method 1 above 22 ERB rate, and

1d speech waves varied by Method 2 above 22 ERB rate.

The LMA filter with 60th-order FFT cepstra was used to synthesize the stimuli, and the duration of each stimulus was approximately 500 ms.

Subjects. The eight listeners (seven males and one female) serving as subjects in the experiment were very familiar with the recorded speakers' voice characteristics. None had any known hearing impairment.

Procedure. The stimuli were presented through binaural earphones at a comfortable loudness level in a soundproof room. Each stimulus was presented to the subjects randomly three times. The task was to identify vowels and speakers, and when the subjects could not identify speakers or vowels they responded with "X."

Results. The speaker identification rates and the vowel identification rates averaged across the subjects for Experiment 1 are shown in Fig. 4. They suggest the following conclusions ($F(1, 14) = 4.60, p < 0.05$).

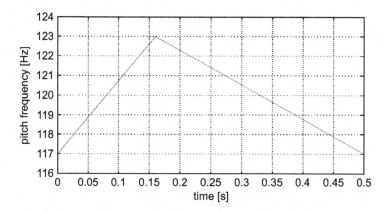

Fig. 2. F0 contour for stimuli used in Experiments 1 and 2

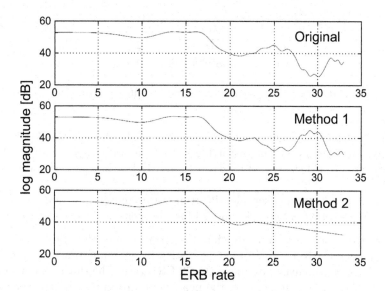

Fig. 3. Spectral envelopes varied by Methods 1 and 2 above 22 ERB rate. The top panel shows the original envelope, the middle one shows the envelope reversed symmetrically with respect to its autoregressive line (Method 1), and the bottom one shows the envelope replaced by its autoregressive line (Method 2).

1. The distortion of the spectral envelopes above the 22 ERB rate does not affect vowel identification but does affect speaker identification ($F(1,14) = 4.51$ between stimuli 1a and 1c for the vowel identification rate, $F(1,14) = 88.90$ between 1a and 1c for the speaker identification rate).

2. The distortion of the spectral envelopes from the 12 to 22 ERB rate affects vowel identification ($F(1,14) = 342.85$ between 1a and 1b for the vowel

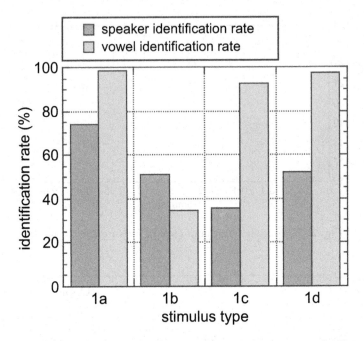

Fig. 4. Speaker (dark bars) and vowel (light bars) identification rates for Experiment 1

identification rate). This distortion has less of an affect on speaker identification rates than does the distortion of the spectral envelopes above the 22 ERB rate ($F(1, 14) = 11.84$ between 1b and 1c for the speaker identification rate).

3. Method 1 affects speaker identification rates more than does Method 2 ($F(1, 14) = 14.32$ between 1c and 1d for the speaker identification rate).

The first two conclusions indicate that the speaker individualities can be controlled independently of vowel characteristics by manipulating the frequency band of the spectral envelopes above the 22 ERB rate. The third conclusion implies that the relationship between the peaks and dips in the spectral envelopes is important in identifying speakers.

2.3   Experiment 2

The results presented in the previous section suggested that speaker individualities in spectral envelopes mainly appear in the high frequency region for the sustained vowels. Experiment 2 was carried out to clarify whether the frequency region is also significant in speaker identification for vowels extracted from a sentence [11].

Stimuli. Speech data were the three Japanese vowels /a/, /i/, and /o/ extracted from the Japanese sentence "shiroi kumo ga aoi yane no ue ni ukande iru" (the white clouds are above the blue roof) uttered by four male native Japanese speakers. The sampling frequency of the data was 20 kHz. The speakers were not instructed as to the prosody of the sentence. F0 of the speech data ranged from 95.7 Hz to 193.2 Hz.

The stimuli were LMA analyzed-synthesized speech waves with fixed power. The LMA filter with 60th-order FFT cepstra averaged for the voiced period of the speech data was used to synthesize the stimuli, and the duration of each stimulus was approximately 500 ms. The F0 contour shown in Fig. 2, after shifting its average to that of each speaker, was used to synthesize the stimuli.

The spectral envelopes of the stimuli were manipulated to study whether the higher frequency region including and above the peak at around the 20–23 ERB rate (1,740–2,489 Hz) illustrated in Fig. 5 is significant for speaker identification. Hereafter, this higher frequency region is referred to as the *higher frequency region* in italics, and the lower frequency region excluding and below the peak is referred to as the *lower frequency region*. Since the spectral peak is significant for voice perception, the *higher frequency region* is defined to include the peak at around the 20–23 ERB rate. The frequency regions are different for the different vowels and speakers. Three types of stimuli were used in Experiment 2.

ORG. LMA speech waves synthesized from the original spectral envelope for the whole frequency region.

LOW. LMA speech waves synthesized from the spectral envelope where the *higher frequency region* was replaced by the average over the speakers, for each vowel. (The original spectral envelope remains for the *lower frequency region.*)

HIGH. LMA speech waves synthesized from the spectral envelope where the *lower frequency region* was replaced by the average over the speakers, for each vowel. (The original spectral envelope remains for the *higher frequency region.*)

It should be noted that the manipulation described above did not affect vowel identification.

Subjects. The eight listeners (seven males and one female) serving as potential subjects in the experiment were very familiar with the recorded speakers' voice characteristics, but they were different from the subjects in Experiment 1. None had any known hearing impairment.

Procedure. The stimuli were filtered through a low-pass filter with a cut-off frequency of 8 kHz to eliminate high-frequency noise, and were presented through binaural earphones at a comfortable loudness level in a soundproof room. Each stimulus was presented randomly five times. The task was to identify the speaker of each stimulus. The subjects were allowed to listen to each stimulus repeatedly.

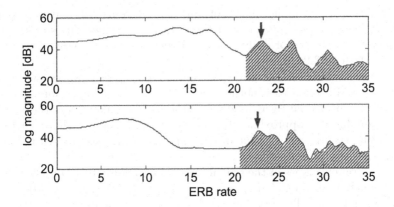

Fig. 5. The *lower* and *higher frequency regions* for the spectral envelopes of the vowels /a/ and /i/. The arrows indicate the peak at around the 20–23 ERB rate (1,740–2,489 Hz), the hatched region shows the *higher frequency region* including and above the peak, and the *lower frequency region* is that excluding and below the peak.

Results. Figure 6 shows the speaker identification results of the psychoacoustic experiment. Two of the subjects whose speaker identification rates for stimulus ORG were less than 40% were excluded from the results. The F test for the speaker identification rate ($F(1,34) = 4.13, p < 0.05$) indicates that speaker individualities mainly exist in the *higher frequency region* of the spectral envelopes for steady periods of the vowels in the sentences, because there is no significant difference between ORG and HIGH ($F(1,34) = 0.51$) whereas there are significant differences between ORG and LOW ($F(1,34) = 52.13$) and between HIGH and LOW ($F(1,34) = 38.00$).

2.4   Experiment 3

Experiment 2 revealed that the *higher frequency region* is crucial for perceptual speaker identification. In Experiment 3, we attempt to evaluate the relative effectiveness of the frequency region for automatic speaker identification or verification. We adopted the simple similarity method [13] to measure the similarity of the shape between two spectral envelopes of specific frequency regions, and demonstrated how the frequency regions contribute to automatic speaker recognition [12].

Speech Data. Five male native Japanese speakers recorded the five Japanese vowels at a sampling rate of 20 kHz with 16-bit resolution. The speakers were instructed to tune the pitch of their voices to the same height as that of a 125 Hz pure tone to avoid the influence of F0 on the evaluation. A steady period of 200 ms duration was extracted from each recorded speech waveform. Five tokens for each vowel were used in the experiment.

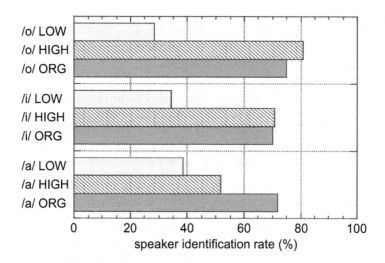

Fig. 6. Speaker identification rate of Experiment 2

Parameter. Spectral envelopes were obtained from 60th-order FFT cepstra averaged over the voiced period of the speech data. The frame length was 25.6 ms and the frame period was 6.4 ms. The 0th-order cepstrum was set to 0 prior to calculating the spectral envelopes to avoid the influence of power.

Simple Similarity Method. The simple similarity method [13] is adopted to evaluate the similarity between two spectral envelopes. The simple similarity value S_s between a reference pattern r and input pattern t is defined as

$$S_s = \frac{(r, t)^2}{||r||^2||t||^2}, \tag{3}$$

where (r, t) is the inner product of r and t, and $||r||$ is the norm of r ($= \sqrt{(r, r)}$).

Averaged Distance. The averaged distance (AD) [14] was adopted to evaluate the performance of speaker classification. When the simple similarity is used as the distance criterion, AD is defined as

$$AD = \frac{\sum_{i}^{N_{sp}} \sum_{j \neq i}^{N_{sp}} \sum_{k}^{N_{set}} \{S_s(r_i, t_{ik}) - S_s(r_i, t_{jk})\}}{N_{sp}(N_{sp} - 1)N_{set}}, \tag{4}$$

where N_{sp} is the number of speakers ($=5$), N_{set} is the number of input patterns ($=5$), r_i is the reference pattern of speaker i ($= 1, \cdots, N_{sp}$), and t_{ik} is the kth input pattern of speaker i. A larger AD value means that the classification method gives higher performance, and the speaker-specific shape of the spectral envelope thus would appear in the region of frequencies that give the higher AD value.

AD values of the following three frequency regions were calculated: 0–33 ERB rate (0–8,000 Hz), 0–20 ERB rate (0–1,740 Hz), and 20–33 ERB rate (1,740–8,000 Hz). The last frequency region is significant for perceptual speaker identification, as shown by the results of Experiment 2.

Results. AD values for the three frequency regions are listed in Table 1. They show that the performance of speaker classification using the frequency region from the 20 to 33 ERB rate is the best. The results indicate that the spectral envelope in that frequency region has essentially a speaker-specific shape, thus making it suitable for speaker classification based on the simple similarity method.

Table 1. Averaged distance value for three frequency regions

frequency region (ERB rate)	vowel				
	/a/	/e/	/i/	/o/	/ɯ/
0 − 33	0.077	0.103	0.110	0.046	0.068
0 − 20	0.033	0.045	0.035	0.017	0.031
20 − 33	0.510	0.351	0.639	0.514	0.579

2.5   Discussions

The results of the psychoacoustic experiments suggested that speaker individualities embedded in static characteristics of spectral envelopes mainly exist in the *higher frequency region*, a result that is consistent with previous psychoacoustic studies [1][2]. The results of Experiment 3 demonstrated that the *higher frequency region* contributes not only to perceptual speaker identification but also to automatic speaker recognition. The effectiveness of selecting or overweighting a higher frequency region for speaker recognition was demonstrated by Hayakawa and Itakura [15], Lin *et al.* [16], and Sivakumaran *et al.* [17].

Most recently, Kitamura *et al.* [18] reported, on the basis of MRI observations, that the shape of the hypopharyngeal cavities is relatively stable regardless of the vowel whereas it displays a large degree of interspeaker variation, and that the interspeaker variation of the hypopharynx affects the spectra in the frequency range beyond approximately 2.5 kHz. They concluded that hypopharyngeal resonance (i.e., the resonance of the laryngeal cavity and the antiresonance of the piriform fossa) constitutes a causal factor of speaker characteristics. It is fair to say that their results provide a biological basis for the experimental results described above.

3   Speaker Individuality in Fundamental Frequency Contours [19]

3.1   Analysis of F0 Contours

To discuss the relationship between F0 contours and the speaker individuality embedded in them, we adopt the Fujisaki F0 model [5] to represent F0 contours in this study. In this model, F0 contours consist of two elements, phrase and accent, which can be controlled independently.

Representation of F0 Contours. The Fujisaki F0 model represents the F0 contour $F0(t)$ as follows [5]:

$$\ln F0(t) = \ln F_b + \sum_{i=1}^{I} A_{pi} G_p(t - T_{0i})$$

$$+ \sum_{j=1}^{J} A_{aj} \{G_a(t - T_{1j}) - G_a(t - T_{2j})\} + A_{pe} G_p(t - T_3) \quad (5)$$

$$\begin{cases} G_p(t) = \alpha^2 t \exp(-\alpha t) \\[2mm] G_a(t) = \min[1 - (1 + \beta t) \exp(-\beta t), 0.9], \end{cases} \quad t \geq 0$$

where F_b is the baseline value of an F0 contour, A_{pi} is the magnitude of the ith phrase command, A_{aj} is the amplitude of the jth accent command, I is the number of phrase commands, J is the number of accent commands, T_{0i} is the timing of the ith phrase command, T_{1j} and T_{2j} are the onset and offset of the jth accent command, and α and β are natural angular frequencies of the phrase and accent control mechanisms, respectively. Parameters α and β characterize dynamic properties of the laryngeal mechanisms for phrase and accent control, and therefore may not vary widely between utterances and speakers [20]. They were thus fixed at $\alpha = 3.0$ and $\beta = 20.0$. The negative phrase command at the end of the utterance was used as T_3. A schematic figure of the model is shown in Fig. 7.

Speech Data. The speech data used for all the experiments were three-mora words with accented second mora: "aōi" (blue), "nagāi" (long), and "niōu" (smell). Each word was uttered ten times by three male speakers: KI, KO, and YO. When recording the speech data, the speakers were instructed to utter the words with the standard Japanese accent and without emotion. Although the speakers are from different areas of Japan, in this study, difference in the speaker's home districts was regarded as one of the causes of speaker individuality.

The speech data were sampled at 20 kHz with 16-bit accuracy, and analyzed using 16th-order LPC in a 30 ms Hanning window for every 5 ms period. The autocorrelation of the LPC residual signal was used to estimate the F0 contours. Equation (5) was then fitted to the contours by the analysis-by-synthesis method.

Analysis Results. To identify physical characteristics that represent speaker individuality in the analyzed parameters, we calculated the F ratio (interspeaker variation divided by averaged intraspeaker variation) for each parameter

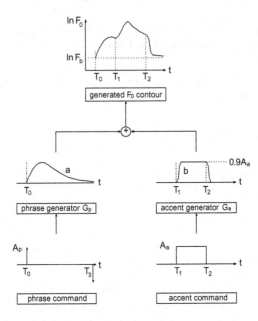

Fig. 7. Schematic figure of Fujisaki F0 model (1 phrase command and 1 accent command). The lower left and right diagrams illustrate a phrase and an accent synthesis, respectively. The upper diagram shows a synthesized F0 contour.

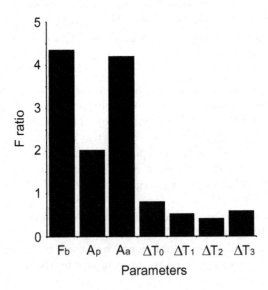

Fig. 8. F ratio of each parameter for the word "aoi." ΔT_i indicates the difference between the command timing and corresponding mora boundary.

of the F0 model. A large F ratio indicates a parameter of which the inter-speaker variation is large and the intraspeaker variation is small, and suggests that the parameter may be significant for speaker identification. Figure 8 shows the F ratio for each parameter of the word "aoi" as an example. In the figure, ΔT_i indicates the difference between the command timing and corresponding mora boundary. The aspects of the F ratio of the other two words are almost the same.

These results indicate that the F ratio of three parameters, F_b, A_p, and A_a, are much larger than those of the other parameters, which suggests that the three parameters are significant for perceptual speaker identification. Parameter F_b is related to the time-averaged F0 and parameters A_p and A_a are related to the dynamic range of the F0 contour. The significance of the time-averaged F0 for speaker perception is well known and it is usually used in automatic speaker identification and verification systems. The dynamics of F0 contours, however, have not been studied in detail.

3.2   Experiment 4

Experiment 4 clarifies whether speaker individuality still exists in the F0 contours when spectral envelopes and amplitude contours are averaged for the three speakers, and when F0 contours are modeled by the Fujisaki F0 model.

Stimuli. The stimuli were original speech waves re-synthesized by the LMA analysis-synthesis system [4]. Four types of stimuli were used for Experiment 4:

4a original speech waves,
4b LMA speech waves without modification of their FFT cepstral data,
4c LMA speech waves with spectral and amplitude envelopes averaged for the three speakers, which we call spectral-averaged LMAs, and
4d spectral-averaged LMA speech waves whose F0 contours were modeled by Eq. (5), which we call F0-modeled LMAs.

Since the time lengths of words uttered by the three speakers were different, a dynamic programming (DP) technique was adopted to shorten or lengthen each word nonlinearly.

The local distance for the DP in this experiment was an linear predictive coding (LPC)-spectrum distance,

$$d(x,y) = \sqrt{2 \sum_{i=1}^{P} (c_i^x - c_i^y)^2}, \tag{6}$$

where c_i^x and c_i^y are ith LPC-cepstra of speakers x and y and P is the LPC order. The LPC-cepstra were analyzed using 16th-order LPC in 30-ms Hanning window for every 5 ms period.

The cepstral and amplitude sequences of each word were time-warped and their duration was normalized by the DP-path of each word. The duration-normalized FFT-cepstral and amplitude sequences were averaged arithmetically in each frame and inversely relengthened or reshortened using each DP-path. The calculated FFT-cepstral sequence was the spectral-averaged FFT-cepstral sequence with the same length as the original. The spectral-averaged LMA (4c) was, thus, resynthesized with the spectral-averaged FFT-cepstral sequence, the averaged amplitude sequence, and the extracted F0 contour. In contrast, the F0-model LMA (4d) was re-synthesized with the spectral-averaged FFT-cepstral sequence, the averaged amplitude sequence, and the F0 contour modeled by Eq. (5).

The stimuli were presented through binaural earphones at a comfortable loudness level in a soundproof room. Each stimulus was presented to each subject six times.

Subjects. The ten male listeners serving as subjects were very familiar with the speakers' voices. No listeners had any known hearing impairment. They also served in Experiments 5 and 6.

Procedures. The task was to identify the speaker of the stimulus. The subjects were allowed to listen to the stimulus repeatedly. Speaker identification rates for the stimuli were averaged for all subjects. This procedure was also used in Experiments 5 and 6.

Results and Discussion. The speaker identification rates are shown in Fig. 9. The results lead to the following conclusions ($F(1, 18) = 4.41, p < 0.05$).

1. Speaker individuality remains in the LMA analysis-synthesis speech. The difference between the speaker identification rates of 99.1% for the original speech waves (4a) and 97.4% for the LMA speech waves without modification (4b) is not significant ($F(1, 18) = 2.82$).
2. The speaker identification rate for the spectral-averaged LMA speech waves (4c) is still large enough to distinguish speakers (92.0%), although the difference between the speaker identification rates for stimuli 4b and 4c is significant ($F(1, 18) = 7.35$). This indicates that there is speaker individuality in the F0 contours, even though both spectral and amplitude envelopes are averaged, and that speaker individuality also exists in the spectral and amplitude envelopes.
3. Speaker individuality still remains in the F0 contours calculated using Eq. 5, because the speaker identification rate for the F0-modeled LMA speech waves 4d is 88.2% and $F(1, 18) = 1.67$ between stimuli 4c and 4d. This result suggests that the Fujisaki F0 model can be used as a basis for controlling speaker individuality.

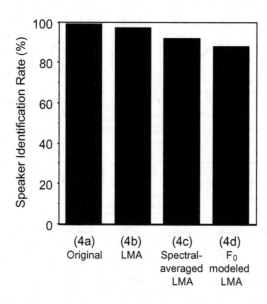

Fig. 9. Speaker identification rates of Experiment 4

3.3   Experiment 5

The aim of Experiment 5 is to clarify whether the three parameters F_b, A_p, and A_a of the Fujisaki F0 model, which were found to have large F ratio values, are effective in identifying speakers, and whether the perceived speaker identity would change upon exchanging parameter values across speakers.

Stimuli. The following types of stimuli were used in Experiment 5:

5a F0-modeled LMA speech waves (same as stimulus 4d), and
5b modified F0-modeled LMA speech waves.

Stimulus 5b was resynthesized with the spectral-averaged FFT-cepstral sequence and the modified F0 contour, of which parameters F_b, A_p, and A_a were exchanged with those of another speaker. The duration of the modified F0-modeled LMA speech waves was the same as that of the original speaker's. We call the speaker who contributes parameters F_b, A_p, and A_a for stimulus 5b the "target" speaker and the speaker who contributes parameters ΔT_0, ΔT_1, ΔT_2, and ΔT_3 for the stimulus the "origin" speaker. Figure 10 is a schematic diagram of the exchange of parameters.

Parameters were exchanged between speakers KI and KO, and YO's speech was used as dummy data. First, the stimuli for 5a were presented two times to the subjects for training and then the stimuli for 5b were presented randomly six times.

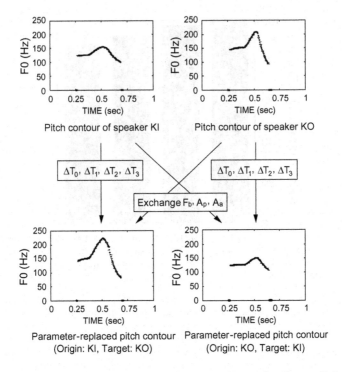

Fig. 10. Schematic diagram of parameter exchange for the word "aoi"

Results and Discussion. The speaker identification rates for Experiment 5 are shown in Fig. 11. The results can be summarized as follows.

1. The speaker identification rate of the target speaker for modified F0-modeled LMA speech waves (5b-2) is 88.9% and $F(1, 18) = 0.05$ between stimuli 5a and 5b-2 in Fig. 11. Note that the speaker identification rate of stimulus 5a is 88.2%.
2. The speaker identification rate of the origin speaker for modified F0-modeled LMA speech waves (5b-1) is 3.4%, and that of the other speaker (5b-3) is 7.8%. These values are much smaller than the rate identified for the target speaker.

These results indicate that the parameters F_b, A_p, and A_a, which describe time-averaged F0 and the dynamic range of an F0 contour, are significant in controlling speaker individuality. The results also suggest that the timings of the commands are not particularly significant in identifying speakers when the perceived speech is a word.

3.4   Experiment 6

The aim of Experiment 6 is to clarify whether speaker individuality can be manipulated by shifting the time-averaged F0. Time-averaged F0 values are often used for automatic speaker identification or verification. This experiment evaluates whether the parameter is also efficient for speaker individuality control.

Stimuli. The following types of stimuli were used in Experiment 6:

6a spectral-averaged LMA speech waves (same as stimulus 4c) and
6b F0-shifted LMA speech waves.

The F0-shifted LMA speech waves were resynthesized with the spectral-averaged FFT-cepstral sequence and the F0 contour whose time average was shifted to equal that of another speaker. Since the time-averaged F0 can be modified by shifting F0 contours, the Fujisaki F0 model is not adopted in this experiment. We call the speaker who contributes all parameters except the time-averaged F0 for stimulus 6b the "origin" speaker and the speaker who contributes the time-averaged F0 for the stimulus the "target" speaker.

F0 contours were shifted between speakers KI and KO, and YO's speech was used as dummy data. First the stimuli for 6a were presented two times to the subjects for training and then the stimuli for 6b were presented randomly six times.

Results and Discussion. The speaker identification rates are shown in Fig. 12. The results indicate that the following.

1. The identification rate of the origin speaker for the stimulus is 50.8% (6b-1 in Fig. 12) and that of the target speaker for stimulus 6b is 37.2 % (6b-2 in Fig. 12). $F(1, 18) = 24.63$ between stimulus 6a and 6b-1, and 67.50 between stimuli 6a and 6b-2.

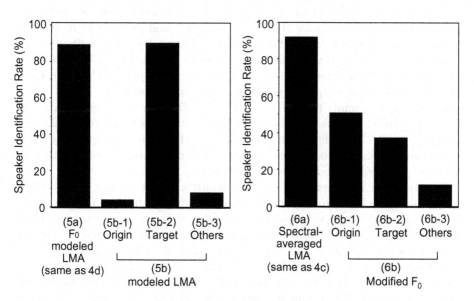

Fig. 11. Speaker identification rates of Experiment 5

Fig. 12. Speaker identification rates of Experiment 6

2. The results indicate no significant difference between the identification rates of the target and origin speakers for the F0-shifted LMA speech waves, because $F(1, 18) = 2.59$ between stimuli 6b-1 and 6b-2.

These results suggest that shifting the time-averaged F0 of one speaker to that of another speaker causes perceptual confusion during speaker identification. Although the time-averaged F0 still contains speaker individuality, these values are not as significant as the dynamics of the F0 contours, which are described by the set of parameters F_b, A_p, and A_a. It is not clear which of F_b, A_p, and A_a is the most significant for speaker identification; however, it is clear that a change in any one of them can cause misperception and all of them are needed to identify speakers.

4   Conclusion

We proposed some physical characteristics related to speaker individuality embedded in the spectral envelopes of vowels and F0 contours of words and investigated the significance of the parameters through speech analyses and psychoacoustic experiments. FFT spectral envelopes of specific frequency regions were directly manipulated, the Fujisaki F0 model was used to extract and manipulate the physical characteristics of F0, and the stimuli were synthesized by the LMA analysis-synthesis system.

Experiments 1 and 2 showed that, under the present experimental conditions, speaker individuality in the spectral envelopes mainly exists including and above the peak at around the 20–23 ERB rate (1,740–2,489 Hz), which is referred to as the *higher frequency region* in this study. The frequency region includes the third and higher formants for the vowels /a/, /o/, and /ɯ/, while it includes the second and higher formants for the vowels /i/ and /e/. The results are supported from a biological viewpoint by the results obtained by Kitamura *et al.* [18], who showed that the shape of the hypopharyngeal part of the vocal tract affects spectra in the frequency region beyond approximately 2.5 kHz. The experimental results imply that speaker individualities can be controlled without influencing vowel identification by manipulating this frequency region. Experiment 3 revealed that the spectral envelope in the *higher frequency region* is more effective for automatic speaker identification or verification than that in the *lower frequency region* or the whole frequency region.

The results of Experiments 4 and 5 demonstrate that speaker individuality exists in both the F0 contours and spectral envelopes and that the parameters F_b, A_p, and A_a, which are related to the dynamics of F0 contours, are significant in identifying speakers. The results suggest that speaker individuality can be controlled when the three parameters are manipulated. Experiment 6 showed that shifting the time-averaged F0 of one speaker to that of another results in a low speaker identification rate. Although the time-averaged F0 is often used as a distinctive feature for automatic speaker identification and verification, these values are not as significant as the dynamics of the F0 contours controlled by the parameters F_b, A_p, and A_a.

In Experiments 4, 5, and 6, the speech data were three-mora words with accented second mora, and the speakers came from different dialectal districts. In future work, we will investigate speaker individuality for different accent or mora words, for sentences, and among speakers of the same dialect. Duration of phonemes and pauses may also be significant in identifying speakers when listening to sentences, and the relative significance of the three parameters, F_b, A_p, and A_a, may change when using speech data of the same dialect.

Acknowledgments. This study was supported in part by Grants-in-Aid for Scientific Research from the Ministry of Education (No. 07680388), from the Japan Society for the Promotion of Science (No. 6157), and from the Ministry of Internal Affairs and Communications through their Strategic Information and Communications R&D Programme (No. 042107002).

References

1. Takahashi, M., Yamamoto, G.: On the physical characteristics of Japanese vowels. Res. Electrotech. Lab. 326 (1931)
2. Furui, S., Akagi, M.: Perception of voice individuality and physical correlates. Trans. Tech. Com. Psychol. Physiol. Acoust. H85-18 (1985)
3. Zhu, W., Kasuya, H.: Perceptual contributions of static and dynamic features of vocal tract characteristics to talker individuality. IEICE Trans. Fundamentals E81-A, 268–274 (1998)
4. Imai, S.: Log magnitude approximation (LMA) filter. IEICE Trans. Fundamentals J63-A, 886–893 (1980)
5. Fujisaki, H., Hirose, K.: Analysis of voice fundamental frequency contours for declarative sentences of Japanese. J. Acoust Soc. Jpn (E) 5, 233–242 (1984)
6. Takeda, K., Sagisaka, Y., Katagiri, S., Abe, M., Kuwabara, H.: Speech database user's manual. ATR Tech. Rep. TR-I-0028 (1988)
7. Glasberg, B.R., Moore, B.C.J.: Derivation of auditory filter shapes from notched-noise data. Hear. Res. 47, 103–138 (1990)
8. Li, K.-P., Hughes, G.W.: Talker differences as they appear in correlation matrices of continuous speech spectra. J. Acoust. Soc. Am. 55, 833–873 (1974)
9. Mokhtari, P., Clermont, F.: Contributions of selected spectral regions to vowel classification accuracy. In: Proc. ICSLP'94, pp 1923–1926 (1994)
10. Kitamura, T., Akagi, M.: Speaker individualities in speech spectral envelopes. J. Acoust. Soc. Jpn (E) 16, 283–289 (1995)
11. Kitamura, T., Akagi, M.: Speaker individualities of vowels in continuous speech. Trans. Tech. Com. Psycho. Physio. Acoust. H-96-98 (1996)
12. Kitamura, T., Akagi, M.: Frequency bands suited to speaker identification by simple similarity method. In: Proc. Autumn Meet. Acoust. Soc. Jpn, pp. 237–238 (1996)
13. Iijima, T.: Theory of Pattern Recognition. Morikita, Tokyo (1989)
14. Kitamura, Y., Iwaki, M., Iijima, T.: Pluralizing method of similarity for speaker-independent vowel recognition. IEICE Tech. Rep. Sp. 95, 47–54 (1996)
15. Hayakawa, S., Itakura, F.: Text-dependent speaker recognition using the information in the higher frequency band. In: Proc. ICSLP'94, pp. 137–140 (1994)

16. Lin, Q., Jan, E.-E., Che, C.-W., Yuk, D.-S., Flanagan, J.: Selective use of the speech spectrum and a VQGMM method for speaker identification. Proc. ICSLP'96, pp 2415–2418 (1996)
17. Sivakumaran, P., Ariyaeeinia, A.M., Loomes, M.J.: Sub-band based text-dependent speaker verification. Speech Commun. 41, 485–509 (2003)
18. Kitamura, T., Honda, K., Takemoto, H.: Individual variation of the hypopharyngeal cavities and its acoustic effects. Acoust. Sci. & Tech. 26, 16–26 (2005)
19. Akagi, M., Ienaga, T.: Speaker individuality in fundamental frequency contours and its control. J. Acoust. Soc. Jpn (E) 18, 73–80 (1997)
20. Fujisaki, H., Ohno, S., Nakamura, K., Guirao, M., Gurlekian, J.: Analysis of accent and intonation in Spanish based on a quantitative model. In: Proc. ICSLP'94, pp. 355–358 (1994)

Speaker Segmentation for Air Traffic Control

Michael Neffe[1], Tuan Van Pham[1], Horst Hering[2], and Gernot Kubin[1]

[1] Signal Processing and Speech Communication Laboratory
Graz University of Technology, Austria
[2] Eurocontrol Experimental Centre, France
`michael.neffe@TUGraz.at`, `v.t.pham@TUGraz.at`,
`horst.hering@eurocontrol.int`, `g.kubin@ieee.org`

Abstract. In this contribution a novel system of speaker segmentation has been designed for improving safety on voice communication in air traffic control. In addition to the usage of the aircraft identification tag to assign speaker turns on the shared communication channel to aircrafts, speaker verification is investigated as an add-on attribute to improve security level effectively for the air traffic control. The verification task is done by training universal background models and speaker dependent models based on Gaussian mixture model approach. The feature extraction and normalization units are especially optimized to deal with small bandwidth restrictions and very short speaker turns. To enhance the robustness of the verification system, a cross verification unit is further applied. The designed system is tested with SPEECHDAT-AT and WSJ0 database to demonstrate its superior performance.

Keywords: Speaker Segmentation/Verification, Air Traffic Control, GMM-UBM, Voice Activity Detection, Quantile Filtering.

1 Introduction

The air ground voice communication between pilots and air traffic controllers is hardly secured. This contribution proposes the introduction of an additional security level based on biological speech parameters. The Air Traffic Control (ATC) voice communication standards have been defined by international conventions in the forties of the last century. These standards do not address security issues in the air ground voice communication. Illegal phantom communication of "jokers" playing the role of pilot or controller has repeatedly been reported. Voice communications from attackers to achieve terrorist goals are possible. In order to address the raised security demands the introduction of security measures for the air ground voice communication is required. The proposed levels of security have to align with the existing technical communication standards. Beside the technical standards for the physical transmission channel, behavioral rules for the users of the so-called party-line channel are established. Party-line communication means that the communication with all aircrafts in a controlled sector handled by an unique controller takes place in a consecutive manner on a

C. Müller (Ed.): Speaker Classification II, LNAI 4441, pp. 177–191, 2007.
© Springer-Verlag Berlin Heidelberg 2007

single shared voice channel. Therefore, pilots have to identify themselves in each voice message with the attributed call sign for this flight. The ATC controller uses the same call sign in any reply to identify the destination of the voice message. Addressing messages by call signs requires permanent attentiveness of all party-line users. Call sign confusion represents an important problem in terms of ATC safety. A recent study showed that call sign confusion can be associated with about 5% of the overall ATC related incidents [1].

In the legacy air ground voice communication system multiple safety and security lacks can be identified. As a consequence, the Eurocontrol Experimental Centre (EEC) proposed the Aircraft Identification Tag (AIT) [2] in 2003, which embeds a digital watermark (e.g., call sign) in the analog voice signal of a speaker before it is transmitted over the Very High Frequency (VHF) communication channel. AIT represents an add-on technology to the existing VHF transceiver equipment, which remains unchanged. The watermarks are not audible for humans. They represent a digital signature of the originator (pilots or controller) hidden in the received voice message. AIT allows the automatic identification of the origin of the voice communication within the first second of speaking as the schematic illustration in Fig.1 shows. As stated previously, spoken call signs are included in each voice message to identify originator and destination of this voice message. Many different reasons like bad technical channel quality, misunderstanding, speaking errors and so on, may make the spoken call sign unserviceable for the destination. This creates supplementary workload and call-sign confusion may affect ATC safety. AIT will help to overcome this safety issue as it can be used to "highlight" the speaking aircraft in real-time. In this manner AIT also introduces some basic level of security for the communication layer.

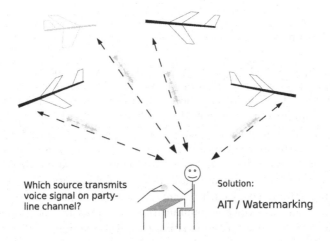

Fig. 1. Illustration of the voice communication between pilots and controller in one control sector. The AIT watermarking system allows the identification of the transmitting source at a particular time.

Our research goes beyond pure AIT watermarking and introduces a new security level for the air ground communication channel by using *behavioral biometric* voice data of the party-line speakers. The idea is that some *behavioral biometric* characteristics can be extracted from the pilots' voices and are automatically enrolled when an aircraft registers the first time to a control sector. At any later occurrence of the same AIT signature the new received speaker voice can be compared with the existing enrolled speaker dependent models to verify whether the speaker had changed as proposed in [3]. Recapitulating, the AIT reduces the problem of distinguishing different speakers on the party-line channel known as the speaker segmentation problem to a binary decision problem of claimant vs. imposter speaker. Note that the use of a speaker segmentation system alone can not satisfy the high security demands needed for ATC. On the one hand because the AIT watermarking is able to determine the beginning of each speaker turn exactly and assigns it to the corresponding aircraft. On the other hand for the verification problem the system has to make a binary decision compared to a one out of N decision for the segmentation task, where N is the number of pilots on the party-line in a certain control sector. Moreover, in ATC only the information of a speaker change in one aircraft is relevant. As one can imagine the error rate for such system is higher compared to a verification system. The enrolled speaker model should be handed over from control sector to control sector to enable a more accurate modeling of speakers with flight progress. This proposed concept secures the up and down link of the ATC voice communication. An illustration of the proposed solutions is shown in Fig.2. Before transmitting a voice message the push-to-talk (PTT) button has to be pressed, which introduces a *click* on the transmission channel. This solution may be considered as a first level of basic security. Using a *click* detector one may determine the start of each talk spurt. On the party-line unfortunately no information of the transmitting source can be gained with such a method. The AIT as shown in the second level in Fig.2 identifies beside the start of each sent voice message also the transmitting source microphone. Base on this framework, the level of security can be improved by embedding a speaker verification system which is depicted as the third level in Fig.2. At the second and third level, the first numerical index determines the aircraft number and the second the speaker. As an example AC22 is not equal to AC21 which is the first speaker assigned to aircraft AC2 when AC2 enters the control sector. Hence AC22 has to be verified as an imposter.

Air traffic communication can be thought of as a special case of the well-studied meeting scenario [4,5]. Here, all participants of the *meeting* communicate over the VHF channel using microphones, whereas the communication protocol is strictly defined. As mentioned before, the communication is highly organized, concurrent speaking is not allowed and no direct conversation between pilots of different aircrafts is allowed, communication is intended only between pilots and controller.

This contribution is organized as follows: Section 2 investigates the restrictions arising from the transceiver equipment and the channel itself and also

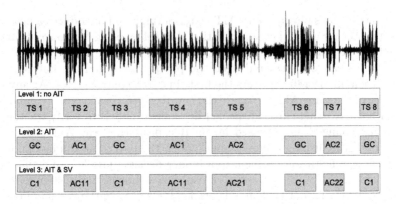

Fig. 2. Coaction of AIT and SV on the party-line in air-ground voice communication. *TS* is the abbreviation for an arbitrary talk spurt, *GC* for a talk spurt originating from the ground (i.e., ground control) and *AC* for a talk spurt originating from a certain aircraft. The first numeric index in level 2 and 3 indicates a specific aircraft and the second index in level 3 the speaker identity. In the first level the nature of the generic speaker segmentation problem is depicted. In level 2 AIT watermarking assigns speaker turns to their source and in level 3 AIT and SV are shown to solve the speaker segmentation task. The waveform at the top shows channel noise between talk spurts.

addresses speaker behaviors. In section 3 the system design is presented with all its processing units. A detailed description of the experimental setup and the databases is given in section 4 where also restrictions discussed in sec. 2 are considered. Experimental results and comparisons are discussed in section 5. The contribution finally ends with some conclusions in section 6.

2 Radio Transceiver Characteristics and Inter-speaker Behaviors

2.1 VHF Transceiver Equipment and Its Limitations

After introducing the ATC security problem the signal conditioning and its effects for speech quality will be analyzed. Speech quality in ATC is impaired in two ways: Firstly, the speech signal is affected by additive background noise (wind, engines) which is not completely excluded by using close talking microphones. Secondly, there is a quality degradation by the radio transmission system and channel which limits the signal in bandwidth and causes distortions. The transmission of the speech signal uses the double sideband amplitude modulation (DSB-AM) technique of a sinusoidal carrier. The system is known to have a low quality. The signal is transmitted over a VHF channel with a 8.3 kHz channel spacing. This yields in an effective bandwidth of only 2200 Hz in the range of 300 − 2500 Hz [6] for speech transmission. The carrier frequency is within a range from 118 MHz to 137 MHz. Dominating effects which are degradating the transmitted signal are path loss, additive noise, multipath propagation caused by reflections, reflection itself, scattering, absorption and Doppler shifts. A thorough description of the degradation of the signal caused by the fading channel can be found in [7,8]. In literature the system proposed in [9] deals with a bandwidth close to but not as narrow as this system has to deal with related to SV.

2.2 Inter-speaker Behaviors

Considering the speaking habits during pilot and controller communication, Hering et al. [2] has shown that one speaker turn is only five seconds on average in length. Training of speaker verification algorithms with speech material of such a short duration is a really challenging task. For comparison Kinnunen [10] used 30 seconds in mean for testing and 119 seconds for training or Chen [11] uses 40 seconds for training.

3   System Design

Considering all these demands, the front-end processing has a main impact on speaker verification performance. First a Voice Activity Detector (VAD) is used to separate speech from silence portions. Based on this segmentation the feature extraction unit performs the transformation from the speech signal in the time domain to a set of feature vectors to yield a more compact, less redundant representation of the speech signal. After the feature extraction, speaker classification is performed. For this application we are only interested in text-independent methods. Main concepts of speaker verification methods in literature are, firstly Support Vector Machine (SVM) where a suitable kernel function has to be found for each speaker. Commonly polynomial and radial basis function are used as kernels. In [12] a slightly worse performance is reported compared to Gaussian Mixture Models (GMM) for clean audio recordings. Secondly, Vector Codebooks [13] have been considered. In this method a vector codebook has to be built using clustering methods to enroll speakers. In the testing phase a distance metric is used to make a decision. They reported a performance slightly lower than GMM but with a lower complexity. Lastly we considered Gaussian Mixture Models. The GMM method had been chosen because firstly the method is well established and understood, secondly is low in complexity and lastly it still delivers state-of-the-art performance. After gender recognition speaker dependent models are derived from a gender dependent Universal Background Model (UBM) which is trained off-line and is assumed to model the whole model space of all existing speakers. For retraining only the speaker dependent model is used. Finally verification of an utterance is done in the verification unit which is shown in the block diagram in Figure 3.

3.1 Pre-processing

Voice Activity Detection: To segregate speech from non-speech, first the short-term log-energy E is extracted from each frame with a length of 16 ms and

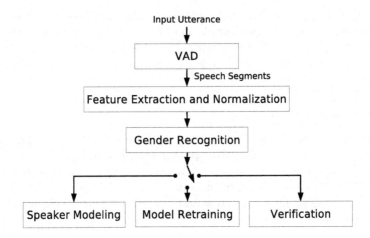

Fig. 3. Processing units of the speaker verification system. Features are extracted only of speech-frames detected by the VAD. After feature normalization gender recognition is performed and only the recognized gender model is further used. Depending on the state of the experiment a new speaker model is derived, a speaker model is retrained or recognition and scoring is done.

8 ms frame shift. Based on the quantile method introduced by Stahl et al. [14] a rough estimate for speech frames is obtained. There a hard threshold is determined experimentally by taking a quantile factor from the range $[0 .. 0.6]$ as produced by Pham et al. in [15]. Because quantile filtering is based on minimum statistics the determined hard threshold adapts over time, leads to high VAD performance [16]. In addition this introduces low complexity because only one parameter, the log-energy, extracted directly from signal domain is needed. By employing a quantile factor of $q = 0.4$, which was selected experimentally to achieve high accuracy of VAD in [16], we expect also high performance for speaker verification. To smooth VAD outputs resulting from fluctuations of non-stationary noise, a duration rule has been applied. In order to adapt with our air traffic speech data, the 15ms/200ms rule as reported in [17] has been modified as 100ms/200ms rule to *bridge* short voice activity regions, preserving only candidates with a minimal duration of 100 ms, and being not more apart than 200 ms from each other. This excludes talk-spurts shorter than 100 ms and re-labels pauses smaller than 200 ms. We propose in the following a new detection method to distinguish between speech signals and consistently high-level noise which results from the transmission channel itself during non-active communication periods.

Long-Term High-Level Noise Detection: To account for long term high-level noise as encountered in air traffic voice communication a new rule is introduced. The 1st discrete derivative of the log-energy values ΔE of all frames which are stored in a buffer of 800 ms are calculated. If the difference between the maximum and minimum values of ΔE in a buffer is below a predefined threshold k as Equation 1 shows, the part is considered as high-level noise and is relabeled as non-speech frame.

$$\max_{i \in \mathbf{Z}}(\Delta E_i) - \min_{i \in \mathbf{Z}}(\Delta E_i) < k \tag{1}$$

The frame index i runs from one to buffer length. Informal experiments showed good performance of long-term noise detection for a threshold of $k = 0.002$ on air traffic recordings provided by EEC [18]. The buffer update rate has been set to 80 ms. In Figure 4 the effect of this method is shown on a true air traffic voice communication recording. The buffer window containing the values of ΔE is slided over the whole signal. As one can easily recognize by visual inspection, the region between second 4.25 and second 7.8 contains only channel-noise. ΔE of these frames depicted as dashed line in Fig. 4 is almost consistent and the difference between $\max \Delta E$ and $\min \Delta E$ is smaller than the predefined threshold in this specified region. This whole section is going to be relabeled as non-speech after applying this rule.

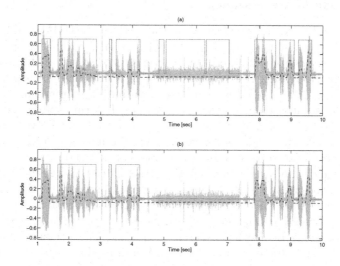

Fig. 4. Example of the effect of the *Long-term high-level noise detection* method on real AT voice communication recordings. The time domain speech signal is depicted as reference, the solid line shows the detected speech of the VAD inside the recording and finally the dashed one the 1st discrete derivative of the log-energy ΔE. The VAD outputs are shown (a) before and (b) after applying the consistent noise detector which eliminates the labeled speech frames of long-term noise duration.

Feature Extraction Unit. Before extracting features as first step mean subtraction and amplitude normalization of the input speech signal is performed. For each speech segment detected by the VAD features are extracted separately. This is necessary to avoid artificial discontinuities when concatenating speech frames. 14 cepstral coefficients are extracted using a linear frequency, triangular shaped filterbank with 23 channels between 300 Hz and 2500 Hz for each frame. Finally the whole feature set comprises these cepstral coefficients calculated in *dB* and the polynomial approximation of its first and second derivatives [19]. Altogether 42 features per frame are used. Performance results using this feature setup but for different frame lengths and frame rates are listed in section 5.

Feature Normalization: In order to reduce the impact of environmental/channel dependent distortions, feature normalization has been carried out. Histogram Equalization (HEQ) [20,21] is known to normalize not only the first and the second moment but also higher-order ones. Experimental results showed that HEQ outperforms the commonly used mean and variance normalization technique [19]. The HEQ method maps an input cumulative histogram distribution onto a Gaussian target distribution. The cumulative histogram distribution is calculated by sorting the input feature distribution into 50 bins. This number has been selected that small to get sufficient statistical reliability in each bin.

3.2   Gaussian Mixture Models

For classification a text-independent statistical method, the GMM-UBM, is used. This method was first introduced by Reynolds et al. [13]. Speaker dependent models are derived from the UBM by *maximum a posteriori* (MAP) adaptation [22]. Our application uses gender dependent UBMs and not merged ones. For training the UBM, the basic model has been initialized randomly and then trained in a consecutive manner by the speech data. To form a speaker dependent model first the log-likelihood of each gender dependent model given the input data is calculated. The gender is determined by selecting the gender-model with the higher score. The corresponding gender dependent UBM is used to adapt a speaker dependent model. For speaker adaptation three EM-steps and a weighting factor of 0.6 for the adapted model and correspondingly 0.4 for the UBM is used to merge these models to the final speaker dependent model. The retraining of an existing speaker dependent model with new speaker data is done in the same manner but with a different weighting. The weighting is directly proportional to the ratio of the total speech length used so far for training and the new utterance length, the model is going to be retrained to. The retraining of the UBM with new data is done in the same way.

Score-Normalization: The score $S(X)$ is calculated using the log-likelihood of the speaker dependent model θ_{SPK} and the UBM θ_{UBM} given the test data X as:

$$S(X) = \log P(X|\theta_{SPK}) - \log P(X|\theta_{UBM}) \qquad (2)$$

For score normalization two methods [23] have been tested. Firstly, the Zero Norm (ZNORM) and secondly the test-normalization (TNORM) was used. For this the mean and variance of the score distribution of ten imposter models have been taken to normalize the score. Both methods have been tested but no increase in performance can be reported.

3.3 Cross Verification

Alternatively to meet the high security expectations in ATC voice communication a cross verification unit is applied as add-on. If an utterance is shorter than a predefined minimum length (i.e., 8 seconds) and the score is not confident enough (positive or negative) the system waits for another utterance and conducts a cross verification. To explain the meaning of cross verification let X_1 and X_2 be the feature vectors of the first and second utterance to be investigated, respectively. Furthermore θ_1 and θ_2 are the adapted speaker models. If the following equation is satisfied

$$S_{\theta_2}(X_1) \,\&\, S_{\theta_1}(X_2) > t \tag{3}$$

i.e., both scores S are above a threshold t and are verified to be from the same gender as defined in 3.2, than it is assumed that both utterances are from the same person and thus are concatenated.

This method shows to increase the robustness of the verification system. Two wrong behaviors may occur in the cross verification unit. Firstly, if two utterances from one speaker are not determined to be uttered from the same speaker, the two utterances are not concatenated and hence are not used together for verification. Furthermore the overall performance stays the same. Secondly, if two utterances from different speakers occupy the same model space e.g., have almost the same "statistical properties" they are verified to be uttered by one speaker and are hence concatenated. But this leaves the score almost the same. Using this method the performance increases because by concatenating two utterances more data for verification are available. This procedure works well in the region of the score distribution where the probabilities of intruders and claimants are almost the same i.e., the confidence is low. Figure 5 shows the region of insufficient confidence in the score distribution histogram. Intruders and true speakers are illustrated separately. The region of low confidence for our experiment as shown in this figure in the white box with dashed borders has been set to -1.8 ± 0.2.

4 Experimental Setup and Data

For development purposes the telephone database from SPEECHDAT-AT [24] was used. It is emphasized that for development, training and testing separate parts of this database are used. Further testings have been carried out on the WSJ0 database [25] where different speakers utter the same text. In order to simulate the conditions of ATC all files were band-pass filtered to a bandwidth from 300 Hz to 2500 Hz and down-sampled to a sampling frequency of 6 kHz. For the experiment a total of 200 speakers comprising 100 females and 100 males were randomly chosen from the entire database. A representative distribution of dialect regions and age was maintained. Background models were trained gender dependent using two minutes of speech material for each of 50 female and 50 male speakers. For training the UBM the speech material of five speakers had been concatenated. This model then was used for training the next five

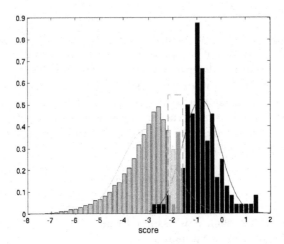

Fig. 5. Histogram and fitted Gauss curves for the score distributions of imposters (left) and true speakers (right). The rectangle with dashed borders illustrates the score region of low confidence.

speakers to yield the UBM in the end. Since also the influence of the number of Gaussian components on the performance is of interest it will be analyzed in more detail in sec. 5. Out of the remaining 100 speakers 20 were marked as reference speakers. Afterwards their speech material was used to train speaker dependent models. Both, for the remaining 99 speakers, known as imposters as well as for the reference speaker, six utterances were used for verification. So each reference speaker was compared to 600 utterances, yielding a total of 12000 test utterances for 20 reference speaker models all together. To match ATC conditions the database was cut artificially in segments of 5 seconds. The experiment was performed twice. Firstly only one segment of 5 seconds was used for training and secondly, which is assumed to be the general case, three segments in a row.

For the tests conducted on the WSJ0 database again the files were pre-processed as for the SPEECHDAT-AT database. The CD 11_2_1 of wsj0 database comprising 23 female and 28 male speakers was used to train the gender dependent UBMs. Since in this database each speaker produces the same utterance 100 seconds of speech were randomly selected from each speaker and used for training. For testing CD 11_1_1 with 45 speakers divided into 26 female and 19 male ones was taken. Here again the speech files for the reference speaker as well as for the claimants were selected randomly. Speech material used for training the reference speaker was labeled and hence excluded from verification. Because the speech files were randomly selected, the experiment was carried out five times. Out of the 45 speakers 24 were labeled as reference speakers, 12 female and 12 male each. As for the SPEECHDAT-AT database the speech files were cut artificially into talk spurts of 5 seconds. Training of the reference speakers was performed by using three segments of speech each of 5 seconds in length. Both, for the remaining 44 speakers, known as imposters as well as for the reference speaker, 12 utterances also five seconds in length were used for verification. So each reference speaker was compared to 540 utterances. Verification had been done for 24 reference speakers which yields in a total number of 12960 test utterances.

Here the influences of mismatches between different microphone types have not been considered, because in general it can be assumed that a pilot does not change the headset during a flight.

5   Results and Discussion

The impacts of front-end processing and model complexity on speaker verification performance are examined. Therefore various numbers of Gaussian mixture components and different frame lengths and frame rates configurations shown in Table 1 are studied. The performance has been measured in terms of equal error rate (EER) and detection cost function (DCF) [26] which are depicted as special points in the detection error trade-off (DET) curve [27]. The DET curve is defined as the plot of the false acceptance (FA) rate vs. the false rejection (FR) rate. In Figure 6 DET curves for both the best system with and without the cross verification unit are shown. As previously expected, the EER for the cross-verification system is lower than for the basic system but with a slight increase of the DCF value. To see the influence of the frame length and frame rate on the one hand and the number of Gaussian components on the other hand experiments on the SPEECHDAT-AT database had been conducted several times using 3 segments of 5 seconds in length for training. Performance results of the various setups measured as EER in [%] and as DCF values are shown in Table 1. We used 16, 38, 64 and 128 Gaussian components (#GC) for the experiments. For feature extraction five different configurations of frame lengths/frame rates (FL/FR) in [ms], (10/5, 20/5, 20/10, 25/5, 25/10), have been examined. Considering only the EER values for the FL/FR ratio of 0.5 e.g., 10/5 ms and 20/10 ms, the EER increases with increasing number of Gaussian components used for modeling. For the remaining FL/FR configurations one can easily recognize a minimum for 38 Gaussian components.

After studying these results one had taken the system with the lowest EER of only 6.51 % as the best setup for this specific application. This system has been finally used for training a speaker model with only one segment of 5 seconds. Here an EER of 13.4 % has still been reached. The recognition rate for the gender recognition is 96 %.

The same experiment for the SPEECHDAT-AT database has been conducted for a background model with 1024 Gaussian mixture components.

The remaining parts have been left untouched. This has been done for comparison reasons since many systems being in place are using GMMs up to this number of components [22]. For this system design and its restrictions an EER of 33 % could be reached. A reason for this result could be the over modeling for speaker dependent models and verification afterwards, using short speaker turns.

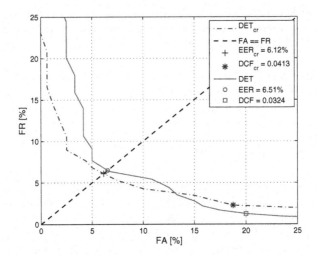

Fig. 6. DET curves with EER point (plus sign and circle) defined as the intersection of the main-diagonal with the DET curve. Additional the DCF values (star and rectangle) for the SPEECHDAT-AT database are depicted. The results are shown for both the normal system and the system with cross-verification. *FA* is the abbreviation for the false acceptance rate and *FR* for the false rejection rate. The results with subscript *cr* are those of the cross-verification system and those without, of the basic system. For training three segments of 5 seconds in length are used.

Table 1. Performance results as a function of the frame length/frame rate and the number of Gaussian components (#GC) tested with SPEECHDAT-AT database. The first value of each table entry is the EER in % and the second one corresponds to the DCF value. For training 3 segments of 5 seconds in length are used.

#GC	Frame Length/Frame Rate [ms]				
	10/5	20/5	20/10	25/5	25/10
16	8.2/0.042	7.4/0.039	6.75/0.042	7.26/0.033	10.14/0.0438
38	9.04/0.042	6.9/0.037	10.83/0.053	6.51/0.0376	8.8/0.0435
64	9.25/0.045	7.88/0.042	11.8/0.054	9.62/0.05	11.2/0.05
128	10.65/0.046	9.15/0.044	13.12/0.068	12.5/0.043	13.6/0.066

At the end the results for the WSJ0 database are presented. As mentioned in sec. 4 the experiment has been done five times. The mean result of the EER and the DCF values are given as follows: The mean EER is 10.8 % and the mean DCF value is 0.0637 with a standard deviation of 0.5823 and 0.0104 respectively. The degradation in performance from SPEECHDAT-AT to WSJ0 database is due to inter-speaker score threshold shifts and not due to false/true speaker discrimination problems. To solve this problem and improve the score for this database in future different score and feature normalization techniques are going to be investigated.

6 Conclusion

A novel speaker segmentation system for voice communication in air traffic control is described for enhancing the security of air traffic voice communication. As a first security level the aircraft identification tag based on watermarking technique is used to assign a talk spurt on the shared communication channel to its source. The air traffic communication safety is then enhanced by applying a speaker verification system based on the optimized front-end processing units for this task. Speaker dependent models are derived in cooperation with gender information for selecting the gender dependent universal background model. Results have been presented with investigation of various numbers of Gaussian components used for modeling. Furthermore its inter-relationship with the frame length and frame rate used to extract features are discussed. Based on *a priori* knowledge of the score distribution a cross verification unit can further reduce the equal error rate. The system has been evaluated on two databases with promising results. As each pilot has to identify its voice message with the call sign a planned extension of the system is the combination of our system with a text-constrained system to get an even higher level of security. Furthermore, the means of the frequency restriction as well as the optimization of VAD for speaker recognition performance will be investigated.

Acknowledgments

We kindly acknowledge the support of the Eurocontrol [18]. The author wishes to thank Franz Pernkopf for fruitful discussions throughout this work.

References

1. Van Es, G.: Air-ground communication safety study: An analysis of pilot-controller occurrences. EUROCONTROL DAP/SAF Ed. 1.0 (2004/04)
2. Hering, H., Hagmuller, M., Kubin, G.: Safety and security increase for air traffic management through unnoticeable watermark aircraft identification tag transmitted with the VHF voice communication. In: Proc. of the 22nd IEEE Dig. Avionics Sys. Conf. DASC '03, vol. 1, 4.E.2–41-10 (2003)
3. Neffe, M., Hering, H., Kubin, G.: Speaker segmentation for conventional ATC voice communication. In: 4th EUROCONTROL Innovative Research Workshop Brétigny-sur-Orge France (2005)
4. Mistral Project, 2005-2006, http://www.mistral-project.at
5. Abad, A., Brutti, A., Chu, S., Hernando, J., Klee, U., Macho, D., McDonough, J., Nadeu, C., Omologo, M., Padrell, J., Potamianos, G., Svaizer, P., Wölfel, M.: First experiments of automatic speech activity detection, source localization and speech recognition in the chil project. In: Proc. of Workshop on Hands-Free Speech Communication and Microphone Arrays, Rutgers University, Piscataway, NJ (2005), http://chil.server.de/servlet/is/101/
6. Airlines electronic engineering committee. Airborne VHF Communications Transceiver, Manual Annapolis, Maryland (2003), https://www.arinc.com/cf/store/catalog_detail.cfm?item_id=493

7. Hofbauer, K., Hering, H., Kubin, G.: A measurement system and the TUG-EEC-Channels database for the aeronautical voice radio. In: IEEE Vehicular Technology Conference, Montreal, Canada (2006)
8. Hofbauer, K., Hering, H., Kubin, G.: Aeronautical voice radio channel modelling and simulation - A tutorial review. ICRAT, Belgrade, Serbia and Montenegro (2006)
9. Reynolds, D.A., Campbell, W., Gleason, T.T., Quillen, C., Sturim, D., Torres-Carrasquillo, P., Adami, A.: The 2004 MIT Lincoln Laboratory Speaker Recognition System. In: Proc. IEEE Int. Conf. on Acoustics, Speech, and Signal Processing (ICASSP), vol.1, pp. 177–180 (2005)
10. Kinnunen, T., Karpov, E., Franti, P.: Real-time speaker identification and verification. IEEE Transactions on Audio, Speech and Language Processing 14(1), 277–288 (2006)
11. Chen, K.: On the use of different speech representations for speaker modeling. IEEE Trans. on Systems, Man and Cybernetics, Part C 35, 301–314 (2005)
12. Wan, V., Campbell, W.M.: Support vector machines for speaker verification and identification. In: Proc. of the IEEE Signal Processing Society Workshop on Neural Networks for Signal Processing X, vol. 2, pp. 775–784 (2000)
13. Reynolds, D.A., Rose, R.C.: Robust text-independent speaker identification using gaussian mixture speaker models. IEEE Transactions on Speech and Audio Processing 3(1), 72–83 (1995)
14. Stahl, V., Fischer, A., Bippus, R.: Quantile based noise estimation for spectral subtraction and Wiener filtering. In: Proc. of the IEEE International Conf. on Acoustics, Speech, and Signal ICASSP '00, vol. 3, pp. 1875–1878 (2000)
15. Pham, T.V., Kubin, G.: WPD-based noise suppression using nonlinearly weighted threshold quantile estimation and optimal wavelet shrinking. In: Proc. Interspeech'05 Lisboa, Portugal, pp. 2089–2092 (september 4-8, 2005)
16. Pham, T.V., Kèpèsi, M., Kubin, G., Weruaga, L., Juffinger, A., Grabner, M.: Noise cancellation frontends for automatic meeting transcription. In: Euronoise Conf. Tampere, Finland, CS.42-445 (2006)
17. Brady, P.T.: A statistical analysis of on-off pattern in 16 conversations. Bell Syst. Tech.J. 47(1), 73–91 (1968)
18. Eurocontrol experimental centre-EEC Brètigny-sur-Orge, France, http://www.eurocontrol.int/eec/public/subsite_homepage/homepage.html
19. Bimbot, F., Bonastre, J.-F., Fredouille, C., Gravier, G., Magrin-Chagnolleau, I., Meignier, S., Merlin, T., Ortega-Garcia, J., Petrovska-Delacretaz, D., Reynolds, D.A.: A tutorial on text-independent speaker verification. EURASIP Journal on Applied Signal Processing, pp. 430–451 (2003)
20. de la Torre, A., Peinado, A.M., Segura, J.C., Perez-Cordoba, J.L., Benitez, M.C., Rubio, A.J.: Histogram equalization of speech representation for robust speech recognition. IEEE Transactions on Speech and Audio Processing 13, 355–366 (2005)
21. Skosan, M., Mashao, D.: Modified segmental histogram equalization for robust speaker verification. Pattern Recognition Letters 27(5), 479–486 (2006)
22. Bimbot, F., Bonastre, J.-F., Fredouille, C., Gravier, G., Magrin-Chagnolleau, I., Meignier, S., Merlin, T., Ortega-Garcia, J., Petrovska-Delacretaz, D., Reynolds, D.A.: Speaker verification using adapted gaussian mixture models. Digital Signal Processing, 10, 19–41 (2000)
23. Auckenthaler, R., Carey, M., Lloyd-Thomas, H.: Score normalization for text-independent speaker verification systems. Digital Signal Processing 10, 42–54 (2000)

24. Baum, M., Erbach, G., Kubin, G.: Speechdat-AT: A telephone speech database for Austrian German. In: Proc. LREC Workshop Very Large Telephone Databases (XLDB) Athen, Greece, pp. 51–56 (2000)
25. Garofalo, J., David Graff, D., Paul, D., Pallett, D.: Continuous Speech Recognition (CSR-I) Wall Street Journal (WSJ0) news, complete. Linguistic Data Consortium, Philadelphia (1993),
    http://ldc.upenn.edu/Catalog/CatalogEntry.jsp?catalogId=LDC93S6A
26. Przybocki, M., Martin, A.: Nist speaker recognition evaluation (1997),
    http://www.nist.gov/speech/tests/spk/1997/sp_v1p1.htm
27. Martin, A., Doddington, G., Kamm, T., Ordowski, M., Przybocki, M.: The det curve in assessment of detection task performance. In: Proc. Eurospeech, Rhodes, pp. 1895–1898 (1997)

Detection of Speaker Characteristics Using Voice Imitation

Elisabeth Zetterholm

Centre for Languages and Literature Linguistics/Language Technology
Lund University
Box 201 SE-221 LUND, Sweden
elisabeth.zetterholm@ling.lu.se

Abstract. When recognizing a voice we attend to particular features of the person's speech and voice. Through voice imitation it is possible to investigate which aspects of the human voice need to be altered to successfully mislead the listener. This suggests that voice and speech imitation can be exploited as a methodological tool to find out which features a voice impersonator picks out in the target voice and which features in the human voice are not changed, thereby making it possible to identify the impersonator instead of the target voice. This article examines whether three impersonators, two professional and one amateur, selected the same features and speaker characteristics when imitating the same target speakers and whether they achieved similar degrees of success. The acoustic-auditory results give an insight into how difficult it is to focus on only one or two features when trying to identify one speaker from his voice.

Keywords: Voice imitation, voice disguise, speaker identification, impersonator, dialect.

1 Introduction

A listener may recognize a voice even without seeing the speaker. There are cues in voice and speech behaviour, which are individual and thus make it possible to recognize familiar voices. We attend to particular features of the person's speech, e.g. fundamental frequency, articulation, voice quality, prosody, dialect/sociolect, studied among others, by Gibbons (2003) [1] and Hollien (2002) [2]. Probably a combination of different features is involved in the recognition process. It is well-known that the individual voice changes throughout lifetime, but also depending on speaking situation, the speaker's health and emotion. Still, in general it is possible to recognize a well-known and a familiar voice. There are studies comparing the difference in performance between listeners who knew the speaker and those who did not know the speaker. Hollien et al. (1982) [3] show that there was a great difference between the two groups of listeners. Those who were familiar with the speaker had a high degree of correct identifications (98 %) compared to listeners who were not familiar with the speaker (32.9 %). Van Lancker

C. Müller (Ed.): Speaker Classification II, LNAI 4441, pp. 192–205, 2007.

(1985a, b) [4] [5] point out that there are different cognitive processes involved in the recognition task, which might be an explanation for the results. For familiar voices pattern recognition is used and for unfamiliar voices feature analysis is involved. In order to mislead the listener, the speaker may disguise the voice in different ways. In the science of forensic phonetics the task is to try to find out how to recognize disguised voices, a speaker identification task. Any aspect of the speech chain, any sound made by the human vocal tract, from production through acoustics to perception and transcription is of interest in forensic phonetics. It is important to understand which features in the human voice to focus on for speaker identification. It is likely that more than one feature is of importance. This article will focus on voice imitation as a kind of voice disguise, and as a tool for detection of speaker characteristics. Imitation can be used both for impersonation of a specific target speaker and the personal identity as well as the imitation of markers of group identity, such as regional and social dialects.

2   Voice Disguise

When trying to hide one's identity the speaker can disguise his/her voice in a number of ways, e.g. changing the pitch, voice quality, prosodic pattern or lip protrusion, clenched jaw or use objects over the mouth. According to Künzel (2000) [6], it seems that falsetto, pertinent creaky voice, whispering, faking a foreign accent, and pinching one's nose are some of the perpetrators' favourites. He also reports that 15-25 % of the annual criminal cases involving speaker identification, involve at least one type of voice disguise (statistics from German Federal Police Office).

Different identification studies with disguised voices show that e.g. whispering and hypernasal voice are effective disguise methods (Yarmey et al. 2001 [7], Reich and Duke 1979 [8]). One might suggest that it is quite easy to recognize a familiar voice even if disguised, but the identification rates drop both for familiar and unfamiliar voices despite of disguise in studies by Hollien et al. (1982) [3]. Imitating a dialect might also be used as a kind of voice disguise. In a study of imitation of some Swedish dialects, Markham (1999) [9] found that there was a wide variation in the ability to create natural-sounding accent readings. Some speakers were able to both hide their own dialect and convince the listeners that they were native speakers of another dialect, other speakers were able to hide their own dialect, but less successful in creating a natural sounding dialect, according to the listeners. He points out that it is important to create an impression of naturalness in order to avoid suspicion.

To investigate the power of dialect in voice identification, a voice line-up with a bidialectal male Swedish speaker was conducted (Sjöström et al. 2006) [10]. Two recordings when reading a fairy tale were made by the speaker in his two different dialects, Scanian and Stockholm dialects. These were used as the familiarization voice in four different voice line-ups. Four more recordings of the fairy tale were made with four mono-dialectal Swedish male speakers, two with a Scanian dialect and two with a Stockholm dialect, used as foils in the line-ups. Different tests were made and the results are quite clear. Each line-up contained the four mono-dialectal male voices and one of the target's two dialect voices. Two tests were created as control tests with the same dialect used both as familiarization and target voice. Two other tests were created as dialect shifting tests, where one dialect was used as a familiarization voice and the other of his dialects was the target voice in the line-up. The control tests are to investigate if the speaker can be recognized among the other voices and if they are recognized to the same degree. Native Swedish speakers from different dialect areas in Sweden participated in the different voice line-ups. A high identification rate was found in both control tests, but the results in the dialect shifting tests show that the target speaker is difficult to identify. Listeners often judged one of the foils speaking with the same dialect as the one presented in the familiarization passage instead of the bidialectal target speaker. Dialect seems to be a strong attribute in an identification task and maybe with a higher priority than other features of the voice.

3   Voice Imitation

Voice imitation can be viewed as a particular form of voice disguise. E.g. in crime, voice imitation can be used to hide one's identity and mislead the police investigation. The purpose of the criminal is to disguise his/her voice, probably not to imitate another specific person, but a group of speakers such as a dialect or a specific pronunciation or intonation of an accent. According to Markham (1997) [11] impersonation is defined as reproduction of another speaker's voice and speech behaviour. When imitating a certain target speaker the impersonator has to select and copy many different features, laryngeal as well as supralaryngeal, to be successful. There are organic differences between speakers, which can not be changed, and it is quite hard to produce very close copies of another speaker's voice and speech. Therefore, mimicry is often a stereotyping process and not an exact copy of the target speaker (Laver 1994 [12]). Despite that, it is shown that high quality voice imitation can mislead the listener (Schlichting and Sullivan 1997 [13]). Imitation for entertainment is often more like a caricature and the impersonator focus on the most prominent features of the target speaker, to strengthen the impression. When listening to a voice imitation, the listeners may have expectations about characteristic features of the target speaker's voice and speech, especially if the impersonator uses words and phrases related to the target speaker. Two experiments with an imitation of a Swedish politician have been done to show the impact of the semantic expectation upon a voice imitation. One imitation with a political speech and one imitation with a cooking passage of the same target speaker, were used in a voice line-up. The results support the hypothesis that listeners' semantic expectations would impact upon the listeners' readiness to accept a voice imitation as the voice of the person being imitated (Zetterholm et al. 2002 [14]). Different studies of professional impersonators and their voice imitations show that it is possible to get close to another speaker's voice and speech behaviour both in perceptual and acoustic analyses, which indicate that the human voice is quite flexible (Zetterholm 2003 [15]).

The study of impersonation may give an insight into centrally important features for speaker recognition and therefore be exploited as a methodological tool to find out which features a voice impersonator picks out in the target voice and which features in the human voice are not changed, which is of high importance in a identification situation. The results are also of forensic interest when trying to detect a disguised voice.

4   Overview of the Present Study

As already mention, the impersonator has to identify and copy the most characteristic features of the target speaker's voice and speech behaviour. If different impersonators imitate the same target speakers – will they select the same features? This might give us an insight into important features of specifically impersonation, but also give us a general understanding of speaker characteristics and speaker identification.

In order to better understand the perceptual impression and the acoustic realization of voice imitation, this study examines whether three Swedish impersonators, two professional and one amateur: select the same features when imitating a set of target voices, and achieve similar degrees of closeness to the target voices in terms of pitch, dialect, speech tempo, voice quality and the features that are generally viewed as the characteristic features of each of the target voices.

5   Material

Voice imitations made by three impersonators were used in this study. Recordings of nine target voices, 22 imitations and the impersonators' own voices were analysed. The impersonators are all native speakers of Swedish, yet live in different dialect areas in Sweden, two of them are professional imitators (Imp I and Imp II) and one is an amateur (Imp III). The recordings of the imitators, both the imitations and with their own voices, are made in their own recording studios or in the recording studio at the Department of Linguistics and Phonetics, Lund University, Sweden. All target voices are well-known male Swedish voices, politicians and TV personalities. These recordings are taken from public appearances. All the texts are different, but all imitations are related to the target speaker's profession. Imp I imitates all the target voices, Imp II imitates seven of the voices and Imp III imitates six of the nine target voices. All target speakers are all older than the three impersonators. It is easier to imitate a speaker at the same age or older, according to an interview with Imp I (Zetterholm 2003:132 [15]).

Imp I has a mixed dialect that is a combination of the western part of Sweden and a more neutral Swedish dialect, Imp II has a dialect from the eastern part of Sweden, and Imp III speaks with a neutral Swedish dialect influenced by the intonation pattern of the south Swedish dialect. As there is a clear distinction in the pronunciation of the phoneme /r/ between the Swedish dialects, it is important to describe the impersonator's usage of this phoneme. All three impersonators use a voiced alveolar trill [r] for the phoneme /r/ and the retroflex variants in the combination /r/ and alveolar consonants.

All target speakers have dialects from the western or eastern part of Sweden, some of them influenced by the dialect of Stockholm, and three speakers have a clear dialect from the Stockholm area.

6   Brief Description of Swedish Dialects

The intention is not to give a complete description of the different Swedish dialects, only a brief presentation, especially of the dialects occurring among the speakers in this study. There is a considerable variation between Swedish regional dialects concerning both phonetics and phonology and some of the most characteristic dialectal markers will be presented. The traditional division of Swedish dialects are as follows: South, West, East, Central and North.

6.1   The Phoneme /r/

There are a number of different forms of the phoneme /r/ in Swedish. The two main types are [r] and [ʀ]. In south Swedish dialects a uvular trill [ʀ] or a uvular fricative [ʁ] is used (Elert 1991 [16]). In central Swedish dialects an alveolar trill [r], a retroflex fricative [ʐ] or an alveolar approximant [ɹ] is used. In some parts of central Sweden a retroflex /r/ is used in the combination with /r/ + alveolar consonant, e.g. /rt/ [ʈ], /rd/ [ɖ], /rs/ [ʂ], /rn/ [ɳ] and /rl/ [ɭ]. In south Swedish dialects the pronunciation of the same combinations are a uvular /r/ and the alveolar consonant (Elert 1991 [16], Markham 1999 [9]).

6.2   The Vowels

Various diphthongizations of long monophthongs are one characteristic dialectal marker of the southern dialects, often with an initial onglide of the target vowel. The short vowels are often monophthongs (Bruce 1970 [17], Elert 1991 [16]). Some of the Swedish dialects have a "damped" i- and y-vowel, called Viby-i and Viby-y. One characteristic of this "damped" vowel is a fricative feature and apical articulation. The acoustic analysis shows that Viby-i has a considerably lowered F2 compared to standard Swedish [i].

6.3   Tonal Word Accents

The two Swedish tonal word accents, accent I and accent II, are represented by a high and a low turning point in all dialects, but the timing in relation to the stressed syllable differs between the dialects. The tonal gesture for accent I always precedes the tonal gesture for accent I independent of the dialect and there is a distinctive difference in timing for accent I and accent II. The tonal gesture is earlier in the dialect of Stockholm compared to South Swedish (Bruce 1998 [18]).

7   Method

An auditory analysis of the recordings was conducted at the Department of Linguistics and Phonetics at Lund University, Sweden. An informal listening test was undertaken by 10 members of faculty and a detailed close auditory analysis done by three of the experienced phoneticians from the department. All the listeners were familiar with the nine target voices. The listeners were asked to comment on the voice imitations and to describe which characteristic features of the target voices the impersonators have selected in their imitations. There was a discussion about the general impression of the imitations as well as specific features.

A second auditory analysis was made by 10 speech therapy students from the Department of Logopedics and Phoniatrics at Lund University. These students carried out a critical and close analysis that focused on the voice quality of the recordings of the target voices, the impersonators own voices as well as the imitations.

The auditory analyses focus on the overall impression, the pitch, the dialect and the voice quality.

To examine how the listeners' impressions, as revealed in the auditory analysis, are evidenced acoustically in the imitations, some specific factors were selected for acoustic analysis. These were: (1) The mean fundamental frequency for the target voices, the voice imitations as well as the impersonators' natural voices was measured in all recordings; the auditory analysis show that this is of fundamental importance to acceptance of an imitation. (2) The formant frequencies of the i-vowel were measured for one of the voices, and the imitations of him, to compare the auditory impression of a damped i-vowel. (3) The frequency of the lower edge of the noise energy plateau in the spectrum of /s/ was measured for two of the voices and the imitations, justified by the auditory analysis. (4) The articulation rate of the target speaker was compared to the imitations in order to find out if that is a strong individual phonetic habit or easily changed when imitating someone else.

8   Auditory Analysis

The general auditory impressions that the different groups of listeners gained was that all three impersonators speak with a normal male pitch, that Imp I has a sonorous but slightly leaky voice, that Imp II has a slightly strained and hypernasal voice quality and that Imp III has a sonorous but slightly creaky voice quality. The listeners further formed the impression that Imp III (the amateur) speaks with a higher speech tempo than Imp I and II.

The phoneticians' general impression of all voice imitations was that the impersonators selected the same primary characteristic features of the target voices. This did not, however, necessarily result in the different voice imitations of the same target speaker sounding similar. In spite of this, the phoneticians concluded that all three of the impersonators succeeded in imitating the target voices with global success.

The listener's detailed descriptions of the imitations focused on phonetic features; pitch level, intonation pattern, with particular reference to dialect, pronunciation of dialectal markers, speech tempo, and individual characteristic features of the target speakers.

8.1   Pitch Level

The auditory impression of the pitch level is that the impersonators are flexible in their different imitations and actively attempt to achieve the target speaker's pitch. This is reflected in some of the listeners holding the opinion that the low pitch in the imitation of HV, by Imp I, is too low and that the high pitch in both imitations of MH is exaggerated. The listeners also commented that Imp III (the amateur) maintained a pitch level that was too high in all his imitations.

8.2   Dialectal and Individual Features

The impersonators captured the different dialect intonation patterns well. They also imitated dialect segmental markers successfully. For example, all three impersonators managed to capture the different pronunciation of the r-segment for each target voice. Exaggeration of dialect targets was noted by the listeners. They commented that in the imitations of CB, the pronunciation of [ʀ], and in the imitations of HV the pronunciation of [ʁ] is exaggerated. The pronunciation of the r-segment is a characteristic of these two speakers. In some of the other voice imitations the impersonators have focused on the regional dialect and exaggerated the characteristic pronunciation of the dialect further than individual features of the target speaker. E.g. in the imitation of LO, Imp I has captured the west Swedish dialect and the clearly downstepped intonation pattern of the target speaker whereas the voice imitation made by Imp II of the same target speaker, LO, is more like an imitation of the west Swedish dialect in general, focusing on both intonation pattern and the pronunciation of the vowels.

At a more detailed segmental level, the listeners commented upon the pronunciation of [ɛː] is more like an [eː] and this is obvious in the imitations of especially CG, by all three impersonators, and the imitations of GP, by Imp I and III. The listeners also noted that the damped i-vowel, that is found in some Stockholm dialects and is a characteristic of IW, was found in the imitations of him.

The listeners also noted audible differences between the pronunciations of the phoneme /s/. In most of the voice imitations it is pronounced [s], but in the imitations of HV and IK, it sounds more like [ʃ]. This is also in accordance with the target speaker's pronunciation of /s/.

8.3   Speech Tempo and Speech Style

It seems that all three impersonators are aware of each targets personal speech tempo; they capture these with a varying degree of success. Imp III has a global imitation problem in that he has a higher speech tempo in his imitations compared to the target speakers. Interestingly, even though in some of his imitations he starts with a slower tempo close to the target speaker, he speeds up after a few seconds.

The auditory analysis also showed that the impersonators try to capture the individual rhetorical speech of each target's voice. In the imitation of CB, the imitators manage to copy his rhetorical speech style and his use of focal words, but perceptually they are less successful in imitating CB's characteristic feature of cutting the end of the last syllable. Perhaps, the listeners did not feel that this affected the quality of the voice imitations. Comments from the listeners indicate that Imp II and III, in particular, had captured the target speaker's speech style; a fast speech tempo with pauses and acceleration on focal words. Imp I, on the other hand, was more successful with his imitation of the hesitation sound that CB makes at the beginning of a phrase. All three impersonators were judged by the listeners to achieve distinct articulation, phrasing and formal speaking style of the target speaker CB.

In the imitations of HV they manage to capture the characteristic speech style with an energetic distinct articulation, engaged speech and a speech rhythm like staccato, which are individual markers and characteristics by this speaker. Further, in the different voice imitations of both LO and MH, the characteristic high speech tempo of the target speakers is achieved. The loudly extensive breath, a speaking manner of GP, is achieved effectively in the imitations of him. More generally the listeners noticed that the prolonging of words, and the many pauses and hesitation sounds that form part of the GP imitations acted to strengthen the acceptance of these voice imitations.

8.4   Voice Quality

The consensus of the listeners was that Imp I and II were successful in altering their voice quality in the direction of the target speakers. This was not the case for Imp III, where the listeners reported that even though he attempted to alter his voice quality, his own voice quality was clearly audible in all his imitations.

One specific observation was that the creaky voice quality perceived in the target speakers CB and HV, was achieved by both Imp I and II. Moreover they were successful in making this quality more obvious at the end of utterances, just like the target speakers. Creaky voice quality is also a feature of the target speaker GB, yet here, only Imp I was successful in capturing this quality.

Another specific observation is related to nasal voice quality. In the dialect of Stockholm, spoken by CG, IW and MH, a slightly nasal voice quality is a dialect marker. Both Imp I and II manage to imitate the nasal voice quality in these imitations.

A tense-breathy voice quality is a kind of personal marker for speaker IK, especially in a speaking style of a political speech. This is exaggerated in the voice imitations of him.

9   Acoustic Analysis

9.1   Mean Fundamental Frequency, F0

With the exception of the target voice IK and IW, there is little variation in the mean F0s of the target voices, see Table 1. When looking at the recordings with the impersonators own voices it is shown that Imp I has the lowest mean F0 (113 Hz), Imp III the highest mean F0 (149 Hz), while Imp II has a mean F0 of 127 Hz. Especially for Imp III his own mean F0 is reflected in the imitations.

A comparison between the target speakers and the imitations show that some of the voice imitations are quite close to the target voices. It is clear that Imp I and II (the professionals) seem to have the same conception about the variation in F0 between the target voices, while there is less variation in the imitations made by Imp III. This corresponds with the findings in the auditory analysis.

Table 1. Mean F0 (in Hz) and std.dev. of the target voices, the imitators' natural voices and the voice imitations

	Target voices		Imp I		Imp II		Imp III	
	Mean F0	Std.dev.	Mean F0	Std.dev.	Mean F0	Std.dev.	Mean F0	Std.dev.
Natural			113	38	127	53	149	32
AS	128	41	133	40	142	35	145	45
CB	135	35	125	23	130	28	157	56
CG	121	21	122	17	103	11	136	26
GP	126	42	96	36	-	-	139	48
HV	135	36	91	14	119	28	145	76
IK	207	33	198	23	255	37	-	-
IW	107	27	99	16	97	15	-	-
LO	149	28	142	39	133	25	-	-
MH	147	31	202	40	-	-	218	25

9.2   The "damped" i-Vowel

The auditory impression of a "damped" i-vowel in the recordings of IW and the imitations of this voice is confirmed in the acoustic analysis (see Figure 1). A considerably lowered F2 compared to standard Swedish /i/ (approximately 2200 Hz) is an acoustic correlate to a "damped" i-vowel. The formant frequencies of F2 are lower than 2000 Hz in all occurrences by both Imp I and II. (NB. There is no imitation of IW by Imp III in this imitation corpus).

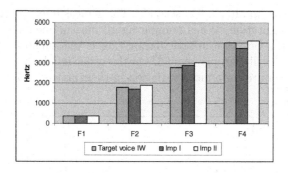

Fig. 1. Mean values (in Hz), of all occurrences, of the formant frequencies of the i-vowel for the target voice IW and the voice imitations by Imp I and II

9.3   The Phoneme /s/

The frequency of the lower edge of the noise energy plateau in the spectrum of the phoneme /s/ was measured for the target voices HV and IK and the imitations of them. Unfortunately, there are no comparable occurrences of words with the same surrounding segments in the recordings, but still a few words for comparison. The acoustic analysis of the target speaker HV, and all the imitations of his voice, confirms the auditory impression of a pronunciation more like [ʃ], with a lowered frequency in the spectrum. In the imitations of IK, Imp I and II solely use [ʃ], whereas IK himself produces /s/ both as [s] and [ʃ]. There is no clear pattern in the context with a preference for one of the phonemes.

9.4   Articulation Rate

The auditory impression is that all three impersonators are aware of the speech tempo of the target speakers and that both Imp I and II manage to imitate that. But concerning the imitations made by Imp III, the listeners comment that he speaks too fast. To get an acoustic correlate to this impression the articulation rate, excluding the silent pauses, were measured in all recordings. Mean articulation rate for Swedish is about 5 syllables per second. In Figure 2 some of the measurements are shown, just to give an insight of the differences between the imitations (there is no imitation of the target speaker LO by Imp III in this corpus). The articulation rate in the recordings with the natural speech with the three impersonators and four of the target voices and the imitations of them are shown in the figure. The tendency is clear; the two professional impersonators (Imp I and II) are more flexible and are closer to the target voices in their imitations than the amateur (Imp III). Concerning the imitations of the target speaker AS, both Imp I and II exaggerates his slow speaking style, more like a caricature.

The auditory impression and the comments from the listeners that Imp III has a high speech tempo in general is not confirmed in the measurements of the articulation rate. One explanation for that might be that he often starts in a slow tempo, but speeds up after a few seconds. These measurements show the mean rate during the imitation. When comparing all measurements of all imitations it is clear that the variation in articulation rate is very small in his imitations, all very close to 5 syllables per second. On the other hand, the two professional impersonators seem to have more variation in their articulation rate.

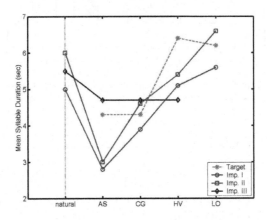

Fig. 2. Articulation rate, syllables per second. The natural voices of the three impersonators, the target voices AS, CG, HV and LO and the imitations of these voices.

10   Concluding Discussion

When listening to a familiar voice, listeners have expectations about the characteristic features of the speaker's voice and speech behaviour. For an imitation on stage, global success and imitation of the most characteristic features may be enough for the listeners to recognize the target voice. But in a critical listening task, by trained phoneticians, there are often passages that reveal the impersonator's identity and voice. On stage, the impersonator can use gestures, body language and other attributes to strengthen the impression of the imitation and distract the listeners from weakness in the voice imitation. Moreover, voice imitations primarily meant for entertainment may be exaggerated, to the extent that they approach a caricature of the target speaker and his/her voice and speech behaviour. This situation permits the impersonator to focus on the most prominent features of the target speaker and build on a situation in which listeners only notice the expected features of the target voice. If a caricature of another speaker is created in a criminal act it would probably be detected to someone who is familiar to the target speaker. To give an impression of naturalness in the imitation it is of importance to convince and deceive the audience (Markham 1999 [9]).

The imitation corpus used in this study show that three different impersonators seems to have different strategies when imitating the same target speakers. It is also shown that the professional impersonators (Imp I and II) are more flexible in their imitations compared to the amateur (Imp III), both concerning voice quality, mean fundamental frequency and articulation rate. Despite that, all listeners agreed that it is clear who they are imitating and that all three impersonators have selected prominent features of each speaker and that they focus on the same characteristics of each target voice. This might indicate which voice features are important for voice recognition and identification by human listeners.

The listeners in this study comment that the impersonators, in almost all imitations, have captured the pitch level, the dialect with different pronunciations and prosody, the speech style, e.g. speech tempo, articulation and individual phonetic habits, such as hesitations sounds, as well as other individual characteristic features of the target speakers. The phoneme /r/ is an important dialectal marker in Swedish and it is obvious that the impersonators focus on this phoneme in some of the imitations. In stressed and focused words and phrases, the impersonators often succeed in imitating vowels, but in unstressed passages the critical listener can hear the impersonators' own voices and their own dialects. Focusing on characteristic features and important passages in the text may be a conscious way of working with and improving the voice imitation.

The anatomically dependent component in the voice quality, such as the size of the vocal tract, is outside of our control, but there is also a component in the speaker's voice quality that is learned. Abercrombie (1967) [19] suggests that most people are probably capable of changing their voice quality and one question is how much of the voice quality is learnt. There is no doubt that a special voice quality characterizes one speaker as well as the members of different groups of speakers. Both laryngeal and supralaryngeal features influence the perception of a speaker's voice quality and a classification is made by Laver (1980) [20]. In spite of these terms we still lack a more complete typology for description of the different voice qualities in normal speech. The categories often described with metaphorical adjectives in a perceptual description, have no consistent acoustic-phonetic correlation. Even though the variation of different voice qualities in normal voices is hard to describe, the listeners were able to tell if the voice quality in the voice imitation is close to the target voice or not. Voice quality is by definition a perceptual matter and according to Hammarberg and Gauffin (1995) [21] perceptual evaluation of voices is subjective and impressionistic, the perceptual aspects are important since they play a role in the listener's acceptance of a voice. It seems to be quite hard to change voice quality, both laryngeal and supralaryngeal features, to make a copy of another speaker's voice quality and to keep the settings for a long period of time. There are passages in this imitation corpus where the natural voice quality of the impersonators is audible. This feature seems to be a strong factor for detecting and acceptance of an imitation and recognition of a voice. When combining the results of the auditory impression and the acoustic measurements, it is shown that it is possible to get quite close to one specific target speaker and that might lead to the conclusion that it is possible to change your own voice and speech in order to disguise your voice. The results of this study also indicate that these three impersonators selected the same features of the target speakers, although to different degrees of success. These features included voice quality, mean F0, dialect with different pronunciation of phonemes and a speaker's individual speech style. With regard to degree of success, as defined by closeness to the target voices, it is clear that the two professional impersonators are able to imitate the voices used in this study with global success, whereas the amateur impersonator is not as successful. In the professional impersonations there were, however, unstressed passages that revealed Imp I and II. The overall impression of the voice imitations made by Imp III is that the imitations are more exaggerated, as if only for entertainment, and that his own voice, both mean F0, articulation rate and voice quality, is reflected in his imitations. The result of the measurements of the articulation rate is of interest when comparing the three impersonators. A professional impersonator is aware of how to change his voice and speech behaviour, both laryngeal and supralaryngeal settings, to get close to the target speaker. There are more variation in the rate in the imitations by the two professional impersonators (Imp I and II) compared to the amateur (Imp III). The question raised before, if the articulation rate is an individual phonetic habit, is not answered, but the results give us an insight into this area and more research is requested. We can thus conclude that the impersonators achieved different degrees of success in getting close to the target voices, both perceptually and acoustically, with all being unsuccessful in one aspect or another. All three impersonators were successful concerning the imitation of the features of the different Swedish dialects.

Applying these findings to a forensic setting that is neither the one of performance or of the critical listener raises a number of issues. It is clear that a caricature would be detected, but will the detailed phonetic errors be noticed by the untrained ear? Equally, it is not clear whether the person familiar with the imitated voice would notice these detailed errors and if so whether they would describe the voice as sounding odd or a similar sounding voice to an investigating officer. Moreover, how well an individual has to know a person to detect these detailed phonetic errors and what parameters interact with knowing a person's voice demands further investigation. All texts in this imitation corpus are within the target speakers' professional life and this would, based on results of Zetterholm et al. (2002) [14], increase the acceptance of a famous voice or by those who are only familiar with the voice in the professional domain.

It is obvious that voice and speech imitation can be used as a method to find out which features the impersonator changes with success and the acoustic correlates of these features, and also to find out which features and characteristics in the human voice are not changed and thus, identify the impersonator rather than the target voice. This is a way of extracting information about individuals from their speech which make it possible to detect speaker characteristics for identification and classification.

Acknowledgements. This study is partly funded by a grant from the Bank of Swedish Tercentenary Foundation Dnr K2002-1121:1-4 to Umeå University for the project "Imitated voices: A research project with applications for security and the law". Thanks to Dr. Frantz Clermont for fruitful discussions and graphs, and of course, thanks to the impersonators and the listeners.

References

1. Gibbons, J.: Forensic Linguistics. Blackwell Publishing, Oxford (2003)
2. Hollien, H.: Forensic Voice Identification. Academic Press, San Diego (2002)
3. Hollien, H., Majewski, W., Doherty, E.: Perceptual identification of voices under normal, stress and disguise speaking conditions. Journal of Phonetics 10, 139–148 (1982)
4. Van Lancker, D., Kreiman, J., Emmorey, K.: Familiar voice recognition: Patterns and parameters. Part 1: recognition of backward voices. Journal of Phonetics 13, 19–38 (1985)
5. Van Lancker, D., Kreiman, J., Wickens, T.D.: Familiar voice recognition: Patterns and parameters. Part 2: Recognition of rate-altered voices. Journal of Phonetics 13, 39–52 (1985)
6. Künzel, H.: Effects of voice disguise on speaking fundamental frequency. Forensic Linguistics 7, 199–289 (2000)
7. Yarmey, D., Yarmey, L., Yarmey, M., Parliament, L.: Commonsense beliefs and the identification of familiar voices. Applied Cognitive Psychology 15, 283–299 (2001)
8. Reich, A., Duke, J.: Effects of selected voice disguises upon speaker identification by listening. Journal of the Acoustical Society of America 66, 1023–1028 (1979)
9. Markham, D.: Listeners and disguised voices: the imitation and perception of dialectal accent. Forensic Linguistics 6(2), 289–299 (1999)
10. Sjöström, M., Eriksson, E., Zetterholm, E., Sullivan, K.: A Switch of Dialect as Disguise. In: Proceedings of Fonetik, Lund University, Centre for Languages & Literature, Linguistics & Phonetics, Working Papers 52, pp. 113–116 (2006)
11. Markham, D.: Phonetic Imitation, Accent and the Learner. PhD thesis, Lund University (1997)
12. Laver, J.: Principles of phonetics. Cambridge University Press, Cambridge (1994)
13. Schlichting, F., Sullivan, K.: The imitated voice – a problem for voice line-ups? Forensic Linguistics 4(1), 148–165 (1997)
14. Zetterholm, E., Sullivan, K., van Doorn, J.: The Impact of Semantic Expectation on the Acceptance of a Voice Imitation. In: Proceedings of the 9th Australian International Conference on Speech Science & Technology, Melbourne, pp. 379–384 (2002)
15. Zetterholm, E.: Voice Imitation. A Phonetic Study of Perceptual Illusions and Acoustic Success. PhD thesis, Travaux de l'institut de linguistique de Lund 44, Lund University (2003)
16. Elert, C.C.: Allmän och svensk fonetik. Norstedts Förlag AB, Stockholm (1991)
17. Bruce, G.: Diphthongization in the Malmö dialect. Working Papers 3, 1–19 (1970)
18. Bruce, G.: Allmän och svensk prosodi. Reprocentralen, Lund (1998)
19. Abercrombie, D.: Elements of General Phonetics. Edinburgh University Press, Edinburgh, UK (1967)
20. Laver, J.: The phonetic description of voice quality. Cambridge University Press, Cambridge (1980)
21. Hammarberg, B., Gauffin, J.: Perceptual and acoustic characteristics of quality differences in pathological voices as related to physiological aspects. In: Fujumura, O., Hirano, M. (eds.) Vocal Fold Physiology. Voice quality control, pp. 283–303. Singular Publishing Group, San Diego (1995)

Reviewing Human Language Identification

Masahiko Komatsu

School of Psychological Science, Health Sciences University of Hokkaido
Ainosato 2-5, Sapporo, 002-8072 Japan
koma2@hoku-iryo-u.ac.jp

Abstract. This article overviews human language identification (LID) experiments, especially focusing on the modification methods of stimulus, mentioning the experimental designs and languages used. A variety of signals to represent prosody have been used as stimuli in perceptual experiments: lowpass-filtered speech, laryngograph output, triangular pulse trains or sinusoidal signals, LPC-resynthesized or residual signals, white-noise driven signals, resynthesized signals preserving or degrading broad phonotactics, syllabic rhythm, or intonation, and parameterized source component of speech signal. Although all of these experiments showed that "prosody" plays a role in LID, the stimuli differ from each other in the amount of information they carry. The article discusses the acoustic natures of these signals and some theoretical backgrounds, featuring the correspondence of the source, in terms of the source-filter theory, to prosody, from a linguistic perspective. It also reviews LID experiments using unmodified speech, research into infants, dialectology and sociophonetic research, and research into foreign accent.

Keywords: Language identification, Human language identification, Speech modification, Source-filter model, Prosody.

1 Introduction

Language Identification (LID) is a process for identifying a language used in speech.[1] Although there have been several reviews of automatic LID by computers ([3][4][5] etc.), there have been no extensive reviews of human, or perceptual, LID research as far as the author knows. As opposed to the well-documented automatic LID research, the research scene of human LID gives the impression that it is not well traffic-controlled and the studies are often sporadic. The backgrounds and motivations of researchers are diverse. Thus, the research into the human capability of LID extends into several disciplines, and the communication seems lacking between disciplines, sometimes even within a discipline.

The cues for identifying languages are classified into two types: segmental and prosodic. The former includes "acoustic phonetics," "phonotactics," and "vocabulary," and the latter corresponds to "prosodics" of the terms in [3]. In the field of automatic LID by computers, much of the research so far has focused on utilizing

---

[1] Part of this article is based on [1][2].

C. Müller (Ed.): Speaker Classification II, LNAI 4441, pp. 206–228, 2007.

segmental features contained in the speech signal, although some research also suggests the importance of incorporating prosodic information into the system ([6][7] etc.). In contrast to this engineering research scene, most of the research on perceptual LID by humans has focused on prosodic information.

Humans' capacity for LID has drawn the attention of engineers, linguists, and psychologists since 1960s. The typical method of research is to conduct perceptual experiments with stimulus signals that are supposed to contain prosodic information of certain languages but not contain segmental information. In other words, the signals are used as the representative of prosody. The modification methods of stimulus signals and the languages used in the experiments have been various and not consistent across researchers. The critical question here is whether the signals used really represent the prosody of language, or more specifically what represents prosody acoustically.

This article aims at giving the reader an overview of the human LID research, discussing the modification methods of speech, the experimental designs, and the relations to the prosodic types of used languages. It also introduces examples from related areas of research. The latter part of the article discusses the acoustic correlates of prosody to advance suggestions for future research.

2   Overview of Human LID Experiments

2.1   LID with Modified Speech

A variety of signals and languages have been used as stimuli in perceptual experiments (see Table A1). All studies listed there have used modified speech that was presumed to represent the prosody of speech, and all of them have concluded that prosody plays some role in LID.

Of the stimuli to represent prosody used in previous experiments, the handiest is lowpass-filtered speech. Atkinson [8] used this signal for the discrimination test of English and Spanish, and showed that these two languages were discriminable and that error rates varied depending on speech styles. The lowpass-filtering technique is still being used (e.g., Mugitani et al. [9], for Eastern and Western Japanese, which have different characteristics of lexical accent).

The most straightforward is a laryngograph signal, which is an indication of variations in glottal electrical resistance, closely related to the glottal waveform. It sounds like a dull buzzing noise, varying in pitch. Maidment [10][11] showed that English and French are discriminable with this signal. Moftah & Roach [12] compared the lowpass-filtered and laryngograph signals and concluded that there was no significant difference in language identification accuracy for Arabic and English.

A synthesized signal was used by Ohala and Gilbert [13]. They made triangular pulse trains that had the same F0 and amplitude as the original speech signal; the amplitude was set to zero where F0 was unavailable, i.e., there was no voicing. The signal simulated the F0, amplitude, and voice timing of the original speech, and sounded like a buzz. They designed the experiments to investigate the relation of prosodic types of languages to explicitly defined acoustic features. They chose three prosodically different languages to test: English (stress-accented, stress-timed), Japanese (pitch-accented, mora-timed), and Chinese (tonal). The results indicated that these languages were discriminated. It also showed that the listeners with prior training performed better than those with no training, that bilingual listeners performed better than trilinguals and monolinguals, and that longer samples were better discriminated than shorter ones. Barkat et al. [14] used sinusoidal signals instead of triangular pulses to test Western and Eastern Arabic, the former of which loses short vowels causing prosodic difference. These two dialects were discriminated by Arabic listeners, but not by non-Arabic listeners.

Application of Linear Predictive Coding (LPC) is comparatively new in the history of research on human LID. LPC separates the speech signal into the source and filter, or spectral, components in terms of the source-filter model. The idea of using LPC can be traced back to Foil's experiment [15], but it was simply a preparatory test for developing an automatic LID system. Foil resynthesized speech by LPC with its filter coefficients constant, resulting in the speech signal that had a constant spectrum all the time, and said that languages were easy to discriminate with this signal. The languages discriminated were not explicitly described.

Navrátil [16] used an inverse LPC filter to remove spectral information of speech; the signal represented prosody of speech. He also made a random-spliced signal, where short segments roughly corresponding to syllables were manually cut out and concatenated in a random order; the resultant signal lost F0 and intensity contours of the original speech and represented syllable-level phonotactic-acoustic information plus duration. He compared the LID results with these signals for Chinese, English, French, German, and Japanese, and concluded that prosody contributes less to LID (see Table 1).

Table 1. Correct identification rates for 6-s excerpts in Navrátil's experiment [16] (Chance level: 20%). Random-spliced speech represnts syllable information, and inverse-LPC-filtered speech represents prosody.

Stimulus	German	English	French	Japanese	Chinese	Overall
Unmodified speech	98.7 %	100.0 %	98.7 %	81.7 %	88.7 %	96.0 %
Random-spliced	79.7 %	98.7 %	79.1 %	54.6 %	57.7 %	73.9 %
Inverse-LPC-filtered	32.1 %	34.3 %	69.4 %	45.3 %	65.9 %	49.4 %

Komatsu et al. [17] used an inverse LPC filter, and further lowpass-filtered the signal with the cutoff of 1 kHz to ensure spectral removal. The resultant signal sounded like muffled speech. They suspected that partial phonotactic information remained in this signal, so they also created the consonant-suppressed signal for comparison, where the amplitude of consonant intervals of the former signal was set to zero to remove possible consonantal effects. In the former signal, the LID for English and Japanese was successful; but in the latter consonant-suppressed signal, it was unsuccessful. Besides, they created signals driven by band-limited white noise. These signals were the replication of what Shannon et al. [18] used for speech recognition experiments. The speech was divided into 1, 2, 3, or 4 frequency bands, the intensity contours of these bands were used to modulate noises of the same bandwidths, and they were summed up altogether. The resultant signals kept only intensity when the number of bands was 1, and broad spectral information increased as the number of bands increased. The correct identification rate increased as the number of bands increased. Comparing the results with all these stimulus types (see Fig. 1), they concluded that LID was possible using signals with segmental information drastically reduced; it was not possible with F0 and intensity only, but possible if partial phonotactic information was also available. The results also suggested the variation due to prosodic difference of languages and listeners' knowledge.

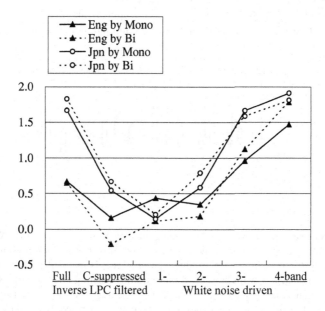

Fig. 1. LID results of English and Japanese in terms of the discriminability index by Komatsu et al. [17]. The index was calculated such that "English" and "Japanese" were scored as +/−2 and "Probably English" and "Probably Japanese" were +/−1, where positive values indicate correct responses and negative, incorrect ones. The graph indicates the results of each stimulus type for English and Japanese samples identified by Japanese monolingual listeners and Japanese-English bilingual listeners, respectively. C-suppressed inverse-LPC-filtered and 1-band white-noise-driven stimuli have only prosodic information (F0, intensity), and the amount of additional information increases when it goes to either side of the graph.

The idea of using LPC was taken a step further by Komatsu et al. [19]. They decomposed the source signal, in terms of the source-filter model, into three parameters, F0, intensity, and Harmonics-to-Noise Ratio (HNR); and created stimulus signals simulating some or all parameters from white noise and/or pulse train. Compared to the previous LPC applications, this method has the merits of the parameterization of the source features and the completeness of spectral removal. They conducted a perceptual discrimination test using excerpts from Chinese, English, Spanish, and Japanese, differing in lexical accent types and rhythm types. In general, the results indicated that humans can discriminate these prosodic types and that the discrimination is easier if more acoustic information is available (see Fig. 2). Further, the results showed that languages with similar rhythm types are difficult to discriminate (i.e., Chinese-English, English-Spanish, and Spanish-Japanese). As to accent types, tonal/non-tonal contrast was easy to detect. They also conducted a preliminary acoustic analysis of the experimental stimuli and found that quick F0 fluctuations in Chinese contribute to the perceptual discrimination of tonal and non-tonal. However, their experiment had a drawback that the number of experimental conditions was too large, which as a consequence had the number of repetitions in each condition too small to run statistical tests. Experiments must be designed to zero in on fewer combinations of acoustic parameters and languages in future.

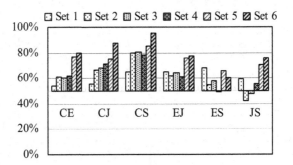

Fig. 2. Correct response rates for each language pair by Komatsu et al. [19]. "C" stands for Chinese, "E" for English, "J" for Japanese, and "S" for Spanish. Set 1 is stimuli made of white noise simulating the intensity of the original speech, Set 2 is made of pulse train simulating intensity, Set 3 is made of white noise and pulse train simulating intensity and HNR, Set 4 is made of pulse train simulating F0, Set 5 is made of pulse train simulating intensity and F0, and Set 6 is made of white noise and pulse train simulating intensity, HNR, and F0. Sets 1-3 represent amplitude-related information, Set 4 represents F0 information, and Sets 5-6 represent both information.

The modification method by Ramus and Mehler [20] is different from others; they are segment-based. They conducted perceptual experiments on English and Japanese, controlling broad phonotactics, syllabic rhythm, and intonation. They segmented the original English and Japanese speech into phonemes and replaced them by French phonemes to exclude the segmental cues to LID. They created four types of stimulus signals differing in the information they contain: "saltanaj", "sasasa", "aaaa", and "flat sasasa". In "saltanaj", all fricatives were replaced by /s/, stops by /t/, liquids by /l/, nasals by /n/, glides by /j/, and vowels by /a/. In "sasasa", all consonants were replaced by /s/, and vowels by /a/. In "aaaa", all segments were replaced by /a/. "Flat sasasa" was the same as "sasasa" but its F0 was made constant. The information that each stimulus contained and the results of LID tests are summarized in Table 2. Ramus and Mehler concluded that syllabic rhythm is a necessary and sufficient cue.

Table 2. Stimuli and LID results of Ramus and Mehler [20]. "+" indicates presence of cue, and "–" indicates absence of cue.

	Intonation	Syllabic rhythm	Broad phonotactics	Result of LID
saltanaj	+	+	+	successful
sasasa	+	+	–	successful
aaaa	+	–	–	unsuccessful
flat sasasa	–	+	–	successful

Although all of these experiments showed that "prosody" plays some role in LID, the stimuli used differ from each other in the amount of information they carry; that is, the acoustic definitions of prosody are not coherent among the studies. An appropriate selection of stimuli is needed for further research.

Experimental procedures in these studies are simple. Participants were provided with a stimulus and instructed to identify a language or dialect. Many experiments simply adopted a multiple choice from two or more language names. Others used somewhat different procedures. In Atkinson's experiment [8], the ABX procedure was used. Ramus and Mehler [20] used a multiple choice from two fictional language names. Maidment [11] and Komatsu et al. [17] used the 4-point scale judgment, e.g., definitely French, probably French, probably English, and definitely English; and Mugitani et al. [9] used the 5-point scale judgment. Komatsu et al. [19] asked the sequential order of the presented stimuli because a multiple choice from four languages would be so difficult to discourage the participants: e.g., participants listened to a Chinese sample and an English sample sequentially and judged whether it was Chinese-English or English-Chinese.

Experimental designs started with a simple one. Discrimination tests were performed for a pair of popular languages: English and Spanish (Atkinson [8]), and English and French (Maidment [10][11]). Mugitani et al. [9] was a pretest for an infants' experiment. Moftah and Roach [12] intended to compare the previously used signals using Arabic and English. Ohala and Gilbert [13] designed their experiment to investigate the relation of prosodic types of languages to explicitly defined acoustic features. They chose three prosodically different languages to test, English (stress-accented, stress-timed), Japanese (pitch-accented, mora-timed), and Chinese (tonal), as well as exploring several other effects. They used conversational speech while preceding studies had predominantly used reading. Barkat et al. [14] focused on the prosodic difference between two Arabic dialects caused by short vowel elision. Navrátil's experiment [16] intended to compare the contributions of prosodic and segmental features, covering five languages. Komatsu et al. [17] compared the LID with segmental features reduced by several methods using English and Japanese. Komatsu et al. [19] parameterized the source features and involved four languages differing in prosodic types (stress-accented, pitch-accented, tonal; stress-timed, syllable-timed, mora-timed) to discuss the relation of the acoustic features to prosodic types. Ramus and Mehler [20] focused on the rhythmic difference of English (stress-timed) and Japanese (mora-timed), which backed up their argument on the acoustic correlates of rhythm.

2.2 LID with Unmodified Speech

It should also be noted that some researchers have used real speech as the stimulus. The purposes and methods of these experiments are different from those using modified speech (see Table A2 for details).

An engineering motivation is the benchmark by humans. Muthusamy et al. [21] did this using 1-, 2-, 4-, and 6-s excerpts of spontaneous speech of 10 languages. The listeners were given feedback on every trial. The obtained results showed that humans are quite capable of identifying languages, but the perceptual cues were not experimentally explored. The cues were sought by Navrátil [16] using two types of modified speech as well as unmodified one, mentioned in section 2.1.

Barkat and Vasilescu [22] sought perceptual cues by two experiments. One is a dialect identification of six Arabic dialects. Endogenous listeners were better at identifying dialects than exogenous listeners. The other used the AB procedure for five Romance languages. The perceptual space was configured by Multi-Dimensional Scaling (MDS) with familiarity and vowel system configuration.

Maddieson and Vasilescu [23] conducted experiments with five languages, combining identification and similarity judgment, and showed that individual variation is poorly explained by prior exposure to the target languages and academic linguistic training.

Bond et al. [24] explored the features that listeners attend using 11 languages from Europe, Asia, and Africa. They used magnitude estimation and MDS techniques and showed that languages were deployed by familiarity, speaker affect (reading dramatic or not), and prosodic pattern (rhythm and F0).

Stockmal et al. [25] challenged to remove the effects of speakers' identity. They did experiments with the AB procedure and similarity judgment for several language pairs using the speech samples spoken by the same bilingual speakers. The results indicated that the listeners discriminated the language pairs spoken by the same speakers and that, in the MDS configuration, they used rhythm information within the context of language familiarity. Stockmal and Bond [26] further eliminated the effect of language familiarity. They replicated the previous experiment only with languages unfamiliar to listeners. The selected languages were all African: all of them are syllable-timed, and all but Swahili were tonal. The results suggested that the listeners discriminated the language pairs using difference in the phoneme inventories.

2.3 Examples from Other Related Areas of Research

Experiments have been conducted with somewhat different perspectives, too. Table A3 listed a few examples of research into infants. Boysson-Bardies et al. [27] showed that the babbling of 8-month-old infants is discriminable by adults. Non-segmental cues such as phonation, F0 contour, and intensity were important. Hayashi et al. [28] and Mugitani et al. [9][29] indicated that infants can discriminate their native language or dialect from others. They used the head-turn preference procedure, which regards the stimulus that infants pay attention longer as preferred, and showed that infants paid attention to their native language or dialect for a longer duration. The original interest of Ramus and Mehler [20], who did the experiment with adults, is in exploring how pre-language infants discriminate languages in bilingual or trilingual environments. See [20][30][31][32] for more literature.

Perceptual experiments have been conducted also for dialectology and sociophonetic purposes (see Table A4 for several recent examples). They seek the perceptual cues of dialect identification and measure the distance among dialects.

Van Bezooijen and Gooskens [33] compared the identification rates between the original speech and monotonized (F0 flattened) speech, representing segmental features, or lowpass-filtered speech, representing prosody, for four Dutch dialects. The results indicated that prosody plays a minor role in dialect identification. A follow-up experiment using only the unmodified signal showed that the difference in the identification rates between spontaneous speech and reading varies among dialects. They also conducted the experiment for five British English dialects, showing that prosody plays a minor role as in Dutch dialects. Gooskens and van Bezooijen [34] adopted a different procedure, 10-point scale judgment of whether dialectal or standard, for six Dutch dialects and six British English dialects. They showed that segmentals are more important, as in their previous experiments, and that the importance of prosody is somewhat larger in English than in Dutch. Gooskens [35] explored 15 Norwegian dialects, and showed that endogenous listeners identify dialects better than exogenous listeners and that prosody is more important than in Dutch dialect identification.

In the United States, Thomas and Reaser [36] did a discrimination test of English spoken by African Americans and European Americans. In order to focus on phonetic characteristics, speech samples were carefully selected to include diagnostic vowels, usually /o/, and subject pronouns, related to intonation variation, but to avoid diagnostic morphosyntactic and lexical variables. European American listeners performed better with monotonized samples than with lowpass-filtered samples; and the detailed analysis indicated that African Americans could not use the vowel quality as a perceptual cue. Thomas et al. [37], who incorporated different techniques, converting all vowels to schwa and swapping F0 and segmental durations, showed that the vowel quality is important although F0 also plays a role and that different listener groups use different cues.

See [36][38] for extensive reviews of the studies in these areas, including experiments with various modification techniques: lowpass-filtering, highpass-filtering, center-clipping; lowpass-filtering vs. monotonization (F0 flattening); bandwidth compression to remove nasality; backward playing, temporal compression; F0 level change of isolated vowels; F2 modification to make vowels front or back; resynthesis of /s/-/ʃ/ to assess the McGurk effect on the perceptual boundary; a synthetic vowel continuum; synthetic vowels; synthetic diphthongs; modification of the intonation and the speaking rate; unmodified, lowpass-filtered, random-spliced, vs. written text.

Table A5 gives examples of the research into foreign accents.[2] Miura et al. [40] and Ohyama and Miura [41] did experiments manipulating a segmental feature (PARCOR[3] coefficients) and prosodic features (F0, intensity, phoneme durations),

---

[2] See also [39].

[3] PARCOR stands for partial auto-correlation.

showing prosodic features contribute more. Miwa and Nakagawa [42] focused on only prosodic features and showed that the sensitivity to such a feature is different between native and non-native listeners.

A confounding factor of perceptual experiments on LID or the naturalness of languages is that prosody is closely related to not only linguistic information but also paralinguistic and nonlinguistic information. Grover et al. [43] found that F0 variation at the continuation junctures of English, French, and German differ significantly, but that the synthetically replaced intonation patterns were regarded by listeners as speakers' variation of emotional attitudes or social classes rather than foreign accents.

Another problem was raised by Munro [44], who investigated the effect of prosody on the perception of foreign accent using lowpass-filtered speech. The results indicated that the foreign-accentedness was recognized in the lowpass-filtered speech. However, they did not show a correlation with the unfiltered, or original, speech, which means that samples regarded as accented when lowpass-filtered may not be regarded as accented when not filtered, suggesting that listeners may use different cues in different conditions.

3   Acoustic Definition of Prosody

3.1   Reviewing Stimulus Signals

The speech modification methods described in section 2.1 may be classified into several groups. The first one is what does not use synthesis or resynthesis techniques: lowpass-filtering and laryngograph output. The second one, which uses synthesis/resynthesis techniques, includes the simple acoustic simulation (triangular pulses, sinusoidal signals, band-limited white noise) and the signal processing based on the source-filter model (inverse LPC filtering, source feature parameterization). Random splicing and phoneme replacing constitute the third group: these modify the signal in segment-based manners, permuting or replacing them, rather than utilizing acoustic processing globally. The second group may be called more "acoustic," and the third group more "phonological."

It is in question whether some of them do properly represent prosody in speech. In lowpass filtering, the cutoff frequency is usually set at 300-600 Hz to make speech unintelligible, but it is reported that speech is sometimes intelligible if repeatedly listened to [45]. In lowpass-filtered speech, some segmental information is preserved under the cutoff frequency, F0 sometimes rises higher than the cutoff, and intensity is not preserved [20]. A perceptual experiment confirmed that, if the cutoff is set at 300 Hz, the filtered signal retains prosodic features and some laryngeal voice quality features but not articulatory features [45]. The laryngograph output is an indication of short-term variations of glottal electrical resistance and virtually uninfluenced by supraglottal resonance and noise source [12][13]. This means that it is not representative of output speech, which we actually hear in usual situations. Due to the loss of resonance and noise source, it does not contain sonority information, which will be discussed in section 3.2. The simple acoustic simulation techniques are close to the source-filter-model-based ones but incomplete because they lack something. The simulation of prosody with pulse or sinusoidal trains does not take the noise source into account. The white-noise driven signal keeps the intensity contour of the original speech but does not have any other information such as F0.

Of the segment-based approaches, random-splicing, of course, destroys prosodic contour information as the experimenter intends. It was reported that speech random-spliced with the segment size between 150-300 ms was unintelligible, and a perceptual experiment confirmed that speech random-spliced with the segment size of 255 ms carries voice quality, some articulatory features, and overall prosodic features (level, range, and variability of pitch, loudness, and sonority) but loses tempo [45]. In Navrátil's experiment [16], segments in length roughly corresponding to syllables were manually cut out.

The processing based on the source-filter model may be the best to represent prosody (see the discussion in section 3.2), but it may have a technical drawback. Inverse LPC filtering does not guarantee the perfect removal of the spectrum. To avoid this problem, Komatsu et al. [17] used a lowpass filter in conjunction with an inverse LPC filter, but still reported that some listeners said they spotted words although it is not clear whether it was true or illusory. On the other hand, the source feature parameterization, in which the stimulus is made of pulses and white noise from scratch, is perfect in the spectral removal but problematic especially in the F0 contour estimation. Komatsu et al. [19] used the MOMEL algorithm [46], originally devised to extract the intonation contour of the intonation languages (i.e., non-pitch-accented, non-tonal). It seems that the algorithm does not only remove microprosody but affects the F0 variation related to pitch accent and tone [47].

To estimate F0 correctly and compare among languages, a model that does not incorporate any phonology of specific languages is desired. For example, modeling by INTSINT [46], which simply encodes F0 patterns, seems more adequate for the present purpose than ToBI [48][49], which describes only F0 variations meaningful in respective languages. Another desirable nature of the model is to divide the contour into components. Although the difference in F0 between languages of different prosodic types have been pointed out [50], local characteristics seem more important than global characteristics [6][51]. Further, three types of F0 characteristics varying across languages have been distinguished: global, recurrent, and local [52]. Although there have been proposed various F0 models [53], not all are adequate for the present purpose. Scrutiny of models is necessary for future research.

The notion of rhythm is also confounding. Since Pike's dichotomy of stress- and syllable-timed rhythms [54], the isochronic recurrence of stress/syllable in speech signal has not been found. This has caused the definition of rhythm to be claimed variously [31][55]. Timing hypotheses argue that there is an isochronic unit or that the length of the higher level unit such as a word can be predictable from the number of lower level units. Rhythm hypotheses argue that the rhythmic difference is the reflection of structural factors, such as syllable structures, phonotactics, etc., rather than timing specifically [30][56][57]. There are also other alternative claims focusing the competence of coordinating units in speech production [58], or the role of the unit in perception [59].

Ramus and Mehler's experiment [20] was to support the rhythm hypothesis. They define "syllabic rhythm" as the temporal alignments of consonant and vowel, which is the reflection of syllable structures, and showed that it is essential to the perceptual discrimination of languages. Here, rhythm is not defined by acoustic features such as the intensity contour but defined by the discrimination of consonant and vowel. This reminds us of the role of broad phonotactics in human LID [17] and automatic LID [60].

The studies pursuant to the rhythm hypothesis are rather phonological than acoustic, because they need phoneme identification. Phonemes must be identified in the stream of speech signal prior to the measurement of durations. However, it seems that this method has been taken as an expedient because appropriate acoustic measures to grasp syllable shapes were not available. Ramus et al. [30] (p. 271 fn) states that "[their] hypothesis should ultimately be formulated in more general terms, e.g. in terms of highs and lows in a universal sonority curve." In another analysis [57], devoiced vowels are treated as consonantal rather than vocalic to reflect more acoustic features. Retesting this hypothesis by calculating sonority in acoustic terms [61] is worth mentioning.

3.2 Correspondence Between Acoustic and Linguistic Features

This section argues that the source at the acoustic level approximately represents prosody at the linguistic level. Fig. 3 shows the simplified correspondence of the articulatory, acoustic, and linguistic models. Note that the figure is simplified for illustration, and that the correspondences of the features in different models are actually not as simple as drawn in the figure. When humans utter speech, especially vowels, the voice source is created at the larynx, is modulated by the vocal tract, and results in the speech sound. This can be modeled by an acoustic model called the source-filter model, in which the source, or the excitation signal, is processed by the filter, resulting in the speech signal. The source consists of three physical elements, F0, intensity, and HNR; and the filter determines the spectral envelope of the sound in the frequency domain. Very naively, prosodic, or suprasegmental, features in the linguistic model seem to involve F0 and intensity of the speech signal controlled by the laryngeal activity: the tone and accent systems seem to involve F0 and intensity, and rhythm seems to involve the temporal variation of intensity. On the other hand, segmental features, i.e. phoneme distinctions, seem to be related to spectral patterns determined by the vocal tract shape, or the movement of articulators. However, their correspondence to each other is actually not so simple. For example, in the recognition of phonemes, it is known that various acoustic cues interact, including not only the spectral pattern but also F0 and intensity. So far, the acoustic contributors to prosodic features have not been thoroughly inquired into. This section discusses whether, or how well, the source elements of the acoustic model approximately represent the linguistic prosody.

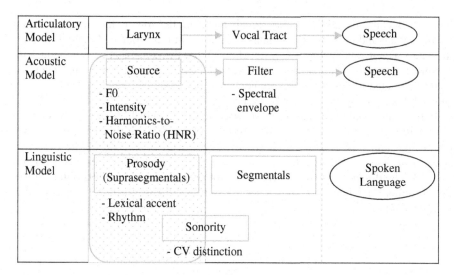

Fig. 3. Approximate correspondence of articulatory, acoustic, and linguistic models. The shaded area indicates the correspondence discussed in this section.

Linguistic features that constitute a prosodic typology include lexical accent (stress accent, pitch accent, and tone), intonation, and rhythm (stress-timed, syllable-timed, and mora-timed). Their acoustic correlates are, basically, F0, intensity, and length. However, assuming that rhythm is, even if partly, the reflection of syllable structures, it follows that acoustic properties that represent sonority contribute to constituting rhythm.

Sonority is a linguistic feature that approximately represents syllable shapes (see Sonority Sequencing Principle [62]). The sonority feature is ambivalently prosodic and segmental by nature. On one hand, it represents syllable shapes, and consequently contributes to rhythm. On the other hand, it is closely related to the articulatory manner of segments, and, as a result, it partially represents some phoneme classes and phonotactics. Consequently, the acoustic properties that represent sonority contain both prosodic and segmental information. Then, it is impossible to completely separate acoustic features corresponding to prosody from acoustic features corresponding to segmentals. The dichotomy of prosody and segmentals are possible in the linguistic model but impossible in the real-world acoustic model.

The important question is, therefore, whether or how the source features at the acoustic level represent sonority, and do not represent segmental features, at the linguistic level. To this end, experiments on Japanese consonant perception were conducted with the LPC residual signal [63]. The identification rate of major classes corresponding to the sonority ranks, i.e., obstruent, nasal, liquid, and glide [62], was as high as 66.4 % while that of phonemes was as low as 20.0 % (chance level: $1/17 = 5.9$ %).

Further, to investigate how sonority is represented in the source, the confusion matrix obtained from this experiment was analyzed with MDS [64]. The analysis showed that sonority can be located in a multi-dimensional perceptual space, and that the dimensions of the space have correspondence to both acoustic and phonological features. Because the LPC residual signals represent the source, the confusion pattern for the signals indicates the consonants' similarities in the source. Although fitting of the data was not satisfactory, the result showed that the consonants can be modeled in a 3-dimensional perceptual space according to their sonority ranks. Its dimensions could be related to acoustic measurements and phonological features. The result also showed that sonority can be mostly defined within the source.

In the perceptual space, consonants with the same sonority rank clearly tended to cluster together. As seen in Fig. 4, voiceless plosives, voiceless fricatives, voiced obstruents, and nasals/glides gathered together. Each dimension of the perceptual space had correspondence to acoustic and phonological features, as shown in Table 3. The dimensions were correlated with acoustic measurements obtained from the stimulus, and had correspondence to some of the sonority-related distinctive features [65].

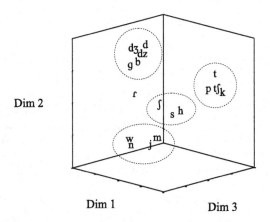

Fig. 4. Three-dimensional analysis of consonant perception in the LPC residual signal (altered from [64])

Table 3. Correspondence of each dimension to acoustic and phonological features

		Lower sonority ⟵⟶	Higher sonority
Dim 1	Acoustics	Lower HNR	Higher HNR
	Phonology	[−voice]	[+voice]
Dim 2	Acoustics	Smaller amplitude Lower F0	Larger amplitude Higher F0
	Phonology	[−sonorant]	[+sonorant]
Dim 3	Acoustics	Shorter duration	Longer duration
	Phonology	[−continuant]	[+continuant]

These results indicate that the source retains sonority information while segmental information, such as cues for phoneme identification, is effectively suppressed. The importance of sonority, or broad phonotactics, has been shown by many previous LID studies, human or automatic.

4 Concluding Remarks

This article started with overviewing human LID experiments, especially focusing on the modification methods of stimulus, also mentioning the experimental designs and languages used (section 2). It was followed by the discussion on what those acoustic features used in human LID experiments mean (section 3). It discussed the acoustic natures of the stimulus signals and some theoretical backgrounds, featuring the correspondence of the source to prosody.

LID is, from a linguistic point of view, a study on the naturalness of a language and the difference from other languages. Language is defined as the pair of the form and meaning. LID research focuses on the form only, and would provide cross-linguistic foundations for the description of the form. Simple manipulations of acoustic features may suffice to engineering purposes; their linguistic meanings have not been inquired into. The author hopes that this article gives the reader some insights into this question.

Acknowledgments. The author is grateful to Dennis R. Preston, Charlotte Gooskens, Erik R. Thomas, and Jiří Navrátil for their kind offer of information.

References

1. Komatsu, M.: What constitutes acoustic evidence of prosody? The use of Linear Predictive Coding residual signal in perceptual language identification. LACUS Forum 28, 277–286 (2002)
2. Komatsu, M.: Acoustic constituents of prosodic types. Doctoral dissertation. Sophia University, Tokyo (2006)
3. Muthusamy, Y.K., Barnard, E., Cole, R.A.: Reviewing automatic language identification. IEEE Signal Processing Magazine 11(4), 33–41 (1994)
4. Zissman, M.A., Berkling, K.M.: Automatic language identification. Speech Communication 35, 115–124 (2001)
5. Navrátil, J.: Automatic language identification. In: Schultz, T., Kirchhoff, K. (eds.) Multilingual speech processing, pp. 233–272. Elsevier, Amsterdam (2006)
6. Thymé-Gobbel, A.E., Hutchins, S.E.: On using prosodic cues in automatic language identification. In: Proceedings of International Conference on Spoken Language Processing '96, pp. 1768–1771 (1996)
7. Itahashi, S., Kiuchi, T., Yamamoto, M.: Spoken language identification utilizing fundamental frequency and cepstra. In: Proceedings of Eurospeech '99, pp. 383–386 (1999)
8. Atkinson, K.: Language identification from nonsegmental cues [Abstract]. Journal of the Acoustical Society of America 44, 378 (1968)
9. Mugitani, R., Hayashi, A., Kiritani, S.: Developmental change of 5 to 8-month-old infants' preferential listening response. Journal of the Phonetic Society of Japan 4(2), 62–71 (2000) (In Japanese)
10. Maidment, J.A.: Voice fundamental frequency characteristics as language differentiators. Speech and Hearing: Work in Progress 2. University College, London, pp. 74–93 (1976)

11. Maidment, J.A.: Language recognition and prosody: Further evidence. Speech, Hearing and Language: Work in Progress 1. University College, London, pp. 133–141 (1983)
12. Moftah, A., Roach, P.: Language recognition from distorted speech: Comparison of techniques. Journal of the International Phonetic Association 18, 50–52 (1988)
13. Ohala, J.J., Gilbert, J.B.: Listeners' ability to identify languages by their prosody. In: Léon, P., Rossi, M. (eds.) Problèmes de prosodie: Expérimentations, modèles et fonctions. Didier, Paris, vol. 2, pp. 123-131 (1979)
14. Barkat, M., Ohala, J., Pellegrino, F.: Prosody as a distinctive feature for the discrimination of Arabic dialects. In: Proceedings of Eurospeech '99, pp. 395–398 (1999)
15. Foil, J.T.: Language identification using noisy speech. In: Proceedings of International Conference on Acoustics, Speech, and Signal Processing '86, pp. 861–864 (1986)
16. Navrátil, J.: Spoken language recognition: A step toward multilinguality in speech processing. IEEE Transactions on Speech and Audio Processing 9, 678–685 (2001)
17. Komatsu, M., Mori, K., Arai, T., Aoyagi, M., Murahara, Y.: Human language identification with reduced segmental information. Acoustical Science and Technology 23, 143–153 (2002)
18. Shannon, R.V., Zeng, F.-G., Kamath, V., Wygonski, J., Ekelid, M.: Speech recognition with primarily temporal cues. Science 270, 303–304 (1995)
19. Komatsu, M., Arai, T., Sugawara, T.: Perceptual discrimination of prosodic types and their preliminary acoustic analysis. In: Proceedings of Interspeech 2004, pp. 3045–3048 (2004)
20. Ramus, F., Mehler, J.: Language identification with suprasegmental cues: A study based on speech resynthesis. Journal of the Acoustical Society of America 105, 512–521 (1999)
21. Muthusamy, Y.K., Jain, N., Cole, R.A.: Perceptual benchmarks for automatic language identification. In: Proceedings of International Conference on Acoustics, Speech, and Signal Processing '94, pp. 333–336 (1994)
22. Barkat, M., Vasilescu, I.: From perceptual designs to linguistic typology and automatic language identification: Overview and perspectives. In: Proceeding of Eurospeech 2001, pp. 1065–1068 (2001)
23. Maddieson, I., Vasilescu, I.: Factors in human language identification. In: Proceedings of International Conference on Spoken Language Processing 2002, pp. 85–88 (2002)
24. Bond, Z.S., Fucci, D., Stockmal, V., McColl, D.: Multi-dimensional scaling of listener responses to complex auditory stimuli. In: Proceedings of International Conference on Spoken Language Processing '98, vol. 2, pp. 93–95 (1998)
25. Stockmal, V., Moates, D.R., Bond, Z.S.: Same talker, different language. In: Proceedings of International Conference on Spoken Language Processing '98, vol. 2, pp. 97–100 (1998)
26. Stockmal, V., Bond, Z.S.: Same talker, different language: A replication. In: Proceedings of International Conference on Spoken Language Processing 2002, pp. 77–80 (2002)
27. Boysson-Bardies, B., de Sagart, L., Durand, C.: Discernible differences in the babbling of infants according to target language. Journal of Child Language 11, 1–15 (1984)
28. Hayashi, A., Deguchi, T., Kiritani, S.: Reponse patterns to speech stimuli in the headturn preference procedure for 4- to 11-month-old infants. Japan Journal of Logopedics and Phoniatrics 37, 317–323 (1996)
29. Mugitani, R., Hayashi, A., Kiritani, S.: The possible preferential cues of infants' response toward their native dialects evidenced by a behavioral experiment and acoustical analysis. Journal of the Phonetic Society of Japan 6(2), 66–74 (2002)
30. Ramus, F., Nespor, M., Mehler, J.: Correlates of linguistic rhythm in the speech signal. Cognition 73, 265–292 (1999)

31. Tajima, K.: Speech rhythm and its relation to issues in phonetics and cognitive science. Journal of the Phonetic Society of Japan 6(2), 42–55 (2002)
32. Hayashi, A.: Perception and acquisition of rhythmic units by infants. Journal of the Phonetic Society of Japan 7(2), 29–34 (2003) (In Japanese)
33. van Bezooijen, R., Gooskens, C.: Identification of language varieties: The contribution of different linguistic levels. Journal of Language and Social Psychology 18, 31–48 (1999)
34. Gooskens, C., van Bezooijen, R.: The role of prosodic and verbal aspects of speech in the perceived divergence of Dutch and English language varieties. In: Berns, J., van Marle, J. (eds.) Present-day dialectology: Problems and findings. Mouton de Gruyter, Berlin, pp. 173–192 (2002)
35. Gooskens, C.: How well can Norwegians identify their dialects? Nordic Journal of Linguistics 28, 37–60 (2005)
36. Thomas, E.R., Reaser, J.: Delimiting perceptual cues used for the ethnic labeling of African American and European American voices. Journal of Sociolinguistics 8, 54–87 (2004)
37. Thomas, E.R., Lass, N.J., Carpenter, J.: Identification of African American speech. In: Preston, D.R., Niedzielski, N. (eds.) Reader in Sociophonetics. Cambridge University Press, Cambridge (in press)
38. Thomas, E.R.: Sociophonetic applications of speech perception experiments. American Speech 77, 115–147 (2002)
39. Gut, U.: Foreign accent. In: Müller, C. (ed.) Speaker classification. LNCS, vol. 4343, pp.75–87, Springer, Heidelberg (2007)
40. Miura, I., Ohyama, G., Suzuki, H.: A study of the prosody of Japanese English using synthesized speech. In: Proceedings of the 1989 Autumn Meeting of the Acoustical Society of Japan, pp. 239–240 (1989) (In Japanese)
41. Ohyama, G., Miura, I.: A study on prosody of Japanese spoken by foreigners. In: Proceedings of the 1990 Spring Meeting of the Acoustical Society of Japan, pp. 263–264 (1990) (In Japanese)
42. Miwa, T., Nakagawa, S.: A comparison between prosodic features of English spoken by Japanese and by Americans. In: Proceedings of the 2001 Autumn Meeting of the Acoustical Society of Japan, pp. 229-230 (2001) (In Japanese)
43. Grover, C., Jamieson, D.G., Dobrovolsky, M.B.: Intonation in English, French and German: Perception and production. Language and Speech 30, 277–295 (1987)
44. Munro, M.J.: Nonsegmental factors in foreign accent: Ratings of filtered speech. Studies in Second Language Acquisition 17, 17–34 (1995)
45. van Bezooijen, R., Boves, L.: The effects of low-pass filtering and random splicing on the perception of speech. Journal of Psycholinguistic Research 15, 403–417 (1986)
46. Hirst, D., Di Cristo, A., Espesser, R.: Levels of representation and levels of analysis for the description of intonation systems. In: Horne, M. (ed.) Prosody: Theory and experiment, pp. 51–87. Kluwer Academic, Dordrecht, The Netherlands (2000)
47. Komatsu, M., Arai, T., Sugawara, T.: Perceptual discrimination of prosodic types. In: Proceedings of Speech Prosody 2004, pp. 725–728 (2004)
48. Venditti, J.J.: Japanese ToBI labelling guidelines. Manuscript, Ohio State University, Columbus (1995)
49. Pierrehumbert, J.: Tonal elements and their alignment. In: Horne, M. (ed.) Prosody: Theory and experiment, pp. 11–36. Kluwer Academic, Dordrecht, The Netherlands (2000)
50. Eady, S.J.: Differences in the F0 patterns of speech: Tone language versus stress language. Language and Speech 25, 29–42 (1982)

51. Komatsu, M., Arai, T.: Acoustic realization of prosodic types: Constructing average syllables. LACUS Forum 29, 259–269 (2003)
52. Hirst, D., Di Cristo, A.: A survey of intonation systems. In: Hirst, D., Di Cristo, A. (eds.) Intonation systems: A survey of twenty languages, pp. 1–44. Cambridge University Press, Cambridge (1998)
53. Shih, C., Kochanski, G.: Prosody and prosodic models. In: Tutorial at International Conference on Spoken Language Processing 2002, Denver CO (2002)
54. Pike, K.L.: The intonation of American English. University of Michigan Press, Ann Arbor (1945)
55. Warner, N., Arai, T.: Japanese mora-timing: A review. Phonetica 58, 1–25 (2001)
56. Dauer, R.M.: Stress-timing and syllable-timing reanalyzed. Journal of Phonetics 11, 51–62 (1983)
57. Grabe, E., Low, E.L.: Durational variability in speech and the Rhythm Class Hypothesis. In: Gussenhoven, C., Warner, N. (eds.) Laboratory phonology 7. Mouton de Gruyter, Berlin, pp. 515–546 (2002)
58. Tajima, K.: Speech rhythm in English and Japanese: Experiments in speech cycling. Doctoral dissertation, Indiana University, Bloomington, IN (1998)
59. Cutler, A., Otake, T.: Contrastive studies of spoken-language perception. Journal of the Phonetic Society of Japan 1(3), 4–13 (1997)
60. Nakagawa, S., Seino, T., Ueda, Y.: Spoken language identification by Ergodic HMMs and its state sequences. IEICE Transactions J77-A(2), 182–189 (1994) (In Japanese)
61. Galves, A., Garcia, J., Duarte, D., Galves, C.: Sonority as a basis for rhythmic class discrimination. In: Proceedings of Speech Prosody 2002, pp. 323–326 (2002)
62. Clements, G.N.: The role of the sonority cycle in core syllabification. In: Beckman, M.E., Kingston, J. (eds.) Papers in laboratory phonology 1, pp. 283–333. Cambridge University Press, Cambridge (1990)
63. Komatsu, M., Tokuma, W., Tokuma, S., Arai, T.: The effect of reduced spectral information on Japanese consonant perception: Comparison between L1 and L2 listeners. In: Proceedings of International Conference on Spoken Language Processing 2000, vol. 3, pp. 750–753 (2000)
64. Komatsu, M., Tokuma, S., Tokuma, W., Arai, T.: Multi-dimensional analysis of sonority: Perception, acoustics, and phonology. In: Proceedings of International Conference on Spoken Language Processing 2002, pp. 2293–2296 (2002)
65. Blevins, J.: The syllable in phonological theory. In: Goldsmith, J.A. (ed.) The handbook of phonological theory, pp. 206–244. Basil Blackwell, Cambridge, MA (1995)

Appendix: Lists of LID Research

Table A1. LID using modified speech

Atkinson (1968) [8]

Language:	English, Spanish
Material:	Poetry, prose, natural speech, nursery rhymes, dramatic dialogues
Modification:	Lowpass-filetered
Method:	Identification (ABX)
Result:	English and Spanish were discriminated. Least error rates in poetry, greatest in prose and nursury rhymes.

Table A1. (*continued*)

Mugitani, Hayashi, & Kiritani (2000) [9]

Language:	Eastern Japanese dialect, Western Japanese dialect
Material:	Elicited speech by a speaker fluent in both dialects
Modification:	(1) Unmodified, (2) Lowpass-filtered (400Hz)
Method:	(1) 5-pt scale (+2=Definitely Eastern, -2=Never Eastern; +2=Definitely Western, -2=Never Western), (2) Identification (Eastern or not)
Result:	(1) Almost perfect, (2) Significant result

Maidment (1976) [10]

Language:	English, French
Material:	Reading
Modification:	Laryngograph waveform
Method:	Identification
Result:	64.5%

Maidment (1983) [11]

Language:	English, French
Material:	Spontaneous speech
Modification:	Laryngograph waveform
Method:	4-pt scale judgment (1=Definitely French, 4=Definitely English)
Result:	74.68% [Calculated such that both "1 Definitely French" and "2 Probably French" counted as French and both "3 Definitely English" and "4 Probably English" counted as English]

Moftah & Roach (1988) [12]

Language:	Arabic, English
Material:	Reading and spontaneous speech
Modification:	(1) Laryngograph waveform, (2) Lowpass-filtered (500Hz)
Method:	Identification
Result:	(1) 63.7%, (2) 65.5%

Ohala & Gilbert (1979) [13]

Language:	English, Japanese, Cantonese Chinese
Material:	Conversation
Modification:	Triangle pulses simulating F0, amplitude, voice timing
Method:	Identification
Result:	56.4% [Chance level: 33.3%]

Barkat, Ohala, & Pellegrino (1999) [14]

Language:	Western Arabic dialects, Eastern Arabic dialects
Material:	Elicited story-telling
Modification:	(1) Unmodified, (2) Sinusoidal pulses simulating F0, amplitude, voice timing
Method:	Identification
Result:	(1) 97% by Arabic listeners, 56% by non-Arabic listeners. (2) 58% by Arabic listeners, 49% by non-Arabic listeners

Table A1. (*continued*)

Foil (1986) [15]

    Language: Unknown (one Slavic and one tonal SouthEast Asian languages?)
    Material: Unknown (noisy radio signals?)
    Modification: LPC-resynthesized with the filter coefficients constant
    Method: Identification
    Result: Easy to distinguish

Navrátil (2001) [16]

    Language: Chinese, English, French, German, Japanese
    Material: Spontaneous speech?
    Modification: (1) Unmodifed, (2) Random-splicing, (3) Inverse-LPC-filtered
    Method: Identification
    Result: (1) 96%, (2) 73.9%, (3) 49.4% [Chance level: 20%]

Komatsu, Mori, Arai, Aoyagi, & Murahara (2002) [17]

    Language: English, Japanese
    Material: Spontaneous speech
    Modification: (1) Inverse-LPC-filtered followed by lowpass-filtered (1kHz), (2) Consonant intervals of (1) suppressed, (3) Band-devided white-noise driven (from 1 to 4 bands)
    Method: 4-pt scale judgment (1=English, 4=Japanese)
    Result: For English, (1) 70.0%, (2) 44.0%, (3) 56.0-95.0% varying over the number of bands; for Japanese, (1) 100.0%, (2) 66.0%, (3) 60.0-96.0% varying over the number of bands

Komatsu, Arai, & Sugawara (2004) [19]

    Language: Chinese, English, Japanese, Spanish
    Material: Reading
    Modification: (1) White noise simulating intensity, (2) Pulse train simulating intensity, (3) Mixture of white noise and pulse train simulating intensity and harmonicity, (4) Pulse train simulating F0, (5) Pulse train simulating intensity and F0, (6) Mixture of white noise and pulse train simulating intensity, harmonicity, and F0
    Method: Judgment on the sequential order by listening to a language pair
    Result: (1) 61.3%, (2) 61.1%, (3) 63.1%, (4) 62.8%, (5) 74.7%, (6) 79.3% [Chance level: 50%]

Ramus & Mehler (1999) [20]

    Language: English, Japanese
    Material: Reading
    Modification: Resynthesized preserving (1) broad phonotactics, rhythm, and intonation, (2) rhythm and intonation, (3) intonation only, (4) rhythm only
    Method: Identification (using fictional language names)
    Result: (1) 66.9%, (2) 65.0%, (3) 50.9%, (4) 68.1%; indicating the importance of rhythm

Table A2. LID using unmodified speech only

Muthusamy, Jain, & Cole (1994) [21]

Language:	10 languages (English, Farsi, French, German, Japanese, Korean, Mandarin Chinese, Spanish, Tamil, Vietnamese)
Material:	Spontaneous speech
Method:	Identification
Result:	With 6-s excerpts, 69.4% (varying from 39.2 to 100.0% over languages) [Chance level: 10%]

Barkat & Vasilescu (2001) [22]

Language:	6 Arabic dialects
Material:	Elicited speech
Method:	Identification
Result:	78% for Western dialicts, 32% for Eastern dialects by Western Arabic listerns; 59% for Western dialects, 90% for Eastern dialects by Eastern Arabic listeners [Chance level: 16.7%]
Language:	5 Romance languages (French, Italian, Spanish, Portuguese, Romanian)
Material:	Reading or story-telling
Method:	AB (same or different)
Result:	MDS configured with familiarity, vowel system complexity

Maddieson & Vasilescu (2002) [23]

Language:	5 languages (Amharic, Romanian, Korean, Morroccan Arabic, Hindi)
Material:	Reading
Method:	(1) Identification, (2) Identification and similarity judgment
Result:	(1) 65% [Chance level: 20%], (2) Partial identification patterns varied among languages

Bond, Fucci, Stockmal, & McColl (1998) [24]

Language:	11 languages from Europe, Asia, Africa (Akan, Arabic, Chinese, English, French, German, Hebrew, Japanese, Latvian, Russian, Swahili)
Material:	Reading
Method:	Similarity to English (magnitude estimation)
Result:	MDS configured with familiarity, speaker affect, prosodic pattern (rhythm, F0)

Stockmal, Moates, & Bond (1998) [25]

Language:	Language pairs (Arabic-French, Hebrew-German, Akan-Swahili, Latvian-Russian, Korean-Japanese, Ombawa-French, Ilocano-Tagalog)
Material:	Each language pair was spoken by the same talker
Method:	(1) AB (same or different), (2) AB (7-pt similarity; 1=very dissimilar, 7=very similar)
Result:	(1) 66.5% and 63.4% depending on the experimental condition [Chance level: 50%], (2) 5.19 for the same-language pairs, 3.45 for the different-language pairs

Stockmal & Bond (2002) [26]

Language:	Language pairs (Akan-Swahili, Haya-Swahili, Kikuyu-Swahili, Luhya-Swahili)
Material:	Reading; each language pair was spoken by the same talker
Method:	AB (same or different)
Result:	71% [Chance level: 50%]

Table A3. Examples of research into infants

Boysson-Bardies, Sagart, & Durand (1984) [27]
Language:	Arabic, Chinese, French
Material:	8- and 10-month-old infants' babbling
Method:	Choice of French from French-Arabic or French-Chinsese pair
Result:	For 8- and 10-month samples respectively; 75.8%, 74.4% (French-Arabic pairs); 69.4%, 31.9% (French-Chinese pairs) [Chance level: 50%]

Language:	Arabic, French
Material:	6-, 8-, 10-month-old infants' babbling
Method:	Choice of French from the pair of babbling
Result:	For 6-, 8-, 10-month samples respectively; 55.5-68%, 67.5-74%, 49-56.9% (varying over experimental conditions) [Chance level: 50%]

Hayashi, Deguchi, & Kiritani (1996) [28]
Language:	Japanese, English
Material:	Spontaneous speech by a bilingual speaker
Method:	Head-tern preference procedure (for infants)
Result:	Infants aged over 200 days preferred the native language Japanese

Mugitani, Hayashi, & Kiritani (2000) [9]
Language:	Eastern Japanese dialect, Western Japanese dialect
Material:	Elicited speech by a speaker fluent in both dialects
Modification:	Unmodified
Method:	Head-turn preference procedure (for infants)
Result:	Greater preference to their native Eastern dialect

Mugitani, Hayashi, & Kiritani (2002) [29]
Language:	Eastern Japanese dialect, Western Japanese dialect
Material:	Elicited speech by a speaker fluent in both dialects
Modification:	Lowpass-filtered (400Hz)
Method:	Head-turn preference procedure (for infants)
Result:	8-month-old infants preferred their native Eastern dialect

---

Table A4. Examples of dialectology and sociophonetic research using modified speech

---

Van Bezooijen & Gooskens (1999) [33]
Language:	4 Dutch dialects
Material:	Spontaneous
Modification:	(1) Unmodified, (2) Monotonized (flat f0), (3) Lowpass-filtered (350Hz)
Method:	Identification of Country, Region, Province, and Place
Result:	(1) Country 90%, Region 60%, Province 40%; (2) Decreased from (1) by 7%, 2%, 4%; (3) Decreased from (1) by 29%, 41%, 32% [Chance levels for Country, Region, Province are 50%, 12.5%, 5.26% respectively; there were almost no answer for Place]

Language:	5 British English dialects
Material:	Spontaneous
Modification:	(1) Unmodified, (2) Monotonized (flat f0), (3) Lowpass-filtered (350Hz), (4) The same as (3) but including typical dialect prosody

Table A4. (*continued*)

Method:	Identification of Country, Region, Area, and Place
Result:	(1) Country 92%, Region 88%, Area 52%; (2) Decreased from (1) by 4%, 10%, 3%; (3) Decreased from (1) by 18%, 43%, 33%, (4) Increased from (3) by 5%, 5%, 1% [Chance levels for Country, Region, Area are 50%, 14.28%, 6.67% respectively; there were almost no answer for Place]

Gooskens & van Bezooijen (2002) [34]

Language:	6 Dutch dialects
Material:	Interview
Modification:	(1) Unmodified, (2) Monotonized (flat F0), (3) Lowpass-filtered (350Hz)
Method:	10-pt scale judgment (1=dialect, 10=standard)
Result:	(1)(2) 4 groups separated, (3) Standard variation and the others were separated

Language:	6 British English dialects
Material:	Interview
Modification:	(1) Unmodified, (2) Monotonized (flat F0), (3) Lowpass-filtered (350Hz)
Method:	10-pt scale judgment (1=dialect, 10=standard)
Result:	(1)(2) 3 groups separated, (3) 2 groups separated

Gooskens (2005) [35]

Language:	15 Norwegian dialects
Material:	Reading
Modification:	(1) Unmodified, (2) Monotonized (flat F0)
Method:	Identification (marking on a map, choosing from 19 countries), Similiarity to the listener's own dialect (15-pt scale)
Result:	(1) 67% by endogenous listeners, 25% by exogenous listeners, (2) 50% by endogenous listeners, 16% by exogenous listeners [Chance level: 5.3%]

Thomas & Reaser (2004) [36]

Language:	English spoken by African Americans and European Americans
Material:	Spontaneous speech (interview)
Modification:	(1) Unmodified, (2) Monotonized (flat F0), (3) Lowpass-filtered (330Hz)
Method:	Identification
Result:	(1) 71.10%, (2) 72.08%, (3) 52.28% by European American listeners [Chance level: 50%]

Thomas, Lass, & Carpenter (in press) [37]

Language:	English spoken by African Americans and European Americans
Material:	Reading
Modification:	(1) Unmodified, (2) Monotonized (flat F0), (3) Conversion of all vowels to schwa
Method:	Identification
Result:	Vowel quality is important; F0 also plays a role

Language:	English spoken by African Americans and European Americans
Material:	Reading
Modification:	Swapping F0 and segmental durations
Method:	Different listener groups use different cues

Table A5. Examples of research into foreign accent using modified speech

---

Miura, Ohyama, & Suzuki (1989) [40]
Language:	English spoken by Japanese speakers
Material:	Reading
Modification:	Substitution of the features of a native speaker's English with those of Japanese English (PARCOR coefficients, F0, Intensity, Phoneme durations)
Method:	Choosing the more natural sample from a pair of samples
Result:	Durations and F0 contribute to the naturalness

Ohyama & Miura (1990) [41]
Language:	Japanese spoken by English, French, Chinese speakers
Material:	Reading
Modification:	Substitution of the features of foreign-accented Japanese with those of a native speaker's Japanese (PARCOR coefficients, F0, Intensity, Phoneme durations)
Method:	Choosing the more natural sample from a pair of samples
Result:	Durations contribute for the speech by English and French speakers; F0 contribute for the speech by Chinese speakers

Miwa & Nakagawa (2001) [42]
Language:	English spoken by native speakers and Japanese
Material:	Reading
Modification:	Resynthesized preserving (1) F0 and intensity, (2) F0, (3) intensity
Method:	Judgment on naturalness (5-pt scale)
Result:	English spoken by native speakers were more natural. Japanese instructors of English were less sensitive to intensity variation than native instructors

Grover, Jamieson, & Dobrovolsky (1987) [43]
Language:	English, French, German
Material:	Reading
Modification:	Replacement of the continuative pattern of F0 with that of another language
Method:	Choosing the more natural sample from a pair of samples
Result:	Not discriminated

Munro (1995) [44]
Language:	English spoken by Mandarin Chinese and Canadian English speakers
Material:	(1) Elicited sentence, (2) Spontaneous speech
Modification:	Lowpass-filtered (225Hz for male speech, 300Hz for female speech)
Method:	Judgment on accentedness (1=Definitely spoken with a foreign accent, 4=Definitely spoken by a native speaker of English)
Result:	For Mandarine speakers (1) 1.8, (2) 2.1; for Canadian English speakers (1) 3.0, (2) 2.8

Underpinning /nailon/: Automatic Estimation of Pitch Range and Speaker Relative Pitch

Jens Edlund and Mattias Heldner

KTH Speech, Music and Hearing, Stockholm, Sweden
edlund@speech.kth.se, heldner@kth.se

Abstract. In this study, we explore what is needed to get an automatic estimation of speaker relative pitch that is good enough for many practical tasks in speech technology. We present analyses of fundamental frequency (F0) distributions from eight speakers with a view to examine (i) the effect of semitone transform on the shape of these distributions; (ii) the errors resulting from calculation of percentiles from the means and standard deviations of the distributions; and (iii) the amount of voiced speech required to obtain a robust estimation of speaker relative pitch. In addition, we provide a hands-on description of how such an estimation can be obtained under real-time online conditions using /nailon/ – our software for online analysis of prosody.

Keywords: pitch extraction; pitch range; speaker relative pitch; fundamental frequency (F0) distribution; online incremental methods; semitone transform; percentiles; speech technology.

1 Introduction

Prosodic features have been used in speech technology and nearby fields for a great many tasks – some directly related to spoken communication, such as segmentation and disambiguation [see e.g. 1 for an overview], and some less obviously communicative in nature, such as speaker identification and even clinical studies [e.g. 2]. Amongst these prosodic features, the fundamental frequency (F0) is perhaps the most widely used. Pitch and intonation has been associated with a large number of functions in speech [see e.g. 3 for an overview] but, in the words of Honorof & Whalen, "its [the pitch's] linguistic significance is based on its relation to the speaker's range, not its absolute value". Others have made similar observations [cf. 4, 5, 6]. It has even been suggested that listeners *must* estimate a base value of a speaker's pitch range in order to recover the information carried by F0 [7]. In other words, what we should be looking for when applying F0 analysis to practical tasks is commonly not the absolute F0 values, but rather an estimation of *speaker relative pitch*. Within speech technology, pitch and pitch range have been used for nearly as many purposes as has prosody in general. In our own studies, we have shown that online pitch analysis can be used to improve the interaction control of spoken dialogue systems [8] and that similar techniques can be used off-line to achieve more intuitive chunking into utterance-like units [9]. Tasks such as these require a

C. Müller (Ed.): Speaker Classification II, LNAI 4441, pp. 229–242, 2007.
© Springer-Verlag Berlin Heidelberg 2007

classification of individuals with respect to the range they produce when they speak. The classification of intonation patterns into for example HIGH and LOW is a more general example of this. The work presented here provides underpinnings for some of the methods used for pitch extraction in /nailon/ – the computer software we maintain and use for the online pitch analysis [10]. In short, we will explore what is needed to create a model that gives a good enough estimation of speaker relative pitch.

Several studies have attempted to investigate how humans accomplish pitch range estimation [e.g. 4, 5, 6]. Here, we are concerned with how to make such estimations automatically for use in real-time in speech technology applications. In other words, we must remain within the constraints set by current technology, which makes some suggested correlates of (position in) pitch range, for example voice quality [e.g. 5], difficult to use. The current work relies on the F0 extraction provided by the ESPS get_f0 function in the Snack Sound Toolkit by Kåre Sjölander (http://www.speech.kth.se/snack/), with online and real-time abilities added by /nailon/. It is worth noting that although get_f0 is more or less the industry standard and very well proven, output from get_f0 is quite noisy. Any model built on automatic extraction of F0 values must be quite resistant to errors in the training data as well as in the test data.

We are primarily interested in how pitch relates to spoken communication, and for F0 patterns to have an effect on communication, they must be perceivable to the interlocutors. It follows that we need a model that is perceptually sound – a model of pitch rather than F0, if you will. It is clear that we lack the know-how to make a model that mimics human perception to perfection, but to the greatest extent we can, we should avoid methods that lie outside the scope of human perception. We may for example want to use a perceptually relevant scale for frequencies, say semitones or ERBs rather than Hertz [11, 12], and we may want to discard differences in frequency that are not discernable to humans.

Furthermore, we want a model that is as general as possible. We will aim at finding a model capable of sensibly estimating speaker relative pitch by searching for (i) the trimmed pitch range of a speaker and (ii) a reasonable description of the distribution of pitch values within that range. Both (i) and (ii) can be approached by searching the F0 distribution for *quantiles*. Quantiles are values dividing an ordered set of observations. Phrased differently, a quantile is the cut-off point under which a certain proportion of the observations fall. Oft-used divisions have specific names, for example quartiles (which divide the observations into four equal parts), quintiles (five parts), deciles (10 parts), and percentiles (100 parts). For (i), the 5th and 95th percentile will be used as targets to achieve a trimming of the data. The exact numbers are ad hoc, although the techniques should not rely on the trimming being set at exactly five per cent at each end. For (ii), we will aim to further divide the speaker's pitch values into four similarly sized parts – the quartiles. In other words, our aim is to find a method to decide if a pitch value is within a speakers range, given that we trim five per cent at the top and at the bottom of the range, and if so, which quartile of a speakers F0 distribution a given pitch value belongs to, with the first and fourth quartiles being somewhat diminished due to the trimming (see Fig. 1). Generally speaking, the task can be formulated as judging a given F0 value against some kind of history of F0 values providing information about whether it belongs to this quantile or the other.

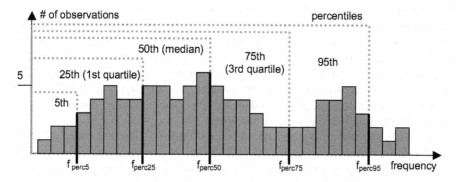

Fig. 1. The aim set forth here is to describe a speaker's pitch range and distribution using percentiles. This histogram over a mock F0 distribution shows the 5th, 25th, 50th, 75th and 95th percentiles.

Given the automatic F0 extraction, this can be done by training a model, or profile, of a speaker's range using pre-recorded speech, or by training a model for a specific group of speakers, say, young male speakers. These methods are burdened with several problems. The first method of pre-compiling user-specific profiles is susceptible to dynamic influences such as emotion and voice fatigue, in that it will fail if a speaker's voice changes from the time of the recording of the training data to the time when the profile is used. It also requires user data to be saved, which has severe implications, both in high demands on storage capacity and on user integrity. Furthermore, both methods require that the speaker be identified, either as a specific speaker or as a member of a class of speakers. In the first case, mistaken identity is a potential problem, and in the second case, there is a corresponding risk that a speaker is associated with the wrong class. Finally, there are naturally a lot of cases where it is simply not practical to pre-compile a model. Notwithstanding these concerns, a model where a large amount of speech from one person has been analysed is, in a sense, the best we can do. We may even say that if we build a model on a set of speech data by ordering all pitch values and counting through them to find our percentiles, this model *is correct*; it constitutes a gold standard as far as the data it is built on is concerned. The descriptive statistics presented in Tables 1 and 2, and in the Appendix, can be viewed as such a gold standard.

If we take the concerns listed above regarding pre-compiled user models seriously – as indeed we should – there is ample reason to look for other solutions. One possibility that eliminates much of the problems with pre-compiled models is to train the models online, when they are needed. In other words, to incrementally train a speaker model as the speaker speaks. As our primary goal is to be able to judge each F0 value with regard to the speakers range online and in real-time, both the time it takes to train the model before it is reasonably reliable and the time it takes to update it with each new data item are major concerns. Unfortunately, there is no efficient way of calculating exact quantiles incrementally – finding range and percentiles in this manner requires us to build a model where the entire set of instances seen is scanned to recalculate the range and percentiles each time a new instance is added. This method places high demands on memory and processor load. The relationship between the methods is the same as the relationship between the mean score and the median (indeed, if we aim to split a user's pitch into two equally sized categories, say HIGH and LOW, then the mean or the median, respectively, would make up the threshold between the categories. All in all, the exact, instance based method of calculation is virtually useless for our purposes.

The stock pile way of getting around process intensive, instance based and exact calculation of distributions is to find some function that describes the data, either more or less exactly, as in the case of the colouring of rabbit youngs, or reasonably well, as when network and database engineers maintain estimated histograms [e.g. 13]. When it comes to the distribution of F0 values in a speaker's speech, we and others have more or less treated them as if they were normally distributed [8, 14]. There are good reasons to suspect this not to be true. Several authors [e.g. 14] have also pointed out that the distribution is indeed not normal.

A quick survey of the F0 distributions of some individuals still leads one to believe that it is somewhere close to normal, and assuming a normal distribution allows us to straightforwardly find the percentiles given the standard deviation and the means, both of which are readily accessible in an incremental manner. The question, then, is: how close to the truth do we get if we assume a normally distributed F0 within a speaker's speech?

Given our intention that the model be perceptually sound it is clear that we can disregard differences that cannot be perceived by humans. It is not entirely easy to say what differences in F0 humans can and cannot perceive, however. Several studies have been made on the subject: 't Hart, Collier, & Cohen [15] gives an overview of studies that place the just noticeable difference (JND) for F0 between 1 Hz and 5 Hz at frequencies around 100 Hz. Unsurprisingly, the higher values come from studies using voice like stimuli, and the lower from studies using sine tones and suchlike. Elsewhere, 't Hart reports the minimum difference between two F0 *movements* for them to be perceived as different. The numbers here are much larger, and range from over one and a half to more than four semitones [16]. For the purposes we have in mind – for example to classify pitch as HIGH, LOW, or MID within a speakers range - we will say that differences of less than one semitone are acceptable.

In the remainder of this chapter, we present the results of analyses of speech from eight speakers with a view to answering the following questions: What does the F0 distribution look like? How does a semitone transformation affect it? What will the error be if we assume normally distributed F0 values to approximate the 5th, 25th, 50th, 75th and 95th percentiles? How much speech is needed to build a reliable model of F0 distribution? Furthermore, we discuss whether there is a need for a decaying model – a model in which the weight of an observation decreases with time – and if so, what the rate of decay should be.

2 Method

2.1 Speech Material

The speech data used for the present study consists of Swedish Map Task dialogues recorded by Pétur Helgason at Stockholm University [17, 18]. The Map Tasks were designed to elicit natural-sounding spontaneous dialogues [19]. There are two participants in a Map Task dialogue: an instruction-giver and an instruction follower. They each have a map, both maps being similar but not identical. For example, certain landmarks on these maps may be shared, whereas others are only present on one map, some landmarks occur on both maps but are in different positions etc. The task is for the instruction-giver to describe a route indicated on his or her map to the follower.

The Swedish Map Task data used in this study consists of recordings of four pairs, or eight speakers including five female (F1-F5) and three male (M1-M3) speakers. Within each pair, each speaker acted as both giver and follower at least once. They were recorded in an anechoic room, using close talking microphones, and facing away from each other. They were recorded on separate channels. The total duration of the complete speech material is about three hours.

2.2 F0 Analysis and Filtering

The Snack sound toolkit (http://www.speech.kth.se/snack/), with a pitch-tracker based on the ESPS tool get_f0 was used to extract F0 and intensity values. The resulting F0 values (in Hz) were transformed onto logarithmic scale (semitones relative to 100 Hz) to enable comparisons between Hertz and semitone data. Although the quality of the recordings was very good, there was some channel leakage. As the pitch-tracker analyzed parts of the leakage as voiced speech, a filter was used to remove the frames in the lower mode of the intensity distribution. Similarly, in case there was a bimodal distribution of F0 values (e.g. due to creaky voice or artefacts introduced by the pitch-tracker), another filter was applied to remove the lower mode of the F0 distribution. No filtering was used for the upper part of the distributions, however.

2.3 Statistical Analysis

SPSS was used to calculate descriptive statistics, percentiles, and Kolmogorov-Smirnov tests of normality for Hertz and semitone data, respectively, in the pre-processed models condition.

For the incremental models condition the percentiles where calculated in two ways. A gold standard was achieved at each calculation interval by sorting and counting every data point, a method that, as already mentioned, is impractical for real, online purposes. Another set of function-based percentiles – which can be achieved realistically under real online circumstances – was then calculated in a two-step process, in which means and standard deviations were first calculated incrementally at regular intervals. The straightforward method of doing this – simply adding up all instances and calculating the means and standard deviation using the sum and the number of instances – is associated with floating point errors which can be quite severe when the number of instances is high. Instead, recursion functions where used, as described by [20]. The percentiles where then looked up in a table of the area under the standard normal distribution [e.g. 21] using the incrementally calculated means and standard deviation. This technique is the same that we use in the /nailon/ software, although /nailon/ uses a very small moving window for get_f0 for performance reasons [10]. Here, in order to make the results of the incremental processing directly comparable to the pre-processed models created in SPSS, we used the same pre-extracted F0 values as input for all methods. An informal test revealed that /nailon/ data would yield very similar results, however.

3   Results: Pre-processed Models

This section presents estimations of pitch range and subdivisions of the distributions using all available F0 data for the speakers. Note that, these estimations in a sense represent a speaker's total pitch range (although the outliers are trimmed) rather than the range to be expected within an individual utterance. Table 1 shows descriptive statistics for F0 distributions calculated from Hertz data for each speaker, and Table 2 the corresponding statistics calculated from semitone data. Unsurprisingly, the male and female speakers differed substantially with respect to means (or medians), and there was considerable individual variation within the male and female groups. The differences in standard deviation (or inter-quartile range IQR), however, and especially those calculated from semitone data, were more modest, and could not be attributed to speaker gender [cf. 14].

Table 1. Descriptive statistics (means, medians, standard deviations SD, interquartile range IQR, skewness, kurtosis, and number of data points N) for F0 distributions based on Hertz data per speaker SP

SP	MEAN	MEDIAN	SD	IQR	SKEW.	KURT.	N
F1	217	213	38	51	0.56	0.27	49070
M1	124	120	30	27	4.72	47.04	79703
F2	194	194	49	67	0.09	-0.16	72554
F3	201	194	45	51	1.47	5.29	77086
F4	238	234	30	31	2.13	11.30	74790
M2	122	119	26	30	1.02	4.03	39064
M3	117	112	33	34	2.85	21.19	63454
F5	200	193	42	59	0.68	0.18	36415

Table 2. Descriptive statistics (means, medians, standard deviations SD, interquartile range IQR, skewness, kurtosis, and number of data points N) for F0 distributions based on semitone data per speaker SD

SP	MEAN	MEDIAN	SD	IQR	SKEW.	KURT.	N
F1	13.1	13.0	3.0	4.1	0.06	-0.10	49070
M1	3.3	3.2	3.5	3.9	0.92	6.53	79703
F2	10.9	11.4	4.7	6.0	-0.72	0.80	72554
F3	11.7	11.5	3.6	4.5	0.32	1.32	77086
F4	14.9	14.7	2.0	2.3	1.12	3.83	74790
M2	3.1	3.0	3.6	4.4	-0.04	0.88	39064
M3	2.1	2.0	4.3	5.2	0.53	2.12	63454
F5	11.6	11.4	3.5	5.2	0.19	-0.37	36415

Although there were considerable individual differences, the F0 distributions were generally asymmetric. For the distributions based on Hertz data, the medians were smaller than the means and the distributions were positively skewed for all speakers. Similarly, the medians were smaller than the means for all but one speaker, and the distributions were positively skewed for six of the speakers in the semitone distributions. Positive skewness values indicate that the tails of the distributions tend to stretch out more on the right (higher) side than in the normal distribution. Furthermore, seven of the distributions based on Hertz data, and six of those based on semitones had positive kurtosis values, indicating that the data were squeezed into the middle of the distributions compared to the normal distribution.

Thus, the F0 distributions in our data generally deviate from the normal distribution, both in terms of skewness and kurtosis. The distributions tend to be clustered more and to have longer tails than in the normal distribution, and those tails tend to be at the right hand side of the distributions [cf. 14]. Kolmogorov-Smirnov tests of normality showed that all distributions, based on Hertz as well as on semitones, differed significantly ($p<0.01$) from a normal distribution.

As it is difficult to assess the effect of these deviations from a normal distribution directly from the skewness and kurtosis values, we calculated the differences (in semitones) between distribution subdivisions based on distance from the mean expressed in standard deviations and the percentiles corresponding to these distances given normally distributed data, as given by a table of the area under the standard normal distribution. That is, the distance between the 5th percentile and the mean minus 1.65 standard deviations; the 25th percentile and the mean minus 0.65 standard deviations; the 50th percentile (the median) and the mean; the 75th percentile and the mean plus 0.65 standard deviations; and the 95th percentile and the mean plus 1.65 standard deviations. The results for Hertz and semitone data are shown in Table 3 and Table 4, respectively. See the Appendix for the percentile values used for these calculations.

Table 3. Distance (in semitones) between distribution subdivisions based on distance from the mean expressed in standard deviations and the percentiles corresponding to these distances in normally distributed data. Distributions based on Hertz data.

SP	5TH PERC	25TH PERC	50TH PERC	75TH PERC	95TH PERC
F1	-0.9	0.3	0.3	0.1	-0.4
M1	-3.6	-0.6	0.5	1.1	1.2
F2	-0.4	0.2	0.1	-0.1	-0.1
F3	-2.3	0.0	0.6	0.6	-0.3
F4	-1.1	-0.1	0.3	0.5	-0.1
M2	-0.8	-0.1	0.4	0.4	-0.2
M3	-3.7	-0.2	0.7	1.0	0.1
F5	-1.7	0.5	0.6	0.0	-0.5

Table 4. Distance (in semitones) between distribution subdivisions based on distance from the mean expressed in standard deviations and the percentiles corresponding to these distances in normally distributed data. Distributions based on semitone data.

SP	5TH PERC	25TH PERC	50TH PERC	75TH PERC	95TH PERC
F1	-0.2	0.1	0.1	0.0	-0.3
M1	-0.9	-0.3	0.1	0.4	0.7
F2	0.5	-0.4	-0.5	-0.2	1.1
F3	-0.7	0.0	0.2	0.1	-0.2
F4	-0.5	0.0	0.2	0.3	-0.1
M2	0.4	-0.2	0.1	0.1	0.1
M3	-0.6	-0.1	0.1	0.3	0.0
F5	-0.6	0.3	0.2	-0.3	-0.2

These calculations showed that a base value for a speaker's pitch calculated as the mean minus 1.65 standard deviations resulted in slightly lower values than the 5th percentile for all speakers in the Hertz distributions, and for 6 of the speakers in the semitone distributions. The differences ranged from 0.4 to 3.7 semitones (average 1.8 ST) in the Hertz distributions, and from 0.2 to 0.9 semitones (average 0.5 ST) in the semitone distributions. Similarly, the mean minus 0.65 standard deviations resulted in values within 0.6 semitones (average 0.3 ST) from the 25th percentile in the Hertz distributions, and within 0.4 semitones (average 0.2 ST) in the semitone distributions. The mean was within 0.7 semitones from the 50th percentile (i.e. the median) in the Hertz distributions, and within 0.5 semitones in the semitone distributions. The mean plus 0.65 standard deviations resulted in values within 1.0 semitone (average 0.5 ST) from the 75th percentile in the Hertz distributions, and within 0.4 semitones (average 0.2 ST) in the semitone distributions. Finally, a top value for a speaker's pitch calculated as the mean plus 1.65 standard deviations resulted in values within 1.2 semitones from the 95th percentile in the Hertz distributions (average 0.4 ST), and within 1.1 semitones (average 0.3 ST) in the semitone distributions.

4 Discussion: Pre-processed Models

The analyses of all F0 data for each speaker support the previous findings that F0 distributions are typically not normally distributed [cf. 12]. There is usually some positive skewness and some positive kurtosis indicating that the distributions lean to the right and are clustered more than the normal distribution. Thus, F0 data typically violate the assumption of normality underlying many statistical procedures, including estimations of pitch range based on means and standard deviations.

Various transformations (including square roots, logarithmic, and inverse transforms) may be used to correct non-normally distributed data [e.g. 21]. Among these, logarithmic transformation (i.e. from Hertz to semitones) makes the most sense here, in that the data are more readily interpreted and perceptually relevant after transformation. For example, semitone transformation makes the pitch ranges of males and females comparable [e.g. 12]. Indeed, our analyses show that such a transformation (N.B. before the calculations of distribution statistics) resulted in lower skewness values for all but one speaker, and in lower kurtosis values for all but two speakers, and hence decreased the deviations from normally distributed data (which has skewness and kurtosis values of zero). Figure 2 shows histograms with superimposed normal distribution curves for one of the speakers (M1) to exemplify this.

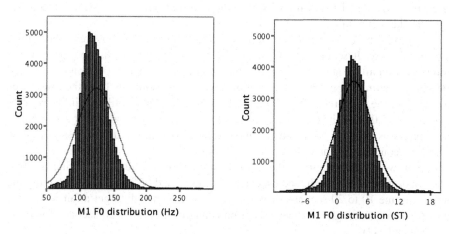

Fig. 2. Histograms showing F0 distributions with superimposed normal distribution curves based on Hertz data (left panel) and on semitone data (right panel) for one of the speakers (M1)

Although our F0 data generally deviated from a normal distribution also after logarithmic transformation, subdivisions based on distance from the mean expressed in standard deviations yielded fairly good estimations of the percentiles. The differences between exact subdivisions and estimations based on the assumption of normality were reduced as a result of the transformation. In the distributions based on semitone data, the differences exceeded one semitone for one subdivision and one speaker only (95th percentile, speaker F2). Given the current knowledge about pitch perception in speech [see e.g. 15 and references mentioned therein] we find it most unlikely that these differences should be perceptually relevant – we consider estimations within one semitone good enough for the kind of classification we are aiming at.

Based on these observations, we argue that use of semitone transformation is advantageous for theoretical as well as for statistical reasons, and hence a sound practice for automatic estimation of pitch range, and furthermore that distribution subdivisions based on means and standard deviations which in turn are based on a fair amount of data, yields a description of the F0 distribution that is good enough to estimate speaker relative pitch in an offline situation. It remains to be shown, however, how much speech is needed to build a reliable model of the F0 distribution in an online situation. This issue will be addressed in the following sections.

5   Results: Incremental Models

As mentioned above, percentiles can either be calculated using an exact method, or, given that a normal distribution can be assumed, from means, standard deviations and a table of the area under the standard normal distribution. Figure 3 shows how percentiles calculated using these two methods differ and evolve over time for the eight speakers in our data.

A comparison of the two methods of calculating percentiles revealed minor differences only. It seems that the method assuming a normal distribution (i.e. the estimation) stabilised at the same rate as the exact method. After 10 seconds of voiced speech, the estimation resulted in percentiles within one semitone from the exact ones in 92.5 % of the cases (counting eight speakers times five percentiles). Similarly, after 20 seconds 95.0 % of the cases differed less than one semitone, and after six minutes, the figure was 97.5 %. There are no large fluctuations in the differences anywhere from 20 seconds to 6 minutes.

As you would have thought, the most drastic changes were found during the first 10 seconds of voiced speech, and especially in the 5th and 95th percentile trimming the outliers. Some 10 to 20 seconds of voiced speech was enough to get a fair model for most speakers, although the models kept on changing well after 20 seconds for some (e.g. M2 and M3).

A rough estimate of the stability of the models can be obtained by comparing the estimated percentiles at 10 and 20 seconds with the percentiles at six minutes (the shortest amount of voiced speech we have available for an individual speaker is six minutes, and we want comparable data sets). After 10 seconds, 80.0 % of the thresholds (again for eight speakers times five percentiles) differed less than one semitone from the thresholds at six minutes, 97.5 % were less than two semitones, and one case exceeded two semitones. Similarly, at 20 seconds, 82.5 % of the thresholds were less than one semitone from the result at six minutes, 97.5 % less than two semitones away, and again; only one case exceeded two semitones.

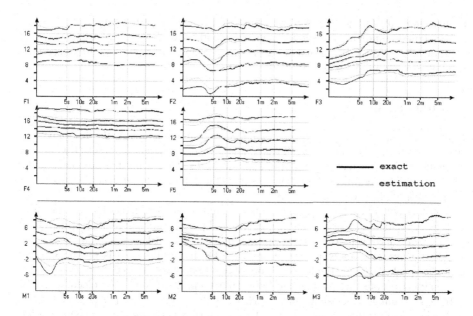

Fig. 3. Cumulative percentiles (in semitones) calculated using an exact method (black lines), and an estimation calculated from incremental means and standard deviations (grey lines) for each speaker. The 5th, 25th, 50th, 75th, and 95th percentiles are shown from the bottom and up. The scale on the time axis is logarithmic.

6 Discussion: Incremental Models

The analysis of the incremental models has shown that a good estimate of the shape of the F0 distribution can be obtained after a short period of time using incrementally calculated means and standard deviations, although it is clear that 10 to 20 seconds is not sufficient to create models of a speaker's *total* pitch range. For most practical purposes, we would not be interested in a speaker's total pitch range, however, but rather of a speaker's *current* pitch range. We need to ask ourselves how stable F0-based models are, how much data needs to go into them, and when they become obsolete.

Fig. 3 may lead us on our way towards the answers. It is important to note that if we use very large quantities of data in a model without putting less weight in old data than in new, we are quickly going to end up with a model that is in effect static, since each new data point has less and less influence on the model as a whole. Such a model will become increasingly burdened with problems typical for static models, most notably susceptibility to dynamic influences. If we on the other hand pay very little attention to older data, the model is going to flutter unpredictably. There are a number of ways to achieve a model that places more weight in new data than in old, for example by letting the weight of data points decrease as they grow older – a decaying model. The question is when the decay should start and at what speed it should proceed. The graphs in Fig. 3 indicate that in several cases, the models are quite fickle before the 10-second line. These fluctuations are not likely to be something we want to capture. At 20 seconds to a minute, the changes are much smaller and slower, and may well be worth modelling. A rough estimate, then, is that decay should commence no sooner than 10 seconds after the data is seen, and should continue slowly over 10 seconds to a minute, perhaps. Further research is needed to test these observations, however.

7 Conclusions

In this contribution, we have examined measurable manifestations of pitch range and speaker relative pitch in the speech signal and provided a hands-on description of how to capture this. The technique works and can be used on a normal computer under real-time online conditions. For concepts like "the centre of the user's F0 range", it is comparable to pre-processed models.

We may conclude that semitone transformation is advantageous for theoretical as well as for statistical reasons, and hence a sound practice for automatic estimation of pitch range, and furthermore that estimations of percentiles based on means and standard deviations yields a description of the F0 distribution that is good enough to estimate speaker relative pitch with errors of less than one semitone.

Finally, we note that somewhere between 10 and 20 seconds of voiced speech is sufficient training material to make such estimations for most speakers, at least in dialogue situations that change at a similar rate to Map Task, and that decaying models are likely to outperform models that grow rigid over large quantities of training data.

Acknowledgements. We thank Anders Eriksson and Rolf Carlson for their comments during this work. Any shortcomings in the present research remain our own. This work was carried out within the CHIL project. CHIL (Computers in the Human Interaction Loop) is an Integrated Project under the European Commission's Sixth Framework Program (IP-506909).

References

1. Shriberg, E., Stolcke, A.: Direct Modeling of Prosody: An Overview of Applications in Automatic Speech Processing. In: Proceedings of Speech Prosody 2004 Nara, Japan, pp. 575-582 (2004)
2. Nilsonne, Å., Sundberg, J., Ternström, S., Askenfelt, A.: Measuring the Rate of Change of Voice Fundamental Frequency in Fluent Speech During Mental Depression. J. Acoust. Soc. Am. 83, 716–728 (1987)
3. Bolinger, D.: Intonation and Its Uses: Melody in Grammar and Discourse. Edward Arnold, London (1989)
4. Carlson, R., Elenius, K., Swerts, M.: Perceptual Judgments of Pitch Range. In: Bel, B., Marlin, I. (eds.) Proceedings of the International Conference on Speech Prosody 2004, Nara, Japan, pp. 689–692 (2004)
5. Honorof, D.N., Whalen, D.H.: Perception of Pitch Location within a Speaker's F0 Range. J. Acoust. Soc. Am. 117, 2193–2200 (2005)
6. Swerts, M., Veldhuis, R.: The Effect of Speech Melody on Voice Quality. Speech Com. 33, 297–303 (2001)
7. Traunmüller, H.: Conventional, Biological and Environmental Factors in Speech Communication: A Modulation Theory. Phonetica. 51, 170–183 (1994)
8. Edlund, J., Heldner, M.: Exploring Prosody in Interaction Control. Phonetica. 62, 215–226 (2005)

9. Edlund, J., Heldner, M., Gustafson, J.: Utterance Segmentation and Turn-Taking in Spoken Dialogue Systems. In: Fisseni, B., Schmitz, H.-C., Schröder, B., Wagner, P. (eds.) Sprachtechnologie, mobile Kommunikation und linguistische Ressourcen. Peter Lang, Frankfurt am Main, Germany, pp. 576–587 (2005)
10. Edlund, J., Heldner, M.: /nailon/ – Software for Online Analysis of Prosody. In: Proceedings of the 9th International Conference on Spoken Language Processing (Interspeech 2006), Pittsburgh, PA, USA, pp. 2022–2025 (2006)
11. Hermes, D.J., van Gestel, J.C.: The Frequency Scale of Speech Intonation. J. Acoust. Soc. Am. 90, 97–102 (1991)
12. Traunmüller, H., Eriksson, A.: The Perceptual Evaluation of F0 Excursions in Speech as Evidenced in Liveliness Estimations. J. Acoust. Soc. Am. 97, 1905–1915 (1995)
13. Lam, E., Salem, K.: Dynamic Histograms for Non-Stationary Updates. In: Proceedings of the 9th International Database Engineering & Application Symposium (IDEAS'05), pp. 235–243. IEEE Computer Society, Los Alamitos (2005)
14. Traunmüller, H., Eriksson, A.: The Frequency Range of the Voice Fundamental in the Speech of Male and Female Adults. [cited 2006-08-01] (1995), Available from http://www.ling.su.se/staff/hartmut/f0_m&f.pdf
15. 't Hart, J., Collier, R., Cohen, A.: A Perceptual Study of Intonation: An Experimental-Phonetic Approach to Speech Melody. In: Cambridge Studies in Speech Science and Communication, Cambridge University Press, Cambridge (1990)
16. 't Hart, J.: Differential Sensitivity to Pitch Distance, Particularly in Speech. J. Acoust. Soc. Am. 69, 811–821 (1981)
17. Helgason, P.: Preaspiration in the Nordic Languages: Synchronic and Diachronic Aspects. PhD dissertation. Department of Linguistics, Stockholm University, Stockholm (2002)
18. Helgason, P.: SMTC - A Swedish Map Task Corpus. In: Working Papers 52: Proceedings from Fonetik 2006. Lund University, pp. 57–60. Centre for Languages and Literature, Lund (2006)
19. Anderson, A.H., Bader, M., Bard, E.G., Boyle, E., Doherty, G., Garrod, S., Isard, S., Kowtko, J., McAllister, J., Miller, J., Sotillo, C., Thompson, H., Weinert, R.: The HCRH Map Task Corpus. Lang. Speech 34, 83–97 (1991)
20. Welford, B.P.: Note on a Method for Calculating Corrected Sums of Squares and Products. Technometrics. 4, 419–420 (1962)
21. Kirk, R.E.: Experimental Design: Procedures for the Behavioral Sciences, 3rd edn. Brooks/Cole Publishing Company, Pacific Grove, CA (1995)

Appendix

Table 5. Percentiles for F0 distributions based on all Hertz data for each speaker

SP	5TH PERC	25TH PERC	50TH PERC	75TH PERC	95TH PERC
F1	162	189	213	239	287
M1	91	108	120	135	162
F2	116	161	194	228	277
F3	146	172	194	223	280
F4	202	220	234	251	288
M2	83	106	119	136	167
M3	78	97	112	131	170
F5	145	168	193	227	276

Table 6. Percentiles for F0 distributions based on all semitone data for each speaker

SP	5TH PERC	25TH PERC	50TH PERC	75TH PERC	95TH PERC
F1	8.3	11	13	15.1	18.3
M1	-1.6	1.3	3.2	5.2	8.4
F2	2.6	8.2	11.4	14.2	17.6
F3	6.5	9.4	11.5	13.9	17.8
F4	12.1	13.6	14.7	15.9	18.3
M2	-3.2	1	3	5.3	8.9
M3	-4.4	-0.6	2	4.6	9.2
F5	6.4	9	11.4	14.2	17.6

Automatic Dialect Identification:
A Study of British English

Emmanuel Ferragne and François Pellegrino

Laboratoire Dynamique Du Langage - UMR 5596 CNRS / Université de Lyon,
France
{Emmanuel.Ferragne,Francois.Pellegrino}@univ-lyon2.fr

Abstract. This contribution deals with the automatic identification of
the dialects of the British Isles. Several methods based on the linguistic
study of dialect-specific vowel systems are proposed and compared using
the Accents of the British Isles (ABI) corpus. The first method examines
differences in diphthongization for the FACE lexical set. Discrimination
scores in a two-dialect discrimination task range from chance to ca. 98 %
of correct decision depending on the pair of dialects under test. Thanks
to the ACCDIST method (developed in [1,2]), the second and third experiments take dialectal differences in the structure of vowel systems into
consideration; evaluation is performed on a 13-dialect closed set identification task. Correct identification reaches up to 90 % with two subsets
of the ABI corpus (/hVd/ set and read passages). All these experiments
rely on a front-end automatic phonetic alignment and are therefore text-
dependent. Results and possible improvements are discussed in the light
of British dialectology.

Keywords: Automatic dialect identification, accents of English, British
Isles, phonetics of vowel systems.

1 Introduction

The specific patterns of pronunciation that are related to speakers' regional
origin or social background greatly contribute to the distinctiveness of their
voices, and therefore to the variability of speech. Dialect – or rather accent[1] –
identification has therefore become an important concern in speech technology.
For instance, it has been shown that automatic speech recognition systems can
perform tremendously better when the training and the test sets are matched
for dialect ([3]). Dialect identification – whether the task be carried out by a
computer or a human expert – also has forensic applications ([4,5]) although, as
is the case with any other component of somebody's voice, the plasticity issue

---

[1] The word *accent* quite often refers to foreign-accented speech, and although it is
appropriate to designate the pronunciation of dialects, the term *dialect* will be used
instead since the present contribution deals exclusively and unambiguously with
pronunciation features.

C. Müller (Ed.): Speaker Classification II, LNAI 4441, pp. 243–257, 2007.
© Springer-Verlag Berlin Heidelberg 2007

(e.g. somebody may alter their accent for sociolinguistic reasons, or in order to deceive) raises daunting challenges for the speech community. Our aim here is to assess to what extent knowledge of the phonetics of dialects can provide an alternative to crude acoustic modelling. A substantial part of this contribution is therefore devoted to some aspects of phonetic vowel variation across dialects. The remainder covers experiments in the automatic classification of the dialects of the British Isles ([6,1,2,7]) with a twofold objective: evaluating classification scores *per se*, and demonstrating how automatic methods can assist researchers in phonetics and dialectology.

2   An Overview of the Dialects of the British Isles

Most of the dialects of the British Isles have been extensively described in the literature; therefore an exhaustive account falls well beyond the scope of this contribution. The reader is advised to consult the following references for thorough information on the phonetic aspect: [8,9,10,11]. However, some features are highlighted in this section because they constitute the necessary background basis for the rest of the discussion. In traditional (areal) dialectology, pronunciation isoglosses, i.e. boundaries demarcating dialects, have commonly been used. The boundaries that delimit differences in vowel systems are of particular interest to us since they are at the heart of the method developed in Experiment 2. By way of example, *gas* does not rhyme with *grass* in the south of England, but it does in the (linguistic) north. Similarly, the vowels of *nut* and *put* are phonologically identical in the north, but a phonemic split caused them to be differentiated in the south. *Good* and *mood* rhyme in Scotland, but not in the rest of the British Isles, while *nurse* and *square* have been reported to have the same vowel in certain speakers from Liverpool and Hull, for example. However, just as surface realization can be affected by sociological factors, vowel systems too may vary within a given location, and speakers sometimes try to "posh up" their accent by adopting the vowel system of a more prestigious variety than their own. This can lead to a phenomenon known as hypercorrection whereby, for instance, a speaker from the north of England (having no distinction between the vowels in *nut* and *put*) tries to imitate a southerner, failing to identify which words should pattern with the southern phonemes of *nut* or *put*, and ends up pronouncing *sugar* with the vowel of *nut* (example taken from [9, page 353]). This may sound trivial, but it has serious consequences on the method we describe in Experiment 2. The question of lexical incidence (roughly speaking: deciding to which phonemic category a vowel token belongs) is indeed crucial here because it suggests extreme caution – and, clearly, expert knowledge – when choosing the key words for creating shibboleth sentences. Suppose a phonetician designs test sentences to elicit the – or the absence of – contrast between *gas* and *grass* or *father* in order to determine whether a speaker is from the north or the south of England. Without prior knowledge of dialectology, he or she may well wrongly infer from the spelling that *mass* patterns with *grass*, or that *gather* rhymes with *father* in southern dialects. Opposing *gather* or *mass* with *gas*, and therefore failing to identify the correct underlying phonological representation of these words, would lead the phonetician to miss the potential contrast under study. We will return to this question further below. Beside systemic differences, dialect variation is also manifested by different phonetic realizations of the same phoneme; this characteristic also plays an important role in Experiment 2, and it is clearly illustrated in Experiment 1, which focuses on diphthongization.

3   Corpus Description

The material comes from the Accents of the British Isles (ABI) corpus ([12]). The database consists of recordings from 14 geographical areas throughout the British Isles. For each variety of English, 20 speakers on average (equally divided into men and women) participated. In the following experiments, two types of data were used: a list of 19 /hVd/ words spoken 5 times by each speaker, and a read passage, containing approximately 290 word tokens, specifically designed to elicit dialect variation. The recordings took place in quiet rooms (e.g. in public libraries) at the beginning of 2003; the participants spoke through a head-mounted microphone that was connected to a PC via an external sound card. The sound files are mono 16 bit 22050 Hz PCM Windows files. Worthy of mention is the total lack of individual information on the participants (age, occupation, etc.), which precludes the inclusion of highly relevant sociolinguistic factors in the study ([5,13,14]). The dialects and the towns where the corresponding recordings took place are listed in Table 1.

Table 1. Dialects of the ABI Database

LABEL	DIALECT	PLACE
brm	Birmingham	Birmingham
crn	Cornwall	Truro
ean	East Anglia	Lowestoft
eyk	East Yorkshire	Hull
gla	Glasgow	Glasgow
ilo	Inner London	London (Tower Hamlet)
lan	Lancashire	Burnley
lvp	Liverpool	Liverpool
ncl	Newcastle	Newcastle
nwa	North Wales	Denbigh
roi	Republic of Ireland	Dublin
shl	Scottish Highlands	Elgin
sse	Standard Southern English	London
uls	Ulster	Belfast

4   Experiment 1: Diphthongization

4.1   Goal

Diphthongization refers to the stability over time of the formant pattern in a vowel. The concept lies at the phonetic level in that it disregards whether a vowel be phonologically termed a diphthong or not. For example the vowels of FLEECE and GOOSE [2] in Standard British English are often described as monophthongs in manuals for foreign learners, but they are clearly diphthongized. Our aim is to come up with an economical and sufficient set of parameters to describe formant stability and then validate the model with a classifier. For the sake of parsimony, and in order to get rid of part of the individual variation, absolute vowel initial and final formant values are discarded (although they are known to be dialect specific) and only dynamic features are considered. In the first experiment we concentrate on the so-called FACE vowel, which occurs in the corpus in the words *sailor, faces, today, takes, same, generations, way, stable, unshakable, faith, later, favour, great, fame, Drake, sail,* and *make*. We posit for practical reasons that all these words belong to the FACE set. Note however that this may be too much of an assumption, and a more cautious approach is taken in Experiment 3 where we no longer consider lexical sets, but individual words instead. Using formant trajectories (i.e. the formant slopes) as a criterion, the FACE vowel has, roughly speaking, three main realizations in the dialects of the British Isles:

1. a long closing diphthong beginning with an open-mid vowel and gliding towards a close front position, e.g. in the south of England (e.g. Figure 1a);
2. a centring diphthong starting from a mid-close (or even closer) quality and gliding towards schwa in Newcastle (e.g. Figure 1b);
3. a rather short front close-mid monophthong, e.g. in Scotland and some dialects of the north of England (e.g. Figure 1c).

It is hypothesized that the slopes of F1 and F2 will adequately model these three types of vowels.

4.2   Method and Results

A transcription at the phonetic level was generated with forced alignment using the Hidden Markov Model Toolkit (HTK) ([15]). The models had been trained on the WSJCAM corpus[3]. Formant values were estimated with the Praat program ([16]) using the Burg algorithm set with default values. Some formant extraction errors occurred (as confirmed by visual inspection of formant tracks); however, in order to keep the procedure as automatic as possible, no attempt was made to manually get rid of outliers. Then the slopes of F1 and F2 were computed with robust linear regression in Matlab. Knowledge of phonetic variation was taken

---

[2] These small capitalized key words stand for lexical sets: [9] popularized this practice in the early 80s, and it is still widely used in British English dialectology nowadays.

[3] We are grateful to Mark Huckvale for kindly providing the HMM models.

into account in order to conceptualize the classification problem. Given that one single linguistic variable (i.e. the FACE vowel) does not allow separability between all dialects, the original task with $C = 14$ classes was approached as $C(C-1)/2 = 91$ separate two-class problems. Another reason for building several two-class models, which would be worth exploring, is to gather an optimal - and therefore presumably different - set of parameters for each pair of dialects. In the absence of any *a priori* reason to the contrary, linear separability was assumed and the classification was performed with a single layer neural net implemented with the Netlab toolbox ([17]). The network has two inputs: the slopes of F1 and F2. For each pair of dialects, all the tokens of all speakers except the speaker under test are passed through the network. This cross-validation procedure is adopted because of the very small size of the dataset. The network is trained with 10 iterations of the iterated re-weighted least squares algorithm. Finally the ouput neuron with a logistic activation function makes a binary decision: the test speaker's tokens either belong to the first or the second dialect of the current pair. A correct classification score is therefore computed for each pair of dialects. In order to save space the 91 scores are not reproduced here; instead, the top and bottom ten pairs are shown in Table 2.

The fourth column shows the geographical distance (in km) between towns. Note how, on average, pairs with high classification scores are farther apart than those with low scores. Actually, a rather low but significant correlation exists between discrimination scores and geographical distances for the 91 pairs ($r = .53$, Spearman rank correlation).

4.3   Discussion

This experiment is the most linguistic-oriented one since the correspondence between formant slope values (the input to the model) and the traditional phonetic vowel quadrilateral facilitates phonetic interpretation. In other words, Experiment 1 not only shows that the method works, but also that the results are directly interpretable in phonetic terms. However, one of the flaws lies in that automatic formant estimation is only partially reliable. Besides, the automatic aspect is quite restricted, and the method described here is therefore very unlikely ever to be implemented in real-life applications. It may however prove a useful tool for testing dialectological hypotheses such as the discriminatory power of a given pronunciation trait.

5   Experiment 2: Vowels in hVd Context

5.1   The ACCDIST Method

In Experiment 2, 19 vowels embedded in /h_d/ consonantal contexts were examined. /hVd/ words have often been used in phonetic studies because the acoustic characteristics of vowels are only slightly affected by these consonants, and keeping the same consonantal context rules out coarticulatory differences.

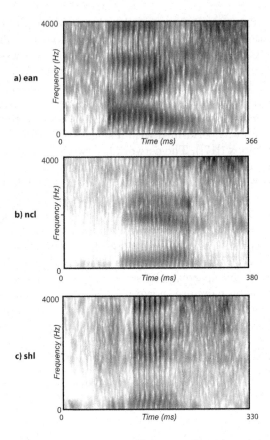

Fig. 1. Spectrograms exemplifying the three realizations of the FACE vowel

In a multi-dialect perspective, the 19 /hVd/ words presumably instantiate all possible phonological contrasts in the dialect that has the biggest inventory[4]. Artificial though the stimuli may seem, they nevertheless give the opportunity to calibrate the system under ideal conditions for subsequent use on data closer to real-life speech (see Experiment 3), and provide a convenient way of studying variation in phonological systems. Prior to the analysis proper, a native English expert phonetician examined the corpus and advised us against including the *ilo* subset on the grounds that the extreme heterogeneity of the speakers could in no way form a single entity (further details are given in section 7.1). More than for any other dialect in the corpus, individual information on speakers would have

---

[4] This again is an oversimplification: to be more accurate, the 19 stimuli exemplify the phonological vowel contrasts of Standard British English, which implies that the other vowel inventories are assessed with reference to that of Standard English, and not to an ideal panlectal representation. Thus, we have no means of knowing whether increasing the number of /hVd/ words would elicit other contrasts in the remaining dialects.

Table 2. Paired-dialect discrimination based on diphthongization. The ten highest and lowest scores are displayed. All scores, unless specified (ns), are significant at the $p = .05$ level (binomial tests).

DIALECT1	DIALECT2	CORRECT DISCRIMINATION (%)		GEOGRAPHICAL DISTANCE (km)
brm	shl	97.8		581
ean	shl	96.8		658
ean	gla	96.3		541
shl	sse	96.1		712
brm	gla	95.9		406
crn	shl	95.5		829
ilo	shl	95.4		712
brm	ncl	95.2		277
lvp	shl	95.0		471
ean	ncl	94.7		354
		...		
lvp	roi	57.3		219
lvp	sse	56.3	ns	285
nwa	roi	52.6	ns	189
crn	lvp	51.4	ns	380
crn	sse	51.3	ns	374
ilo	sse	51.3	ns	0
gla	ncl	51.0	ns	193
lvp	nwa	49.7	ns	38
eyk	lan	49.3	ns	125
crn	ilo	42.5		374

been essential. ABI comes complete with a word-level segmentation; assuming – although this is not totally accurate – that voiced frames corresponded to vowels, automatic pitch detection with the Snack Sound Toolkit ([18]) was employed to estimate vowel boundaries. 12 MFCC and one energy feature were computed at 25 %, 50 %, and 75 % of the duration of the vowel, and the duration itself was included to form a vector of 40 features. The computation was done with the melfcc routine from the rastamat toolbox ([19]); the options were those that the author recommends to duplicate HTK's MFCC, except that the window length and the analysis step were set to 20 ms and 10 ms, respectively. After removing the speakers from *ilo* and two participants who did not complete the whole set of test words, we were left with 261 speakers. The rationale for the classification method was first introduced, as far as we know, by [6], and it was later adopted by [1,2], who devised the ACCDIST method (Accent Characterisation by Comparison of Distances in the Inter-segment Similarity Table), which is central to this section[5]. Speaker normalization is a critical issue in phonetics: differences in

---

[5] [1,2] also used the ABI corpus; he however worked on a different part of the database, namely, a set of shibboleth sentences.

individual acoustic spaces, either due to physiological constraints or habit, have
to be factored out. [6] and later [1,2] got round the problem by representing
vowels with reference to a speaker's vowel space structure, and not to average
stored values. One way to do this is to compute distances between each pair of
vowels. As mentioned above, a vector of size 40 was computed for each vowel.
For a given speaker, the values for the five repetitions of each /hVd/ type were
averaged. Then, distances were calculated between the 19 vowel types, yielding,
for each speaker, a 19×19 symmetric distance matrix. Quite a few distance
measures for continuous variables are available in the literature (see for example
[20], for a discussion of the properties of some of them), and the choice of the
appropriate one depends on the particular kind of data. Central to the problem
is the issue of variable weighting: in our $n \times p$ matrices (where n are the 19 vowels
of a speaker and p the 40 spectral and duration features), the ranges and scales
of the p variables differ substantially. It is common practice to standardize each
variable to zero mean and unit variance (i.e. computing a so-called z-score); yet,
we assumed that, given that the computation of MFCC is based on an auditory
filter bank, the differential weightings induced by differences in scales and ranges
reflected perceptually relevant attributes of the spectrum, and should therefore
be preserved. A good choice in such cases is to use a family of distance metrics
whose general form is the Minkowski distance:

$$d_{ij} = \left(\sum_{k=1}^{p} |x_{ik} - x_{jk}|^r \right)^{\frac{1}{r}} \tag{1}$$

where r must be superior or equal to 1. As the chosen r value increases, the
differential weighting of the p variables also increases: large differences are given
relatively more weight than small ones. Bearing in mind what has just been
said about the perceptual relevance of our feature space, we want to avoid dis-
torting it by using high exponents and will therefore stick to low values such
as $r = 1$ and $r = 2$, which correspond to the Manhattan and Euclidean distances, respectively. So, once a 19×19 distance matrix has been computed for
each speaker, the classification method described in [1,2] is carried out: 13 dialect matrices are obtained by getting the mean of the individual matrices for
each dialect. The validation procedure goes as follows: the dialect matrix of the
speaker under test is re-computed without her/his individual matrix and then
each individual matrix is compared to the 13 dialect matrices. Matrix similarity
is estimated with a matrix correlation coefficient: the two matrices, i.e. the test
speaker and the dialect matrix (or rather: either the upper or lower triangular
part, since they are symmetric) are unfolded onto a row vector, then the Pearson
product-moment correlation is computed. The speaker under test is classified as
belonging to the dialect whose correlation with her/his matrix is highest. Percent
correct identification scores are 86.6 %, 89.0 %, and 89.7 %, for men, women, and
both sexes respectively, using the Euclidean distance. Slight improvements are
obtained in both sexes condition with the Manhattan distance: 90.0 %. Correlation measures are insensible to scale magnitude, which solves the question of

Table 3. Confusion Matrix: /hVd/ words; all subjects; Manhattan distance. Overall correct score: 90.0 %.

TEST DIALECT	BRM	CRN	EAN	EYK	GLA	LAN	LVP	NCL	NWA	ROI	SHL	SSE	ULS
brm	18	-	1	-	-	1	-	-	-	-	-	-	-
crn	-	16	-	-	-	-	-	-	-	1	-	3	-
ean	1	-	15	-	-	-	-	-	-	-	-	3	-
eyk	2	-	-	22	-	-	-	-	-	-	-	1	-
gla	-	-	-	-	18	-	-	-	-	-	-	-	2
lan	-	-	-	-	-	21	-	-	-	-	-	-	-
lvp	-	-	-	-	-	-	19	-	-	-	-	-	-
ncl	-	-	-	-	-	-	-	18	1	-	-	-	-
nwa	1	-	-	-	-	-	1	-	18	-	-	-	-
roi	-	-	-	-	-	-	-	-	1	19	-	-	-
shl	1	-	-	-	1	-	-	-	-	-	19	-	1
sse	1	1	2	-	-	-	-	-	-	-	-	12	-
uls	-	-	-	-	-	-	-	-	-	-	-	-	20

speaker normalization. Note incidentally that the method is unaffected by sex differences.

5.2   Gaussian Modelling

An alternative classification using Gaussian modelling was carried out with the Netlab ([17]) toolbox. The model takes z-scored individual distance matrices as input and estimates one Gaussian model $N(\mu, \sigma)$ per dialect. As before, the test speaker is excluded from the training set; in other words, for each speaker, a new model is trained on all the data minus this speaker's matrix. The estimated dialect identity is then given according to the Maximum Likelihood decision. This statistical decision yields a non significant improvement over the previous method: for the both sexes condition, the model achieves 90.4 % correct classification.

5.3   Discussion

Both methods seem to perform equally well, which might indicate that a ceiling has been reached for this particular corpus. This question will be addressed more in depth in Section 7.1. A close examination of Table 3 suggests that linguistic explanations can often justify some of the misclassifications. For example, the historical link between *ean* and *sse* may account for the 3 *ean* speakers being classified as *sse*, and the 2 speakers of *sse* being classified as *ean*. The fact that 2 speakers of *gla*, and 1 from *shl* were identified as *uls* could be accounted for by saying that the 3 dialects belong to a common super region, namely, the Celtic countries. The high scores were of course facilitated by the absence of

co-articulatory variation; yet, it is worth pointing out that even /hVd/ words
– whose weaknesses are constantly condemned – contain essential information
about dialect. And, what is more, they probably constitute the quickest and
most convenient way to form an opinion about the linguistic quality of a corpus,
or the feasibility of a classification task.

6   Experiment 3: Dialect Classification with Read Passages

The ACCDIST procedure is then applied to the read passage part of the
corpus. The segmentation was obtained through forced-alignment as in Experiment 1. The number of words uttered by all speakers amounted to 61. When
words were polysyllabic, only the stressed syllable was kept for the classification.
The same spectral and duration parameters as in Experiment 2 were computed.
One option would have been to classify the vowels according to the lexical set
they belonged to. However, this would have artificially reduced the diversity of
coarticulatory phenomena, possibly leading to poor performances, and it would
have necessitated the intervention of a human expert in order to infer lexical set
membership of the stressed vowel in a given word. This would in turn have led to
a manifold increase in the tedium and the time to carry out the classification, not
to mention the questionable theoretical validity of such inferences. A sounder approach that by-passes such linguistic hypotheses was therefore adopted: instead
of vowel types, distances were computed between vowel tokens. Note here that
264 speakers are included. The 61×61 individual distance matrices were then
classified with the same correlation-based procedure that was used for the /hVd/
words. 89.6 %, 87.6 %, and 90.5 % correct classification are obtained for men,
women, and both sexes respectively with the Euclidean distance. The Manhattan
distance yields 87.4 % and 89.4 % for men and women; there is no improvement
for the third condition.

7   General Discussion

7.1   Guidelines to Assess Classification Scores

One of the questions underlying these experiments is how good a 90 % correct
classification score is with respect to the data that has been analysed. A fundamental conceptual discrepancy between language identification and dialect
identification should help us come up with a tentative answer. Except for a
few borderline cases – including code-switching–, language sets are in principle
mutually exclusive; in other terms, a speaker either speaks language A or language B, and certainly not a mixture of the two. Matters get more complicated
for dialect corpora: dialect membership for a speaker does not mean that the
speaker produces all the phonetic features of that particular dialect, nor does
it mean that s/he does not use features from other dialects. And as the distance (however it is measured) of a speaker from its dialect prototype increases,

so does the risk of this speaker being associated (by a naive listener, an expert phonetician, or the machine) with another dialect. In other words, it is undoubtedly more adequate to view dialect classes as fuzzy sets, and language classes as hard sets, although quite circularly, depending on linguistic denominations, we may come across borderline cases: if we use the linguistic criterion of mutual intelligibility, some entities traditionaly termed "languages" can overlap (see the case of Danish, Swedish, and Norwegian) while others called "dialects" may be rather distinct (possibly the case for distant dialects of Arabic). Translating this into figures, it could be said that language identification scores must be judged against the maximal achievable score (i.e. 100 % in almost all cases) whereas, there is no simple way to estimate this figure for dialects. There probably exists a floor (above chance level) below which the scores of an automatic dialect identification system can be considered bad; this floor could be given by classification carried out by naive listeners. And there certainly is another threshold around which scores can be deemed excellent. We tried to estimate the value of the latter threshold with an informal experiment: a native speaker expert phonetician was asked to listen to one third of all the passages spoken by men in the ABI corpus. The experiment was actually divided into 14 (one per dialect) separate verification tasks. In each task, the expert had to listen to a stimulus and say whether it had been uttered by a speaker of the dialect of the current task or not. We will not go into too much detail since this is beyond the scope of the present research, suffice it to say that the expert scored 89.6 %. Of course, proficient though the expert may have been, his degree of acquaintance with dialects probably varied from one to the next, but this is the closest we can get to estimating the highest possible classification score. Ceiling effects in classification accuracy are also suggested by a statement in the documentation of the corpus acknowledging that some speakers, particularly in *crn* and *nwa*, have an accent that might not be regarded as typical.

7.2   Descriptive Scope

Part of the descriptive task of the phonetician is to come up with linguistically interpretable visual representations from multidimensional raw numerical data. Graphical displays, particularly vowel plots, have been frequently used to illustrate phonetic phenomena. This section exemplifies how the methods employed in Experiment 2 for classification can be used as a descriptive tool. The dendrograms in Figures 2 and 3 display the output of hierarchical clustering computed with the single linkage algorithm implemented in Matlab for a selected set of vowels in two female speakers from *eyk* and *shl* respectively. The first tree clearly shows the relative proximity of *hood* and *Hudd*, exemplifying the well-known absence of phonemic split in the north of England we discussed in Section 2. The second tree illustrates the phonemic merger in Scotland involving the vowels of *hood* and *who'd*.

Figure 4 shows the scatter of women from six selected dialects based on individual 19 × 19 distance matrices computed with the /hVd/ words. Each individual matrix was z-scored and dimensionality was reduced with principal

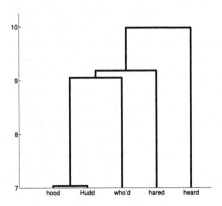

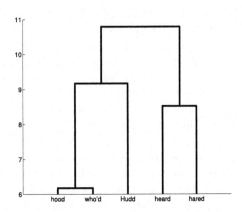

Fig. 2. Dendrogram illustrating the absence of *hood* vs. *Hudd* phonemic split in *eyk*

Fig. 3. Dendrogram illustrating the phonemic merger involving *hood* and *who'd* in *shl*

component analysis. The plane is defined by the first two principal components, which account for approximately 35 % of the variance of the original data. High though the distorsion may be, meaningful patterns can still be identified on the graph: an imaginery oblique line separates the dialects of England (*ean,lan*, and *ncl*) from those of the Celtic countries (*gla*, *roi*, and *shl*). Then, within the English group, an almost geographical picture emerges: *ean* in the south east, *lan* in the north west, and *ncl* in the north east. In the Celtic group, Scotland and Ireland are neatly split, with *roi* being distinct from *gla* and *shl*[6]. Finally, within the Scottish subset, the situation looks more fuzzy (but this may simply be a consequence of dimensionality reduction), although there is a tendency for *gla* speakers to cluster near the bottom of the graph, and *shl* speakers above the latter. Whatever the goodness of the final display, the efficiency of inter-segment distance matrices to capture dialect specifities is confirmed by the bidimensional map whose interpretation in linguistic and geographical terms makes perfect sense.

7.3   Suggested Improvements

We now turn to the question of how to improve the classification scores. Consider the $n \times p$ matrix where n refers to the 261 speakers and p to the $19(19-1)/2 = 171$ distances (i.e. the unfolded 19×19 individual symmetric matrix) between pairs of vowels. It is very unlikely that all distances possess equal discriminatory power: some may be extremely relevant, e.g. those between two vowels that can be either merged or not depending on the specific vowel system, others may have only slight discriminatory power, for example those implying minute phonetic differences, and others may be irrelevant altogether. In addition, measurements

---

[6] Note however that one speaker from *roi* ended up with the *ncl* cluster.

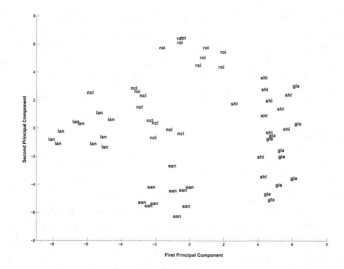

Fig. 4. Female speakers from 6 dialects along 1st and 2nd principal components derived from the distance matrices on /hVd/ words

on continuous scales contain noise, hence we could consider dichotomizing some of the quantitative variables. For example, phomenic mergers, or the absence of phonemic splits, could be regarded as binary events: on a continuous scale, the distance between *hood* and *Hudd* in northern English dialects is never equal to zero, although it should be in systemic (phonological) parlance. Besides, it may vary between speakers despite their producing exactly the same target vowel in these two words. The varying distances between *hood* and *Hudd* in a set of speakers having no *hood* vs. *Hudd* contrast is linguistically irrelevant, and it adds noise to the system. So there must be a threshold in the distance measured on a continuous scale below which the two vowels can be regarded as identical; and above this threshold, the two vowels can be considered different. Feature selection (recall that the features are the $p = 171$ distances between vowel pairs) would be desirable for at least three reasons. Firstly, it would rid the system of noisy variables, possibly improving classifications scores and reducing computational cost. Secondly, some modelling techniques require a subtle balance between the number of examples to classify and the size of the feature space (the n and p dimensions in the matrix respectively); given the small size of n in our data, reducing p is imperative. Thirdly, and most interestingly, special cases of feature selection such as feature ranking and feature weighting can provide explanatory principles: such methods as K-means partitioning may be used to assess the relative weight of each feature ([21]). This assessment could in turn validate linguistic hypotheses on the discriminatory power of each feature. All these methods work *a posteriori* in that they need the data first; another possible improvement would be to include linguistic knowledge prior to data analysis. [6] applied such a procedure to increase the potential differentiation of dialects: for example, if the distance between the vowels of *father* and *after* is smaller than that between *cat* and *after*, then strong evidence for a southern English dialect is obtained, whereas this weighs against northern English dialects, and neither favours nor disfavours Scottish dialects. So [6] came up with an *a priori* trivalent weight system which somewhat enhances the discrimination on the basis of phonological knowledge after the raw numerical evidence has been accumulated.

7.4   Perspectives

The classication method presented here is text-dependent: what is being said must be known beforehand, and the words of the training and test sets must match. Besides, it is based on phonetic and phonological knowledge of dialect differences, and we must bear in mind that the stimuli (/hVd/ words and read passages) were precisely designed to elicit dialect variation, and therefore facilitate discrimination. So this approach can be termed shibboleth-based. Now, how good would the performance be with a randomly chosen text? More specifically, how could one deal with mismatches between the vowels of the training datasets and those of the test speaker set? Another challenge is the transposition of the method to spontaneous speech. Future research will focus on text-independency and include other phonetic cues such as consonants and suprasegmentals.

Acknowledgements

We are grateful to Mark Huckvale and Francis Nolan for their help. This study was supported by a Eurodoc grant from the Région Rhône-Alpes.

References

1. Huckvale, M.: ACCDIST: a metric for comparing speakers' accents. In: Proceedings of Interspeech 2005, Jeju, Korea, pp. 29–32 (2004)
2. Huckvale, M.: ACCDIST: an accent similarity metric for accent recognition and diagnosis. In: Müller, C. (ed.) Speaker Classification. Lecture Notes in Computer Science / Artificial Intelligence, vol. 4343, Springer, Heidelberg (2007) (this issue)
3. Yan, Q., Vaseghi, S.: A comparative analysis of UK and US english accents in recognition and synthesis. In: Proceedings of ICASSP, Orlando, Florida, USA, pp. 413–417 (2002)
4. Ellis, S.: The Yorkshire Ripper enquiry: Part I. Forensic linguistics 1, 197–206 (1994)
5. Jessen, M.: Speaker Classification in Forensic Phonetics and Acoustics. In: Müller, C. (ed.) Speaker Classification. LNCS(LNAI), vol. 4343, Springer, Heidelberg (2007) (this issue)
6. Barry, W.J., Hoequist, C.E., Nolan, F.J.: An approach to the problem of regional accent in automatic speech recognition. Computer Speech and Language 3, 355–366 (1989)
7. Huang, R., Hansen, J.: Advances in word based dialect/accent classification. In: Proceedings of Interspeech 2005, Jeju, Korea, pp. 2241–2244 (2005)

8. Orton, H., Sanderson, S., Widdowson, J.: The linguistic atlas of England. Croom Helm, London (1978)
9. Wells, J.: Accents of English. The British Isles, vol. 2. Cambridge University Press, Cambridge (1982)
10. Foulkes, P., Docherty, G.: Urban Voices. Accent Studies in the British Isles. Arnold, London (1999)
11. Kortmann, B., Schneider, E.W.: A Handbook of Varieties of English. Mouton de Gruyter, Berlin, Germany (2004)
12. D'Arcy, S.M., Russell, M.J., Browning, S.R., Tomlinson, M.J.: The Accents of the British Isles (ABI) corpus. In: Proceedings of MIDL 2004 Workshop, Paris, France, LIMSI-CNRS, pp. 115–119 (2004)
13. Schötz, S., Müller, C.: A Study of Acoustic Correlates of Speaker Age. In: Müller, C. (ed.) Speaker Classification. Lecture Notes in Computer Science / Artificial Intelligence, vol. 4343, Springer, Heidelberg (2007) (this issue)
14. Schötz, S.: Acoustic Analysis of Adult Speaker Age. In: Müller, C. (ed.) Speaker Classification. Lecture Notes in Computer Science / Artificial Intelligence, vol. 4343, Springer, Heidelberg (2007) (this issue)
15. Young, S., Evermann, G., Kershaw, D., Moore, G., Odell, J., Ollason, D., Povey, D., Valtchev, V., Woodland, P.: The HTK Book (for HTK version 3.2). Cambridge University Engineering Department, Cambridge (2002)
16. Boersma, P., Weenink, D.: Praat. Doing Phonetics by Computer. version 4.4.22 (2006)
17. Nabney, I.T.: Netlab. Algorithms for Pattern Recognition. Springer, London (2002)
18. Sjölander, K., Beskow, J.: Wavesurfer - an open source speech tool. In: Proceedings of ICSLP 2000, Beijing, China, pp. 464–467 (2000)
19. Ellis, D.P.W.: PLP and RASTA (and MFCC, and inversion) in Matlab, online web resource (2005)
20. Gower, J.C., Legendre, P.: Metric and Euclidean properties of dissimilarity coefficients. Journal of Classification, 5–48 (1986)
21. Makarenkov, V., Legendre, P.: Optimal variable weighting for ultrametric and additive trees and K-means partitioning: Methods and software. Journal of Classification, 245–271 (2001)

ACCDIST: An Accent Similarity Metric for Accent Recognition and Diagnosis

Mark Huckvale

Phonetics and Linguistics, University College London,
Gower Street, London, WC1E 6BT, U.K.
M.Huckvale@ucl.ac.uk

Abstract. ACCDIST is a metric of the similarity between speakers' accents that is largely uninfluenced by the individual characteristics of the speakers' voices. In this article we describe the ACCDIST approach and contrast its performance with formant and spectral-envelope similarity measures. Using a database of 14 regional accents of the British Isles, we show that the ACCDIST metric outperforms linear discriminant analysis based on either spectral-envelope or normalised formant features. Using vowel measurements from 10 male and 10 female speakers in each accent, the best spectral-envelope metric assigned the correct accent group to a held-out speaker 78.8% of the time, while the best normalised formant-frequency metric was correct 89.4% of the time. The ACCDIST metric based on spectral-envelope features, scored 92.3%. ACCDIST is also effective in clustering speakers by accent and has applications in speech technology, language learning, forensic phonetics and accent studies.

Keywords: Accent, accent recognition, accent similarity, speaker normalisation.

1 Introduction

1.1 What Is an Accent Similarity Metric?

An accent similarity metric is a formal procedure to estimate the distance between the accents of two speakers as measured from recordings of their speech. An accent metric would analyse the phonetic content of two transcribed recordings and deliver a single number that would reflect the extent that the two speakers had similar accents in their production of that material. If it were possible to create such a metric, then it would be useful in a number of ways in both speech science and speech technology. For example, an accent metric could be used to cluster speakers into accent groups automatically, which could be used to track the spread and development of accents according to regional or social class parameters. The identification of accent groups would be useful for building speech recognition systems that adapted to a speaker's accent. An accent metric might be used as part of a forensic investigation to help

C. Müller (Ed.): Speaker Classification II, LNAI 4441, pp. 258–275, 2007.

identify characteristics of an unknown speaker. It might be used to evaluate the "nativeness" of the accents of second language learners. Or an accent metric could be used in a diagnostic way to discover the most significant differences between accents.

However the development of an accent metric is not without problems. At the same time as a metric has to be sensitive to accent, it must also be insensitive to the individual characteristics of the speakers themselves: to whether the speaker is a man or a woman, young or old, large or small. Ideally an accent metric would also be insensitive to characteristics of the recording, to differences in the speaking styles exhibited by the speakers, to the emotional states of the speakers involved, and to the exact content of the phonetic material analysed. These, however, are aspirations, while insensitivity to speaker characteristics is essential.

This article describes some fundamental investigations into the practicality of an accent metric using a database of 275 speakers representative of 14 regional accents of the British Isles. It explores the effect of speaker characteristics, such as speaker sex, on the ability to assign a sample speaker to an accent group. It explores the effect of different acoustic parameter sets, different speaker normalisation procedures, and different pattern classification functions.

The rest of this section provides an overview of accent measurement: including what to measure, how to measure, how to normalise measurements and how to compare measurements. Section 2 demonstrates the application of the currently most effective technique based on normalised formant frequencies and Linear Discriminant Analysis (LDA) using the British Isles data. Section 3 presents an alternative normalisation strategy for accent similarity measurement called ACCDIST which finesses the speaker normalisation problem and which provides demonstrably superior performance.

1.2 What to Measure?

Research has uncovered phonetic variation across social groups at all levels of speech: phonological organisation, prosody, phonetic quality and effect of context on phonetic realisation. In terms of *phonological variation*, for example, Northern and Southern accents of the British Isles differ in the lexical distribution of the PALM and TRAP vowels, so that words like "bath" and "ask" have the PALM vowel in one accent the TRAP vowel in another [1]. Another example is that English accents vary according to "rhoticity" - whether orthographic "r" letters are realised as an /r/ segment pre-consonantally or in syllable final position. In terms of *prosodic variation*, differences have been observed across accents in the realisation of intonational tunes – for example the use of either a low-falling or a low-rising pattern for the ends of declarative sentences [2]. In terms of *phonetic quality*, the preferred articulation of vowels and consonants can vary across accents. Some London accents notoriously exploit a glottal stop for /t/ in some syllable-final environments, while Liverpool accents use a softer fricated form of /t/ in syllable-initial position [1]. Similarly the realisation of /l/ in different syllable contexts varies markedly across accents: some like Irish-English use a fronted "clear" [l] in all positions, while many American-English accents use a palatalised "dark" [l], and an ongoing trend in Southern England is for a syllable-final /l/ to be realised as a back rounded vowel [1].

Although there are all these differences between accents, the greatest body of research has concentrated on the quality of vowels. For example, the extensive study by Labov and his colleagues into American-English accent variation across regions of the U.S. has looked almost exclusively at vowels [3]. Various reasons could be proposed for this: (i) vowels are readily mutable by speakers and changes in vowel quality are widely exploited to establish differences between accents, (ii) vowels are common, acoustically intense and relatively stable, which makes them easy to record and identify, and (iii) vowels are easy to describe using a few acoustic parameters, such as average formant frequencies. This article will also focus on measurements of vowel quality, but the reader should keep in mind that vowel quality is only one part of the phonetic difference between accents.

Even when the choice is to concentrate on vowel quality, it is also important to realise that any operation which averages across vowel qualities realised in different words makes an implicit assumption about the phonological distribution of vowels in the lexicon. While it may make sense in an American accent to average the measurements of the vowels in "tack" and "task", this does not make sense in a Southern British accent where these are phonologically different vowels. Similarly, it may be satisfactory to average vowels measurements across "palm" and "harm" in Southern British accent, but it would not be for a rhotic accent, where the /r/ in "harm" will have some effect on the vowel quality used. In summary, we need to take care when averaging measurements across words, since this might obscure differences in phonological distribution which are themselves characteristics of accents.

1.3 How to Measure?

How can we measure the quality of vowels? It is worth asking what properties we require of the measurements? These include: (i) that the numbers are of low dimensionality, but that (ii) they capture all the important perceptual differences between vowels, that (iii) the parameters are easy to measure and are robust to variation caused by the recording equipment or the recording environment. Finally we need measurements (iv) which can be processed in such a way as to remove characteristics of speaker identity while preserving characteristics of accent identity.

There are really two main approaches: firstly to estimate the centre frequencies of the main concentrations of spectral energy in the acoustic form of the vowel – this provides estimates of the vocal tract resonant frequencies used in vowel production; or secondly to estimate the shape of the spectral envelope of the whole vowel sound. Typically the former gives rise to between 2 and 4 numbers representing the formant frequencies of the vocal tract, while the latter gives rise to between 10 and 20 spectral coefficients representing the shape of the spectral envelope.

Historically, formant frequencies have been most widely used in the field of experimental phonetics, and have been the basis for the influential studies by Labov, quoted above. Spectral envelope measures, on the other hand, have been most widely used in speech technology to represent phonetically relevant properties of the speech signal. The advantages of formant measures are that a significant fraction of perceptual vowel quality can be captured (and indeed plotted) using just two numbers; and that a frequency-based measure leads to a natural means for speaker normalisation through frequency scaling (see next section). Disadvantages of formant measures are that they are difficult to extract from the signal automatically (and measurement errors have a non-normal distribution), and that they ignore the perceptual effect of changes in formant amplitude. On the other hand while spectral measures capture all the perceptually important aspects of vowel quality, they use far more parameters which are also less amenable to normalisation. The effect of acoustic representation on accent similarity is investigated in section 2.

Finally, it is also worth noting that vowels are not static sound objects: not only are some vowels intrinsically dynamic in nature (diphthongs) but also vowels are affected by the syllabic context in which they occur, particularly the influence of the consonantal environment. Thus measurements of a vowel realisation need to be sensitive to any changes in vowel quality that occur across time.

1.4 How to Normalise?

Since any speaker could speak any accent, knowledge of the physical characteristics of the speaker as expressed in their speech is of no help in identifying the speaker's accent. Thus we need some process by which the influences of the physical characteristics of the speaker on the recorded speech need to be removed or ignored. Typically these will be aspects of the signal related to a speaker's vocal anatomy and physiology, for example: voice quality, pitch range and vocal tract size.

The extensive use of formant frequencies to characterise vowels in experimental phonetics research has led to a lot of work into the best means by which formant frequencies can be normalised to remove the influence on the measurements caused by the speaker's vocal tract size. Fortunately, a recent study by Adank [4] has compared a large number of these normalisation techniques specifically within the domain of accents research. Adank tested a number of formant frequency normalisation procedures on vowels from a number of speakers within a task which assessed whether the normalisation also affected the discriminability of their accents. The best method turned out to be very simple: the conversion of formant frequencies into z-scores using the distribution of individual formant frequencies by each individual speaker. Thus a normalised formant frequency describes the relative position of the formant within the range of its frequencies typically used by the speaker, and independently from other formants.

Normalisation in the frequency domain seems sensible since the major effect of a change of vocal tract size is indeed to scale its resonant frequencies. However it is harder to see how frequency domain normalisation can be as simply applied to measures of the spectral envelope. For example, a spectral envelope estimate extracted by measuring the energy across a number of fixed frequency bands has confounded measures of energy and frequency. No simple scaling using the distribution of measurements from a speaker would produce an envelope that would be independent of vocal tract size.

There are ways of representing the spectral envelope which appear to make it less sensitive to speaker characteristics: firstly to put the spectral envelope on a logarithmic frequency axis, and secondly to use "cepstral" coefficients which are more sensitive to shape than to absolute position within the spectrum. The most common form of spectral envelope measure used in speech recognition approximates these two suggestions, and is called mel-scaled frequency cepstral coefficients (MFCC) derived from the work of [5].

Although the use of MFCCs to represent the spectral envelope does seem to reduce the sensitivity of speech recognisers to speaker characteristics, it is not clear whether the use of MFCC also affects the measurement of accent. This is investigated in section 2.

Recently a radical alternative approach to normalisation was proposed by Nobuaki Minematsu [6] and formulated explicitly for the purpose of comparing accents by Huckvale [7]. The idea is to represent the accent of a speaker not in terms of the absolute quality of the sounds realised in their speech, but to represent them in terms of the relative similarity of those sounds. That is, to describe a speaker through a set of vowel-to-vowel distances rather than in terms of the individual vowel qualities. The advantage of such an approach is that a speaker's vowels are only compared with each other, not with the vowels of other speakers. Thus normalisation is implicit in the representation, rather than added in a post-processing step. The effectiveness of this kind of representation will be investigated in section 3.

1.5   How to Compare Accents?

Given a set of normalised measurements of vowels for a speaker, how can we compare speakers to get a measure of accent similarity? If we assume that we have matched phonetic material, the task comes down to finding the distance between two instances of the same vowel, represented by two vectors of acoustic measurements: x_1 and x_2. There are a number of ways in which we might approach this: (i) *Correlation* distance: this gives the similarity between the two vectors independently from the mean and variance of each vector; (ii) *Unweighted Euclidean* distance: this gives the root mean square difference between the means of the vectors independently from their variance; (iii) *Weighted Euclidean* distance: this gives the root mean square difference between the means but each squared acoustic parameter difference is first weighted by the variance of that parameter calculated across all data from all speakers; (iv) *Mahalanobis* distance: this gives the difference between the vectors weighted by the covariance between all acoustic parameters calculated across all data from all speakers; and (v) *Linear Discriminant Analysis* distance: this is the Mahalanobis distance where the covariance is calculated separately for each accent group.

Which of these distance functions to use will depend on the nature of the acoustic parameter set and on the amount of data available. If the acoustic parameters are independent and identically distributed, then the Euclidean metric is appropriate. However if the acoustic parameters are independent but have different variances, then the Weighted Euclidean metric should be used. If the acoustic parameters are not independent, then the Mahalanobis distance is best. However, the use of the full covariance matrix in the distance measure relies on that matrix itself being well estimated from the data – a poorly estimated covariance matrix may perform worse than the Euclidean metric. Linear Discriminant Analysis is a powerful way of making use of the fact that the covariance matrix may well be different in the different accents – however it requires that speakers be assigned to accent groups in the first place. The benefits of the different distance measures are explored in section 2.

2   Accent Data and Baseline Performance

2.1   Speech Data

Speech material was extracted from the Accents of the British Isles (ABI) corpus recorded by the University of Birmingham under contract to 2020 Speech Ltd. Nominally, ten male and ten female speakers from 14 accent areas (see Table 1) spoke the same set of 20 short sentences (see Table 2). However, there are some gaps in the database and the material in fact totalled 275 speakers and 5208 sentences.

Table 1. The 14 accents areas in the ABI corpus

Code	Accent	Code	Accent
brm	Birmingham	lvp	Liverpool
crn	Cornwall	ncl	Newcastle
ean	East Anglia	nwa	North Wales
eyk	East Yorkshire	roi	Dublin
gla	Glasgow	shl	Scottish Highlands
ilo	Inner London	sse	South East
lan	Lancashire	uls	Ulster

Table 2. List of sentences used for the experiments

1	Kangaroo Point overlooked the ocean
2	where were you while we were away
3	the high security prison was surrounded by barbed wire
4	an official deadline cannot be postponed
5	few people live to be a hundred
6	co-operation and understanding go a long way to alleviate dispute
7	they often go out in the evening
8	glucose and fructose are natural sugars found in fruit
9	help celebrate your brother's success
10	young children should avoid exposure to contagious diseases
11	the oasis was a mirage
12	comedies never have enough villains
13	cement is measured in cubic yards
14	I itemise all accounts in my agency
15	a young mouse scampered across the field and disappeared
16	Gary attacked the project with extra determination
17	a good attitude is unbeatable
18	her auburn hair reminded him of autumn leaves
19	after tea father fed the cat
20	father cooked two of the puddings in batter

The designers of the ABI corpus [8] selected the first 18 of the 20 sentences to achieve 'phonetic balance' of British English, while the last two sentences were adapted from the 'accent revealing' sentences used by Barry *et al* [9]. A phonological transcription was generated for each sentence using Southern British English pronunciations, and phonetic segmentation was performed by forced alignment using the HTK Hidden Markov Modelling toolkit[1]. All subsequent analysis was made using only the vowel segments in the 20 sentences including diphthongs but excluding schwa. This gave up to 145 vowel measurements per speaker. The fact that a single phonological transcription was used for annotating all accents was not important as only the segment boundaries, rather than the phonetic identity was used in subsequent analysis. After segmentation, phones were effectively referred to as, e.g., the "first vowel in 'after'" rather than as /ɑː/ or /æ/.

2.2 Accent Recognition with a Formant Frequency Distance Metric

To assess how well formant frequency measurements can be used in a metric to compare speakers' accents, an accent recognition experiment was constructed using the accent labels in the ABI corpus as the 'correct' answer for each speaker. Accent classification accuracy can then be used as a benchmark for different accent similarity metrics, assuming that the corpus was well designed. Formant frequency estimation was performed using the FORMANAL program of the Speech Filing System[2]; this delivers spectral peak estimates each 10ms using LP analysis. The automatic segmentation labels were used to divide each vowel into two halves by time, and an average frequency of each formant in each half was calculated from the trimmed mean of the frame values. The trimmed mean was calculated using values from the 20th to 80th percentiles. Average formant frequency values from each half were then abutted to create a single vector for classification.

Accent recognition performance was evaluated by comparing each speaker to the data set remaining (274 speakers) once that speaker had been removed ("leave one out"). Only vectors from identical vowels in identical word positions were compared across speakers. The overall distance was computed from the mean of the individual matched vowel distances. Evaluation was done for two, three and four formants. Similarity was judged using an unweighted Euclidean metric, a weighted Euclidean metric (using the diagonal elements of the covariance matrix) and with a Mahalanobis metric (using the full covariance matrix) on the hertz values. Performance was also evaluated in three gender conditions: (i) where the unknown speaker is compared only to speakers of the same sex, (ii) where comparison is made to speakers of either sex, and (iii) where comparison is made only to speakers of the opposite sex. The idea here was to expose any sensitivity of the comparison to the physical characteristics of the speakers. The recognised accent was chosen from the accent of the most similar individual speaker. Results are shown in Table 3.

---

[1] htk.eng.cam.ac.uk
[2] www.phon.ucl.ac.uk/resource/sfs/

Table 3. Baseline accent recognition rate for held-out speaker using 2, 3 or 4 formant frequency data, in 3 covariance conditions (U=unweighted, D=diagonal, F=full-covariance) across 3 gender match conditions. N=275.

%Correct	Same Sex			Any Sex			Opposite Sex		
#Formants	U	D	F	U	D	F	U	D	F
2	68.2	69.0	70.4	67.5	68.2	67.2	38.0	37.2	41.2
3	55.5	58.8	60.2	55.1	58.8	60.6	29.9	31.8	34.7
4	51.1	55.8	57.7	52.9	59.5	57.3	28.1	32.1	33.6

The best recognition performance without taking notice of speaker sex came from using just two formants and a weighted Euclidean metric, but at only 68.2%, performance is not very good. More formants do not mean better performance. The matching to speakers of the opposite sex to the test speaker makes performance considerably worse (37.2%), showing the sensitivity of formant frequency values to speaker characteristics. Matching to speakers of the same sex only makes performance a little better (69.0%), because speakers tend to match best to speakers of the same sex in any case. The use of the weighted distance measures shows a small improvement, particularly for 4 formants, where it accommodates the fact that the variance of the formant frequencies are not equal.

2.3 Formant Frequency Normalisation

It is expected that a normalisation procedure should improve accent recognition performance when formant frequencies are used. Firstly, the reduction in inter-speaker variability means the speakers will form more compact clusters within an accent group. Secondly, the sex of the test speakers should have less effect. Thirdly, the difference in variance of the formant frequencies (when expressed in hertz) should be removed, making the frequencies more compatible with the unweighted metric. To evaluate the importance of normalisation, the formant frequencies for each speaker were normalised to a unit normal distribution (i.e. to z-scores) using all the vowel frames for each speaker and each formant independently. Accent recognition performance using the normalised formant frequency values is shown in Table 4:

Table 4. Accent recognition rate for held-out speaker using normalised 2, 3 or 4 formant frequency data, in 3 covariance conditions (U=unweighted, D=diagonal, F=full-covariance) across 3 gender match conditions. N=275.

% Correct	Same Sex			Any Sex			Opposite Sex		
# Formants	U	D	F	U	D	F	U	D	F
2	76.3	76.3	74.1	79.2	79.2	74.5	66.8	67.2	63.9
3	67.5	70.8	68.2	69.3	72.6	72.6	50.0	57.3	54.0
4	57.3	66.1	64.6	56.6	66.4	65.0	39.1	47.4	48.5

Normalisation has a large effect, with the best performance in the any sex condition now reaching 79.2%. Most noticeable, however is that there is much less variation across the three gender matching conditions. Matching only to the opposite sex still has a somewhat lower performance, showing that even formant frequency normalisation does not remove all speaker characteristics from the data. The normalisation of the formant frequencies from hertz to z-scores reduces the benefit of the weighted distance functions as expected, although there is still a benefit for 3 and 4 formants.

2.4 Pooling of Speakers and Contexts

So far we have chosen the accent group for our test speaker by finding the accent of the single closest known speaker in the training data. However, if we assume that the training speakers are drawn from a population of speakers of that accent, we might do better to compare our unknown speaker to the estimated mean of that accent population. Individual speakers with idiosyncratic pronunciations then have less effect. On the other hand, this assumes that the speakers in the database do indeed form a homogeneous group with a single, meaningful average. Another advantage of pooling speakers into a mean is also that separate acoustic parameter covariance matrices for each accent can be computed. Any differences in covariance between accents can then be exploited in recognition. Table 5 shows the results for accent recognition against the accent group means using Linear Discriminant analysis.

Table 5. Accent recognition rate for held-out speaker against accent group means using normalised 2, 3 or 4 formant frequency data, in 3 covariance conditions (U=unweighted, LD=diagonal LDA, LF=full-covariance LDA) across 3 gender match conditions. N=275.

% Correct	Same Sex			Any Sex			Opposite Sex		
# Formants	U	LD	LF	U	LD	LF	U	LD	LF
2	84.3	85.4	84.3	85.8	85.4	86.9	74.5	72.3	74.5
3	83.6	85.0	84.7	82.1	86.9	89.4	71.9	71.9	79.2
4	75.2	83.9	85.0	70.4	83.6	88.7	59.9	71.2	76.6

Even with the unweighted distance measure, performance is markedly improved by matching to the accent group means rather than to individual speakers. This shows that the means are indeed a reasonable way of describing the accents in this data. The use of LDA has very little effect when only two formants are used, but there is increasing benefit of using LDA with 3 or 4 formants. In other words there are differences in the covariance matrix across accents which are useful to accent recognition. The best performance overall now comes from 3 normalised formant frequencies with full covariance LDA at 89.4%.

If pooling speakers into groups helps, then it is worth considering whether it would also help to pool vowels together across words. For example, it would be possible to average instances of /i/ vowels across all words in the sentences that a dictionary would indicate contain /i/. Pooling in this way, of course, assumes that the phonological units used by the speakers in the words are known, and as we have seen in section 1.2, this is not necessarily the case. Accents vary in terms of the inventory and distribution of phonological units in the lexicon. Table 6 shows the effect of averaging vowel measurements within each speaker using a Southern British English dictionary as the phonological model.

Table 6. Accent recognition rate for held-out speaker against accent group means using normalised 2, 3 or 4 formant frequency data pooled across phonologically matched vowels, in 3 covariance conditions (U=unweighted, LD=diagonal LDA, LF=full-covariance LDA) across 3 gender match conditions. N=275.

% Correct	Same Sex			Any Sex			Opposite Sex		
# Formants	U	LD	LF	U	LD	LF	U	LD	LF
2	79.9	65.0	63.9	79.6	67.2	66.1	65.0	58.8	56.9
3	80.3	69.7	69.0	74.1	65.3	65.7	64.2	58.0	55.5
4	74.5	70.8	71.5	65.7	65.0	67.2	48.5	48.2	53.6

In comparison with Table 5, these results show that pooling over word contexts makes performance considerably worse, with the best performance in the any sex condition falling from 89.4% to 79.6%. Although pooling over contexts may be of benefit when the number of measurements is small - so as to obtain more robust measures of the average form of a vowel - when there is enough data, the disadvantage of pooling becomes significant. When the data is normalised, it is beneficial to average across speakers when the word contexts are the same, but that doesn't extend to averaging across vowel contexts, even within the same speaker.

2.5 Spectral Envelope Measures

Formant frequencies are just one way of characterising the important features of the sound spectrum for speech. Although vowel quality can largely be preserved through formant frequencies alone, there are other aspects of the spectrum which affect our perception of timbre, and these may be significant in judging vowel similarity across speakers [10]. Another problem with formant frequencies as acoustic parameters is the difficulty of measuring them in situations of noise, poor voice quality, nasality or high pitch. Finally, accent variation is not solely restricted to vowels, and formant frequencies are unlikely to be appropriate for consonantal speech sounds. An alternative feature set comes from parameterisation of the spectral envelope directly, expressing how energy is distributed across frequency without trying to estimate the location of spectral peaks. Such representations have been shown to correlate well with the perception of timbre, including vowel quality [11]. Speech coding and speech recognition systems use such features extensively, and many designs have been evaluated. Here we will choose two possible candidates for comparison with the formant metric: an approach used for speech coding that uses an auditory filterbank, and an approach widely used in speech recognition using cepstral coefficients.

The auditory filterbank analysis was adapted from the 19-channel vocoder of Holmes [12], which has 19 band-pass channels between 180 and 4000Hz with bandwidths based roughly on auditory filter widths. The output of the filterbank was rectified, smoothed by a low-pass filter at 50Hz, downsampled to 100 frames/sec then logarithmically compressed. The mean of each frame was then subtracted and added as a 20th parameter. This makes the vector more compatible with an unweighted Euclidean metric by encoding the difference in overall energy once rather than 19 times. For each vowel, the mean frame was calculated from each half, and the two means abutted to create a 40 dimensional classification vector. Processing was performed using the voc19 program of SFS.

The second spectral envelope feature set was based on the mel-frequency scaled cepstral coefficient (MFCC) approach described by Davis & Mermelstein [5]. 25ms windows of signal are transformed by DFT into spectral energies which are collected into 20 triangular channels spaced according to a mel-scaling of the frequency space. The first 12 coefficients of the cosine transform of the channel energies are then calculated, and these are then weighted with a sinusoidal cepstral lifter. Finally the overall window energy is added in bels as a 13th coefficient. The cosine transform makes the parameters relatively independent (i.e. diagonal covariance), while the cepstral liftering and the energy coding gives the parameters more equal variance. Both of these make the feature set more suitable for use with an unweighted Euclidean metric. Frames are computed every 10ms. For each vowel, the mean cepstral coefficients were calculated from each half, and the two means abutted to create a 26 dimensional classification vector. Processing was performed with the MFCC program of SFS. For comparison with the normalised formant frequency metric, accent recognition was performed as in Table 5, with pooling across speakers and using LDA, see Table 7.

Table 7. Accent recognition rate for held-out speaker against accent group means using spectral envelope parameters, in 3 covariance conditions (U=unweighted, LD=diagonal LDA, LF=full-covariance LDA) across 3 gender match conditions. N=275.

% Correct	Same Sex			Any Sex			Opposite Sex		
Features	U	LD	LF	U	LD	LF	U	LD	LF
voc19e	79.2	79.9	77.4	70.1	75.2	71.9	52.6	54.4	44.9
MFCC	65.3	70.8	82.1	56.2	63.5	78.8	26.6	44.5	55.5

It is clear that the switch back to un-normalised parameters has produced a significant drop in recognition performance compared to formants. The best performance is now only 78.8% in the any sex condition, compared to 89.4% for 3 normalised formants. The spectral envelope parameters, like the unnormalised formant frequency parameters also show a great sensitivity to speaker sex, with very poor performance in the opposite sex condition. Linear Discriminant Analysis has a large benefit for the MFCC feature set, but less so for the vocoder feature set, possibly because the latter had a far larger covariance matrix to be estimated. Overall the MFCC feature set gave better performance, perhaps related to the smaller number of parameters.

2.6 Summary

In this section, a large and difficult accent recognition task has been used to evaluate some different ways to measure accent similarity. We have shown that (i) normalisation has a large effect on recognition performance, not only because it reduces the differences between speakers, but because it enables (ii) pooling of speakers into accent groups, which in turn allows for (iii) the use of independent covariance matrices per accent in the LDA distance measure. However, only formant frequencies are amenable to speaker normalisation: through a mapping from hertz to z-scores. In the next section, we will show how the ACCDIST approach allows speaker normalisation to be performed using any acoustic parameter set.

3 ACCDIST Metric

3.1 Construction

Section 2 has shown the importance of speaker normalisation to the construction of an accent similarity metric. The real challenge is how to create a representation of the pronunciation preferences of a speaker which is sensitive to all the significant phonetic quality of his or her speech without it being affected by the physical characteristics of his or her particular vocal apparatus. There are two previous studies in the area which give hints on how this might be achieved. The first is the work of Barry, Hoequist and Nolan [9] who developed a regional accent classification technique based on acoustic comparisons made within known sentences. Speakers were asked to record specific sentences which were automatically segmented. Formant frequencies were estimated for particular vowels in particular words in those sentences. Then, relative values of the formant frequencies were compared across vowels within a sentence and a set of threshold rules were used to assign the speaker to one of four English regional accents. By this means the system was able to classify accents with an accuracy of 74%. Although the work is conventional in that it used prior knowledge about what vowel realisations characterise a particular accent, it was unconventional in that was based on comparisons between vowels spoken by one speaker rather than on comparisons to some absolute norm. Thus the procedure was self-normalising: the influence of the speaker's physical characteristics, and the influence of the linguistic context on the realisation of vowels were made irrelevant by the use of only relative measurements within a fixed set of sentences.

The idea to quantify the effect of accent through the relationship between the realisations of known segments rather than their absolute spectral quality was recently advanced further by the work of Nobuaki Minematsu [6]. In a study of Japanese learners of English as a second language, Minematsu trained a set of HMM phone models for individual speakers and then performed hierarchical cluster analysis on the pair-wise similarity of those phone models. The cluster analysis produced a dendrogram which he then used to compare individual learners with a native English speaker. The analysis showed, for example, that /l/ and /r/ were more similar (clustered lower in the dendrogram) for Japanese speakers of English compared to native speakers. Here too, the use of comparisons within a speaker were used to address the speaker variability problem: it is the results of cluster analysis of phone similarities that are compared rather than the characteristics of the phones themselves.

The work of Barry and Minematsu suggests that we can gain information about the pronunciation preferences of a speaker solely through comparisons made within that person's speech. So, for example, we could compare the vowel a speaker uses in "after" with the vowel he/she uses in "cat" rather than comparing it to some reference pronunciation. We would expect these two vowels to be more similar in some accents and less similar in others. Providing we know what pairs of vowels to study and what typical similarities occur in different accents, then this process can be used to recognise accents – indeed this is the basis of the Barry *et al* system. But we need to generalise the idea if we want to work with unknown accents where we know neither which vowels are important nor what is the typical variation.

The first part of the solution is to make comparisons between all pairs of segments spoken by the speaker, to create an inter-segment distance table. This must be based on individual instances of tokens, since we do not know yet which words share the same phonological units. The following example analyses the pairwise similarity between the vowels in the utterance "after tea father fed the cat" spoken by a Birmingham speaker and a South East British speaker from the ABI corpus. Table 8 shows the unweighted Euclidean distance between MFCC measurements for all pairs of vowels (following the procedure described in 2.3).

The distance table shows very clearly that the vowel in "after" for the Birmingham speaker is most similar to the vowel he uses in "cat", while for the South East speaker it is most similar to the vowel he uses in "father". Thus the distance tables capture a significant aspect of the known accent difference without comparing speakers' vowels with each other. Table 8 also shows a potential problem, however, in that the absolute size of the similarities also varies from speaker to speaker. Consider the difference between "fed" and "tea" which is twice as large for the South East speaker than for the Birmingham speaker. This variation may be due to the articulatory quality of the speakers, and have nothing to with their accent.

Table 8. MFCC distances between vowels in the sentence "after tea father fed the cat" spoken by a Birmingham and a South East speaker

BRM_M_01	father	cat	fed	tea
after	66.6	11.0	31.5	58.4
father		81.4	63.9	78.9
cat			28.5	56.4
fed				24.2

SSE_M_01	father	cat	fed	tea
after	16.1	62.7	76.9	97.0
father		73.9	69.3	83.9
cat			42.1	105.0
fed				68.0

How can we use the distance tables to compare speaker's accents? Effectively Minematsu performed cluster analysis on such distance tables (although his tables were based on HMM phone models, not single tokens) to derive a dendrogram which he called a 'phonetic tree' [6]. However, it is hard to compute the similarity between trees while it is relatively easy to consider ways to compute the similarity between raw distance tables. Since the absolute values in the distance tables vary from speaker to speaker, a correlation measure seems the most appropriate. This overall strategy is called ACCDIST for Accent Characterisation by Comparison of Distances in the Inter-segment Similarity Table [7].

3.2 Accent Recognition Performance Using the ACCDIST Metric

To establish the validity of the basic approach, the accent recognition experiments described in sections 2.4 and 2.5 were repeated using the ACCDIST metric to compare speakers to accent group means. The process was as follows: firstly the 140 or so vowel instances of each speaker were used to create the distance table for the speaker. This distance table was based on either the raw formant frequency parameters or the spectral envelope parameters, and pairwise similarities were calculated using an unweighted Euclidean metric. Then accent recognition was performed by taking each speaker in turn and comparing his or her distance table to the mean distance table of each accent group (excluding the speaker under test) by a correlation distance measure. Since not all vowels were spoken by all speakers, the correlations only took into account those vowel-pairs which occurred in the test speaker. The tests were run under three sex conditions as before, and the results for the formant parameters and the spectral envelope parameters are shown in Table 9.

Table 9. Accent recognition rate for held-out speaker against accent group means using the ACCDIST metric based on formant and spectral envelope parameters across 3 gender match conditions, compared with the best previous score from Tables 3-7 for that condition. N=275.

% Correct	Same Sex		Any Sex		Opposite Sex	
Features	BEST	ACCDIST	BEST	ACCDIST	BEST	ACCDIST
2 formant	85.4	82.5	86.9	86.5	74.5	78.1
3 formant	85.0	82.1	89.4	85.8	79.2	73.7
4 formant	85.0	78.8	88.7	84.7	76.6	70.8
voc19e	79.9	88.3	75.2	89.4	54.4	82.1
MFCC	82.1	85.8	78.8	92.3	55.5	84.3

For formant parameters, there is no advantage to using the ACCDIST metric rather than z-score normalisation. In only one case, in the opposite sex condition for 2 formants, did ACCDIST outperform LDA. However, there is considerable advantage to using the ACCDIST metric with the spectral envelope features. Large performance gains were seen in all conditions, particularly in the opposite sex condition. This shows that ACCDIST does indeed perform effective speaker normalisation, even on spectral envelope features. Overall, ACCDIST with MFCC features has the best performance of all, with 92.3% accent recognition, a reduction in the error rate of over 25% compared to LDA on normalised formant frequencies. A graphical summary of the important results of all the experiments is shown in Figure 1.

The best score of over 92% is very impressive for a database of regional accents, where differences are often hard for a human listener to judge. However there are no published human classification performance statistics on these data, so we are not sure what the ultimate performance might be. Subjects were chosen for the corpus on geographical grounds rather than on an analysis of their accent.

In summary these results are promising for the utility of the ACCDIST metric. Regional accent classification performance is high even without a better means to compare segment realisations than an unweighted Euclidean distance, and with the use of only the vowel segments in the sentences.

3.3 Example Use in Clustering and Diagnosis

Accent recognition is not the only possible application of an accent similarity metric. A particularly interesting application is for the discovery of accent groups "bottom up", from a corpus of recordings of unlabelled speakers. A metric would help find the significant clusters of speakers, and could be used in a diagnostic sense to discover which aspects of the speech are most significant for each cluster.

To demonstrate this, we shall cluster the speakers in the British Isles database and compare the contents of the clusters with the known accent labels. If the clustering is successful, we should see that the speaker clusters are also accent clusters, that is that all speakers of one accent group should end up in one cluster.

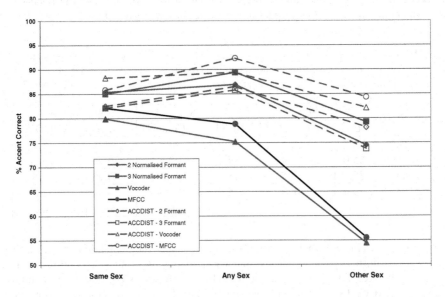

Fig. 1. Summary of accent recognition performance as function of feature set, similarity measure and gender condition

Figure 2 is a graphical representation of k-means clustering of the 275 speakers into four clusters based on the vowel data used in section 2, and with three different accent similarity measures. In the figure, the source accent group for each speaker is shown across the top, while the four output clusters are shown down the side. The area of each filled circle represents the proportion of speakers of the accent group that ended up in each cluster.

Figure 2a shows that clustering based on spectral envelope does not create clusters which have any useful relationship to the underlying accent groups. Indeed the likely result is that one cluster contains mainly male speakers and one cluster mainly female speakers! Figure 2b shows clustering based on normalised formant frequencies, the kind of similarity measure used by Labov [3]. Here it is possible to see that the output clusters do indeed relate to the underlying accent groups. However, there is also a significant amount of imprecision, whereby the members of some accent groups are divided across clusters. Figure 2c shows clustering using the ACCDIST metric on MFCC features. There is a very clear clustering behaviour seen, where all the members of an accent group end up almost completely in a single output cluster. Furthermore, investigation of how the accents are assigned to clusters, brings out four geographical areas of the British Isles: cluster 0 is Northern England, cluster 1 is Ireland, cluster 2 is Southern England, and cluster 3 is Scotland.

	brm	crn	ean	eyk	gla	ilo	lan	lvp	ncl	nwa	roi	shl	sse	uls
c0	•	•	●	•	●	•	●	·	●		●	●	·	•
c1	●	●	●	●	●	●	●	●	●	●	●	●	●	●
c2	•	•	·	•	●	•	·	•	●	·	•	●	•	·
c3	●	●	●	●	•	●	●	●	•	●	●	●	●	●

(a) Euclidean distance on MFCC features

	brm	crn	ean	eyk	gla	ilo	lan	lvp	ncl	nwa	roi	shl	sse	uls
c0	●	●	•	●		●	•	•	●	●	●		●	•
c1		•	•	●		·	●	●	●	●	●	•		
c2	●	●	●	•		●		•		•	•	●		
c3				●							●			●

(b) Euclidean distance on 2 normalised formant frequencies

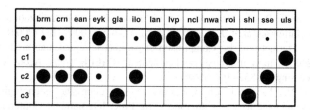

	brm	crn	ean	eyk	gla	ilo	lan	lvp	ncl	nwa	roi	shl	sse	uls
c0	•	•	·	●		•	●	●	●	●	•		•	
c1		•									●			●
c2	●	●	●	•		●						●		
c3				●							●			

(c) ACCDIST distance on MFCC features

Fig. 2. K-means clustering of 275 speakers into 4 clusters using different feature set and distance measures. The area of the circles represents the proportion of each ABI accent group assigned to each cluster.

In summary, ACCDIST can also be used to help uncover the identity of accent groups among a corpus of speakers. This could be useful in accents research for studying how accents develop and change, and does not require a priori knowledge about which accents exist. Once accent groups have been established, these can then be used as the basis for training accent-specific speech recognition systems. Or the ACCDIST metric can help in uncovering what particular characteristics of the speech of the members of a cluster are most significant in defining the group.

4 Conclusions

An accent similarity metric could be very useful in both speech technology and socio-phonetic research. In this article we have discussed the main problems and shown how a high-performance accent similarity metric can be constructed. In a difficult regional accent recognition task, the ACCDIST metric assigned a held-out speaker to the correct accent group 92.3% of the time. The ACCDIST metric had 25% fewer errors than the best linear discriminant analysis using normalised formant frequencies. Also ACCDIST gave a cleaner clustering of speakers into accent groups.

There are many ways in which this work could be extended. One particular area would be to extend measurements to consonants as well as vowels, and to pitch and timing as well as to segmental quality. Fortunately, the inter-segment distance tables used by ACCDIST can readily incorporate such information.

Acknowledgements. I would like to thank Nobuaki Minematsu for introducing me to the idea of speaker-dependent phone clustering; and Paul Iverson for suggesting correlation as a suitable similarity measure. Thanks to 2020Speech Ltd for making the ABI corpus available for research purposes.

References

1. Wells, J.C.: Accents of English. Cambridge University Press, Cambridge (1982)
2. Cruttenden, A.: Intonation, 2nd edn. Longman, London (1997)
3. Labov, W.: Principles of Linguistic Change: Internal Factors, vol. 1. Blackwell, Oxford (1994)
4. Adank, P., Smits, R., van Hout, R.: A comparison of vowel normalization procedures for language variation research. J. Acoust. Soc. Am. 116(5), 3099–3107 (2004)
5. Davis, S.B., Mermelstein, P.: Comparison of parametric representations for monosyllabic word recognition in continuously-spoken sentences. IEEE Trans. Acoustics, Speech and Signal Processing 28, 357–366 (1980)
6. Minematsu, N.: Pronunciation assessment based upon the phonological distortions observed in language learner's utterances. In: Proc. International Conference on Spoken Language Processing, Korea, pp. 1669–1672 (2004)
7. Huckvale, M.: ACCDIST: a metric for comparing speakers' accents. In: Proc. International Conference on Spoken Language Processing, Korea, pp. 29–32 (2004)
8. D'Arcy, S.M., Russell, M.J., Browning, S.R., Tomlinson, M.J.: The Accents of the British Isles (ABI) Corpus. In: Proc. Modélisations pour l'Identification des Langues, MIDL Paris, pp. 115–119 (2004)

9. Barry, W.J., Heoquist, C.E., Nolan, F.J.: An approach to the problem of regional accent in automatic speech recognition. Computer Speech and Language 3, 355–366 (1989)
10. Ito, M., Tsuchida, J., Yano, M.: On the effectiveness of whole spectral shape for vowel perception. J. Acoust. Soc. Am. 110, 1141–1149 (2001)
11. Bladon, R.A., Lindblom, B.: Modeling the judgement of vowel quality differences. J. Acoust. Soc. Am. 69, 1414–1422 (1981)
12. Holmes, J.N.: The JSRU channel vocoder. In: Proceedings IEE, Pt. F, vol. 127, pp. 53–60 (1980)

Selecting Representative Speakers for a Speech Database on the Basis of Heterogeneous Similarity Criteria

Sacha Krstulović[1,*], Frédéric Bimbot[1], Olivier Boëffard[2], Delphine Charlet[3], Dominique Fohr[4], and Odile Mella[4]

[1] IRISA/METISS, Campus de Beaulieu, 35 042 Rennes Cedex, France
sacha@dfki.de, olivier.boeffard@univ-rennes1.fr
[2] IRISA/CORDIAL, 6 r. Kerampont, BP 80518, 22 305 Lannion Cedex, France
bimbot@irisa.fr
[3] France Télécom R&D, 2 ave. Marzin, 22 307 Lannion, France
delphine.charlet@francetelecom.com
[4] LORIA, Campus Universitaire, BP239, 54 506 Vandoeuvre Cedex, France
{dominique.fohr,odile.mella}@loria.fr

Abstract. In the context of the NEOLOGOS French speech database creation project[1], a general methodology was defined for the selection of representative speaker recordings. The selection aims at providing a good coverage in terms of speaker variability while limiting the number of recorded speakers. This is intended to make the resulting database both more adapted to the development of recently proposed multi-model methods and less expensive to collect.

The presented methodology proposes a selection process based on the optimization of a quality criterion defined in a variety of speaker similarity modeling frameworks. The selection can be achieved with respect to a unique similarity criterion, using classical clustering methods such as Hierarchical or K-Medians clustering, or it can combine several speaker similarity criteria, thanks to a newly developed clustering method called Focal Speakers Selection.

In this framework, four different speaker similarity criteria are tested, and three different speaker clustering algorithms are compared. Results pertaining to the collection of the NEOLOGOS database are also discussed.

Keywords: speech database,minimization,speaker selection, speaker clustering, optimal coverage, multi-models, speech and speaker recognition, speech synthesis.

1   Introduction

General goals – The state of the art techniques in the various domains of Automatic Speech Processing (be it for Automatic Speaker Recognition,

---

* Now at DFKI GmBH, Stuhlsatzenhausweg 3, D-66123 Saarbrcken, Germany.
[1] The NEOLOGOS project was funded by the French Ministry of Research in the framework of the TECHNOLANGUE program.

C. Müller (Ed.): Speaker Classification II, LNAI 4441, pp. 276–292, 2007.

Automatic Speech Recognition or Text-To-Speech Synthesis) make extensive use of speech databases. Nevertheless, the problem of optimizing the contents of these databases to make them adequate to the development of a considered speech processing task has seldom been studied (see e.g. [1]). The usual definition of speech databases consists in collecting a volume of data that is supposed sufficiently large to represent a wide range of speakers and a wide range of acoustic conditions [2,3]. Nevertheless, identifying and omitting some redundant data may prove more efficient with respect to the development and evaluation costs as well as with respect to the performances of the targeted system [1]. Alternately, the most recently developed speech recognition and adaptation algorithms tend to make use of several specialized models instead of a unique general model, and hence require an important volume of data to guarantee that the variability of speech is accurately modeled. Similarly, the most recent advances in Text-To-Speech synthesis (TTS) require the availability of a wide range of speakers to investigate the degradation of quality which is still noticeable in the synthetic voices. Hence, The above-mentioned developments require a much larger quantity of data per speaker than the traditional databases can offer (e.g. Speech-Dat[2]). Nevertheless, the increase in the collection cost for such newer and larger databases should be limited as much as possible.

Thus, the present work, partly led in the framework of the NEOLOGOS project[3], focuses on optimizing the contents of the speech databases in order to control the diversity of the recorded speech, both at the segmental and suprasegmental levels. In addition to this scientific objective, it addresses the practical concern of reducing the collection costs for new speech databases.

Proposed database design – The starting point of this work is to consider that the variability of speech can be decomposed along two axes, one of speaker-dependent variability and one of purely phonological variability. The classical speech databases [4,3] seek to provide a sufficient sampling of both variabilities by collecting few data over many random speakers (typically, several thousands). Alternatively, we propose to optimize explicitly the coverage in terms of speaker variability, prior to extending the phonetic coverage by collecting a lot of data over a reduced number of *reference speakers.*

In this framework, the reference speakers should come out of a selection process which guarantees that their recorded voices are non-redundant but keep a balanced coverage of the speech space. Thus, we propose to lead the collection of the corpus along a three stage process:

1. a bootstrap database is collected by recording a first set of many different speakers. This database should provide a wide and potentially redundant

---

[2] See http://www.elda.org/ for the specifications of the currently available Speech-Dat databases.

[3] The following public, academic and industrial partners have participated in the NEOLOGOS project, funded by the French Ministry of Research in the framework of the TECHNOLANGUE program: the ELDA agency, the ENSSAT lab, the France Telecom R&D company/lab, the IRISA lab, the LORIA lab and the Telisma company.

sampling of the speaker space, balanced in turn by a limitation of the quantity of recorded speech per speaker. To compensate for this limitation and to ensure an acceptable sampling of the phonological variability, the phonological coverage is maximized through a careful specification of the linguistic contents of the prompted text. For NEOLOGOS, 1,000 speakers were recorded over the fixed telephone network, and the recorded utterances were a set of 45 phonetically balanced sentences, optimized using techniques derived from TTS methods [5], identical for all the speakers and recorded in one call;

2. a reduced subset of reference speakers (200 for NEOLOGOS) is selected through a clustering of the voice characteristics of the bigger set of bootstrap speakers, and the reduction of the clusters to their most representative element;

3. the final database, comprising only the reference speakers' voices, is collected (for this project, the database of 200 speakers was called IDIOLOGOS). The reference speakers are requested to pronounce a larger corpus of specific sentences. For NEOLOGOS, the 200 reference speakers were requested to pronounce 450 sentences, identical for all the speakers, in 10 successive telephone calls that had to be completed in a short period of time to avoid shifts in the voice characteristics.

The use of clustering methods supposes the definition of a distance, or similarity metric, between the objects to be gathered and their prototype. In our context, the definition of speech clusters represented by a centroid speaker relies on the definition of some similarity metrics between speaker voices. The choice of a relevant speaker similarity metrics has to be made *a priori* among the range offered by the state of the art speech and speaker recognition techniques, since a direct optimization of this choice is infeasible in practice. To avoid a restriction of this choice to a particular *a priori* metrics, we define a framework where several sets of reference speakers can emerge from various speaker similarity metrics, and where the extracted sets can be cross-validated with respect to a metrics different from their metrics of origin (this is detailed in section 2). This approach preserves a diversity of criteria for the selection of reference speakers, and it allows to combine these criteria in order to keep a certain level of generality in the definition of the speech coverage brought by the reference speakers.

Working hypotheses, claims and limitations – As mentioned above, the context of this work (the NEOLOGOS project) is limited by a strong practical constraint : a large number of speakers, namely 1000, are recorded in one single session to set up the bootstrap database. From this initial step, a fraction of this population, namely 200 "typical" speakers, are selected for a complete data collection of 50 sessions, and only these speakers are recorded extensively. This prevents a full-loop evaluation of the proposed speaker selection method, which ideally would consist in comparing the performance of speech applications trained with the 50 recordings of the 200 typical speakers as opposed to those obtained with the 50 recordings from random selections of 200 speakers. As a result of this situation, the work and observations reported in this chapter can not be used to support the universal effectiveness of the proposed methodology, nor the absolute superiority of the NEOLOGOS/IDIOLOGOS corpus thus obtained.

However, the approach adopted and the results reported in this chapter suggest a number of relevant properties of the proposed framework. They are bound to contribute and be reused in other contexts of speech-corpus construction, especially in what concerns:

- the formulation of the "loss of quality" of a sub-corpus with respect to a reference corpus;
- the definition and relevance of several metrics between speech utterances;
- a number of approaches for speech utterances clustering and selection;
- the principle and consistency of the "focal speakers" reference selection approach;
- a framework for combining heterogeneous criteria for utterance selection.

It is also important to note that the present work is based on the assumption that one recording provides an accurate representation of a speaker's voice, which is clearly a simplistic view with respect to reality.

Overview of the chapter – Section 2 exposes our speaker selection methodology and the design of the related corpus. Section 3 proposes and discusses some particular instances of speaker similarity metrics. Section 4 presents some clustering methods for the selection of the reference speakers. Section 5 discusses an evaluation of the clustering methods in the framework of the NEOLOGOS database, while section 6 is dedicated to some conclusions and perspectives.

2   Methodology and Corpus

This section exposes the different steps in the formulation of our general approach, applicable to any database, and ends with more specific details about the linguistic contents of NEOLOGOS.

Reference speakers – Let M be a large number of speakers x_i, $i = 1, \cdots, M$, among which we want to choose $N \ll M$ reference speakers. Let $L = \{x_j; j = 1, \cdots, N\}$ be a given list of N speakers x_j.

Let $d_A(x_i, x_j)$ be a function able to measure the distance, or dissimilarity, between two speakers x_i and x_j. $d_A(\cdot, \cdot)$ depends upon a similarity modeling method A. The lower the distance, the more representative x_j is of x_i in the sense of the similarity modeling method A.

Let $\mathrm{ref}_A(x_i|L)$ be a function able to find out, among the list L, the reference speaker which best represents the speaker x_i in the similarity modeling framework A. Given the above definitions, it can be obtained as:

$$\mathrm{ref}_A(x_i|L) = \arg \min_{x_j \in L} d_A(x_i, x_j) \qquad (1)$$

Quality of a list of reference speakers – Given the ability to represent every speaker x_i of the initial set by a reference speaker issued from a given list L, then:

$$Q_A(L) = \sum_{i=1}^{M} d_A\left(x_i, \mathrm{ref}_A(x_i|L)\right) \tag{2}$$

measures the total cost, or total loss of quality[4], that occurs when replacing each of the M initial speakers by their reference among the N speakers listed in L, according to the similarity modeling method A. The smaller this total loss, the more representative the reference list.

For instance, if the distance $d_A(x_i, x_j)$ is defined as a crossed likelihood between the model of speaker x_i and the data of speaker x_j, then $Q_A(L)$ measures the loss of likelihood that occurs when replacing each speaker x_i with its reference speaker $\mathrm{ref}_A(x_i|L)$. Nevertheless, this interpretation goes beyond the likelihood-based distances: for any distance $d_A(\cdot, \cdot)$, $Q_A(L)$ measures a loss of information, or loss of diversity, relevant to the characteristics of speaker variability modeled by $d_A(x_i, x_j)$.

Optimal list of reference speakers – In turn, finding the optimal list L^A of reference speakers with respect to the similarity modeling method A translates as:

$$L^A = \arg\min_{L} Q_A(L) \tag{3}$$

Due to the dimensions of the databases, solving this optimization problem by an exhaustive search across all the possible combinations of N speakers taken among M speakers is infeasible in practice, due to the huge number of combinations:

$$C_M^N = \binom{M}{N} = \frac{M!}{N!(M-N)!} \tag{4}$$

Nevertheless, clustering methods such as Hierarchical Clustering or K-means clustering can provide locally optimal solutions. The application of these clustering methods to our problem will be developed in section 4.

Comparison of reference lists – Within equation (2), the quality of any reference list L can be measured. In particular, L can be a list L^B issued from an optimization in a similarity modeling framework B:

$$L^B = \arg\min_{L} Q_B(L) \tag{5}$$

In this case, the reference speakers can be attributed from L^B *with respect to an alternate similarity modeling framework* A:

$$\mathrm{ref}_A(x_i|L^B) = \arg\min_{x_j \in L^B} d_A(x_i, x_j) \tag{6}$$

---

[4] Although $Q_A(L)$ actually denotes the loss of quality, it will be mostly referred to as 'quality' in the rest of the chapter, for the sake of simplicity.

It follows that the quality of a selection of reference speakers L^B made in the framework of the similarity modeling method B can be evaluated in the scope of the similarity modeling method A:

$$Q_A(L^B) = \sum_{i=1}^{M} d_A(x_i, \mathrm{ref}_A(x_i|L^B)) \tag{7}$$

This case illustrates the fact that the quality defined by equation (2) brings a general answer to the problem of comparing various reference lists, even when the reference lists are issued from selections made with respect to different speaker similarity criteria. Defining the similarity of the lists in the space of the qualities is more general than trying to implement a direct comparison of the lists' contents. For example, if the method B would form a list L^B by replacing every speaker of a list L^A by its nearest neighbor in the sense of A, then L^A and L^B would have no speaker in common, whereas L^B may still have a good quality in the sense of A.

Moreover, the cross-quality defined by equation (7) suggests a way to find out reference speakers which satisfy a combination of similarity criteria. Let $\mathcal{L}^A = \{L_k^A;\ k = 1, \cdots, K\}$ be a set of K lists of reference speakers which keep a good quality with respect to the similarity modeling method A. Then it is possible to find a set of speakers $L^{A|B}$ so that:

$$L^{A|B} = \arg \min_{L \in \mathcal{L}^A} Q_B(L) \tag{8}$$

This corresponds to finding the best list, in the sense of the similarity criterion B, among a set of lists which are good in the sense of A. For a sufficiently large K, the list $L^{A|B}$ represents a compromise of quality across the similarity criteria A and B (in a non-commutative way), though it may represent a sub-optimal solution within each of these individual frameworks.

The formulation of the loss of quality based on a framework A assumes that A involves a unique type of metrics between speakers. Therefore, L^A refers to a list optimized with respect to the loss of quality Q_A (involving the similarity modeling framework A) and $L^{A|B}$ refers to a list optimized with respect to the loss of quality Q_B (involving the framework B, among lists that are "good" with respect to A). Conversely, regarding the notations, the sets $\mathcal{L}^X$ can contain any type of lists, possibly issued from optimizations in several frameworks. Hence, the superscripts X of the sets $\mathcal{L}^X$, in this passage and the rest of the chapter, will refer to the contents of the set and will not necessarily map back to a unique similarity modeling framework. (See, e.g., note 5 below.)

Calibration of the measure of quality – For the quality of a reference speaker selection to be interpretable and comparable across several modeling criteria, it is necessary to *calibrate* it. This is done by ranking Q_A with respect to an estimate of the distribution of qualities, computed over a "big enough" selection of K random reference lists $\mathcal{L}^{\mathrm{rand}} = \{L_k^{\mathrm{rand}};\ k = 1, \cdots, K\}^5$. In a non-parametric framework, the values of $Q_A\,(\mathcal{L}^{\mathrm{rand}})$ are simply sorted in decreasing order, i.e., from the worst random list to the best. To evaluate a particular list L, we rank $Q_A(L)$ against the sorted qualities and divide the result by the total number of random lists. This normalized rank is called a Figure Of Merit (FOM) and is very easily interpretable: $\mathrm{FOM}_A(L) = 80\%$ means that the list L is better, in the framework of A, than 80% of the random lists in $\mathcal{L}^{\mathrm{rand}}$. The closer to 100%, the better the list.

Linguistic contents – For NEOLOGOS, Some financial and methodological considerations have suggested to limit the volume of the bootstrap database to 50 sentences per speaker, to be recorded in one call. The definition of the linguistic contents of these 50 sentences uses the following method: from a set of 50,000 short sentences (from 5 to 15 words) taken from the French newspaper "Le Monde", a greedy algorithm is applied to find a minimal subset which would cover 99 classes of diphones for at least two times [5]. The obtained set of 76 sentences is further reduced to 36 sentences through a manual sorting of the groups of words carrying the highest count for the diphones classes of interest, and their re-arrangement in a way which preserves natural semantics. Finally, 9 sentences which duplicate already represented diphone classes are added to fit TTS synthesis assessment purposes, and 5 utterances of various natures (numbers and spellings) are added for further control and testing purposes. The definition of the linguistic contents for the 450 sentences of the final NEOLOGOS/IDIOLOGOS database uses a generalized version of the same method, the main difference being that all the diphones can be covered; it will not be detailed in the present chapter. (See [6] for more detail.) A phonetic alignment of the speech data has been generated using standard HMM-based techniques (see [6]).

3   Modeling the Speaker Similarity

Many inter-speaker metrics have been studied in the context of clustering applications (e.g., [7], [8], [9], etc.; see [6] for more). For NEOLOGOS, we have chosen to apply four methods which focus on a variety of speech modeling aspects.

Canonical-Vowels metrics – The Canonical-Vowels (CV) metrics, defined as a Kullback-Leibler distance between Gaussian models of the canonical vowels, accounts for physiological differences between speakers, related to their vocal tract dimensions, in a maximum likelihood modeling framework.

In practice, speaker-dependent mono-Gaussian models are estimated for the three cardinal vowels /a/, /i/ and /u/, on the basis of Mel-Filterbank Cepstral Coefficients (MFCCs). Denoting by $p_i^\alpha = \mathcal{N}(\mu_i, \sigma_i)$ the mono-Gaussian model of a phoneme α for speaker x_i (in a D-dimensional space), the Kullback-Leibler

---

[5] Here, the rand superscript of set $\mathcal{L}^{\mathrm{rand}}$ does not refer to any particular similarity modeling framework. $\mathcal{L}^{\mathrm{rand}}$ contains uniformly distributed random lists.

divergence between the models of two speakers x_i and x_j can be expressed through the phoneme α as:

$$KL(p_i^\alpha \| p_j^\alpha) = \sum_{d=1}^{D} \frac{1}{2} \left(\log \frac{\sigma_j^{d2}}{\sigma_i^{d2}} + \frac{\sigma_i^{d2}}{\sigma_j^{d2}} + \frac{(\mu_i^d - \mu_j^d)^2}{\sigma_j^{d2}} - 1 \right) \tag{9}$$

and a symmetrical speaker dissimilarity can be defined as:

$$d_\alpha(x_i, x_j) = KL(p_i^\alpha \| p_j^\alpha) + KL(p_j^\alpha \| p_i^\alpha) \tag{10}$$

The distance $d_\alpha(x_i, x_j)$ can be computed for $\alpha = $ /a/, /i/, /u/, and a global distance d_{CV} can be defined as a simple sum of the phoneme-dependent distances:

$$d_{\mathrm{CV}}(x_i, x_j) = d_{/a/}(x_i, x_j) + d_{/i/}(x_i, x_j) + d_{/u/}(x_i, x_j) \tag{11}$$

Dynamic Time Warping metrics – The Dynamic Time Warping (DTW) metrics makes minimal modeling assumptions, provides a "direct" comparison of the speech signals, and is affiliated with classical speech recognition techniques.

In practice, the standard DTW algorithm is applied between the 160 breath groups making the original sentences, as determined by phonetician experts from the orthographic transcripts. Standard symmetrical left-to-right displacement plus matching boundaries constraints are applied to the DTW graph. The local distance that we use is the Hamming distance between feature frames based on cepstral coefficients:

$$d(Y(t_j), X(t_i)) = \sum_{c=1}^{N_c} |Y^c(t_j) - X^c(t_i)| \tag{12}$$

where N_c is the number of coefficients in the acoustic vector and $Y^c(t_j)$ is the cth coefficient of the cepstral features vector $Y(t_j)$. The total distance between two speakers is given by the average distance over correct pronunciations of the matching breath groups. Denoting by $\{\mathrm{bg}_k^i\}_{k=1,..,K}$ (resp. $\{\mathrm{bg}_k^j\}_{k=1,..,K}$) the set of $K = 160$ breath groups pronounced by speaker x_i (resp. x_j), and defining $\delta_k(i,j) = \left\{ 1 \text{ if } \mathrm{bg}_k^i \text{ and } \mathrm{bg}_k^j \text{ are correctly pronounced; } 0 \text{ otherwise} \right\}$ we have:

$$d_{\mathrm{DTW}}(x_i, x_j) = \frac{\sum_k \delta_k(i,j) \cdot D_{\mathrm{DTW}}(\mathrm{bg}_k^i, \mathrm{bg}_k^j)}{\sum_k \delta_k(i,j)} \tag{13}$$

where $D_{\mathrm{DTW}}(\mathrm{bg}_k^i, \mathrm{bg}_k^j)$ is the Hamming distance cumulated along the optimal path. Given the used displacement constraints, this distance is symmetrical.

Gaussian Mixture Models metrics – The Gaussian Mixture Models (GMM) metrics estimates a Kullback-Leibler distance between the GMM speaker models that are employed in state-of-the-art speaker recognition.

For this distance, speaker-dependent GMMs are trained using MAP adaptation [10] from a common Universal Background Model [11], a method currently acknowledged as state of the art in the domain of Automatic Speaker Identity Verification. For the GMMs, no analytic expression is available for the Kullback-Leibler distance, but it is possible to approximate it by using an unbiased estimator based on a suitable number of random observations, according to the Monte-Carlo method [12]:

$$KL(p_i \| p_j) = \frac{1}{N} \sum_{n=1}^{N} \log \frac{p_i(o_n^i)}{p_j(o_n^i)} \qquad (14)$$

where: p_i (resp. p_j) models the speech of speaker x_i (resp. x_j), in the form of a GMM; o_n^i (resp. o_n^j), $n = 1 \cdots N$, is a sample of speech data (or *observations*) spoken by speaker x_i (resp. speaker x_j) or synthesized by random draws according to the probability law modeled by p_i (resp. p_j). In our experiments, a limited number of synthetic MFCC vectors was used instead of the natural speech data, in order to speed up the computation of equation (14).

The $KL(p_i \| p_j)$ quantity measures the loss of likelihood of the data of speaker x_i with respect to the model of speaker x_j. This measure is not symmetrical, but the sum of $KL(p_i \| p_j)$ and $KL(p_j \| p_i)$ is; hence, a symmetrical speaker dissimilarity metric can be defined as:

$$d_{\text{GMM}}(x_i, x_j) = \frac{1}{N} \sum_{n=1}^{N} \log \frac{p_i(o_n^i)}{p_j(o_n^i)} + \frac{1}{N} \sum_{n=1}^{N} \log \frac{p_j(o_n^j)}{p_i(o_n^j)} \qquad (15)$$

where p_i and p_j correspond to the Gaussian mixture models of the speakers x_i and x_j.

Hidden Markov phoneme Models metrics – The Hidden Markov phoneme Models (HMM) metrics uses crossed likelihoods between complete sets of speaker-dependent, 3-states, left-right hidden Markov models of French phonemes, and is affiliated with state-of-the-art speech recognition.

This metrics is defined as:

$$
\begin{aligned}
d_{\text{HMM}}(x_i, x_j) = {} & \Lambda\left(\Omega_i \,|\, \Theta_i^{\text{HMM}}\right) - \Lambda\left(\Omega_i \,|\, \Theta_j^{\text{HMM}}\right) \\
& + \Lambda\left(\Omega_j \,|\, \Theta_j^{\text{HMM}}\right) - \Lambda\left(\Omega_j \,|\, \Theta_i^{\text{HMM}}\right)
\end{aligned}
\qquad (16)
$$

where $\Lambda\left(\Omega_i \,|\, \Theta_j^{\text{HMM}}\right)$ is a crossed likelihood between the data of x_i and the models of x_j, defined as: $\Lambda\left(\Omega_i \,|\, \Theta_j^{\text{HMM}}\right) = \sum_\alpha \frac{1}{\nu_i^\alpha} \Lambda\left(\omega_i^\alpha \,|\, \vartheta_j^\alpha\right)$, with: ω_i^α the set of frame sequences (or *observations*) corresponding to the phoneme α pronounced by the speaker x_i; ν_i^α the number of frames in the set ω_i^α; ϑ_j^α a 3-states, left-to-right hidden Markov model of the sequences of frames corresponding to the phoneme α, and estimated from the sample ω_i^α; Ω_i the phonetic data of any speaker x_i, and $\Theta_j^{\text{HMM}} = \{\vartheta_j^\alpha, \forall \alpha\}$ the set of all the phoneme models for a speaker x_j. This distance is symmetrical by definition.

Each metric has been developed independently by the labs participating in this work, according to their know-how and on the basis of their standard tools.

In particular, the choice of the acoustic features used in each metric is related to the observed performances of the local systems. These features are mostly based on Mel-scale cepstra, which are known to provide a good model for the speech spectral envelope, often followed by Cepstral Mean Subtraction, which is known to compensate roughly for some channel distortions. Greater detail about the implementation of each of these metrics can be found in [6].

4   Speaker Clustering and Selection

Finding a global optimum by an exhaustive search among every possible combination of speakers is not tractable in practice, due to the high number C_M^N or $\binom{M}{N}$ of possible combinations (e.g., for NEOLOGOS: $C_{1000}^{200} = \binom{1000}{200} = 6.6172 \cdot 10^{215}$). Nevertheless, this optimization problem can be understood as a clustering task. Classical clustering algorithms, able to find locally optimal solutions, include the K-Means algorithm (or a K-Medians variant) and the Hierarchical Clustering algorithm. In addition, we propose a new method called the Focal Speakers selection.

The K-Means/K-Medians algorithm − The K-Means algorithm [13] aims at grouping data in classes by locally minimizing the following criterion:

$$Q = \sum_{i=1}^{M} d(x_i, \mathrm{ref}(x_i|C)) = \sum_{n=1}^{N} \left(\sum_{x_i \in \mathcal{C}_n} d(x_i, c_n) \right) \tag{17}$$

where C is a list of N classes $\mathcal{C}_n$ in which the data x_i will be clustered, and $c_n = \mathrm{ref}(x_i|C)$ indicates the position of the centroid which abstracts the class $\mathcal{C}_n$. In our framework, the centroids have to be ultimately assimilated to real speakers. Besides, if this assimilation is made at each iteration, a lot of computation can be saved, because the distances between the centroids and the speakers can be read from a pre-computed matrix of inter-speaker distances. The corresponding discretized version of the K-Means algorithm, called the K-Medians, uses the following steps:

1. computation of the matrix of speaker similarities for the considered modeling method (CV, DTW or GMM);
2. random initialization, by a uniform draw of N reference speakers among the $M > N$ initial speakers;
3. assignation of each speaker to the cluster characterized by the closest reference speaker;
4. for each new cluster, determination of the reference speaker as the median speaker, i.e. the one for which the sum of the distances to every other speaker in the cluster is minimum:

$$c_n = \arg \min_{x_j \in \mathcal{C}_n} \sum_{x_i \in \mathcal{C}_n} d_A(x_i, x_j) \tag{18}$$

5. iteration of steps (3) and (4) until the N clusters stabilize.

At step 3, the assignation is done so that each of the $d\left(x_i, \text{ref}(x_i | C)\right)$ terms of the sum in equation (17) diminishes or stays the same; then, at step 4, the upgrade of c_n for each class C_n minimizes the $\sum_{x_i \in C_n} d(x_i, c_n)$ term explicitly, so that the second expression in equation (17) is further minimized. Therefore, the final solution will get a quality better than or equal to that of the list used for the initialization at step 2.

As a matter of fact, the result of the K-Medians is very dependent on the initialization, and the degree of quality of a locally optimal solution is undefined a priori. A solution consists in realizing a great number of runs of the algorithm, with different initializations, and to keep the local solution which reaches the best quality.

Hierarchical clustering – The Hierarchical Clustering algorithm [13] proceeds by establishing a typology of the data which can be described by a tree, or *dendrogram*, where each node describes a group of observations, characteristic of a particular class of data. The building of the tree can be operated in two manners:

– *agglomerative hierarchical clustering*: The classes described in the parent nodes are determined by merging the characteristics defined in the child nodes. The nodes to merge are chosen so that they minimize the following criterion:

$$\Delta(\Theta_i, \Theta_j) = \sum_{x_k \in \pi_{i \cup j}} d_A\left(x_k, \Theta_{i \cup j}\right) - \sum_{x_k \in \pi_i} d_A\left(x_k, \Theta_i\right) - \sum_{x_k \in \pi_j} d_A\left(x_k, \Theta_j\right) \quad (19)$$

where π_j is the population of the cluster/node represented by Θ_j, and $\pi_{i \cup j}$ is the union of the π_i and π_j populations. It can be shown that this criterion corresponds to a direct optimization of the quality Q_A within the constraints of the dendrogram construction. After each merge, a new representative speaker is chosen as the centroid of the merged population.

– *divisive hierarchical clustering*: The child nodes inherit from the characteristics of their parent, but are further divided so that they refine the taxonomy of the data. The node to divide is chosen so that it minimizes the criterion (19). For each node splitting, the speaker assignments and the centroids for the two child nodes are determined by the local application of a 2-classes K-Medians on the population of the parent node. Since this K-Medians is repeated over all the parent nodes to minimize the criterion (19), the divisive version is significantly heavier than the agglomerative one.

In any case, the tree-building procedure is stopped when the number of nodes reaches the requested number of clusters (200 for NEOLOGOS). The list of reference speakers obtained by the Hierarchical Clustering procedure can be used as an initialization for the K-Medians.

The Focal Speakers method – This method is based on empirical considerations. It starts from the hypothesis that speaker subsets with a good quality are

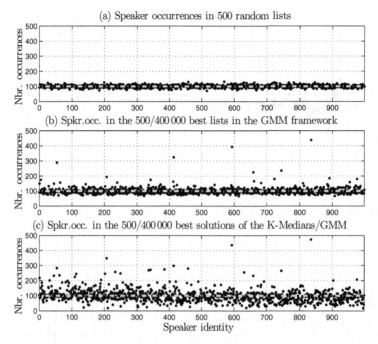

Fig. 1. Number of speaker occurrences for various compositions of $\mathcal{L}_{500}$

more likely to contain some speakers of the global optimum. If this hypothesis is true, the reference speakers of the global optimum should appear more often than others in a set $\mathcal{L}_K = \{L_k; \ k = 1, .., K\}$ made of a union of locally good speaker lists. To verify this, we computed the number of occurrences of each of the M initial speakers x_i among: (a) K random lists of N speakers; (b) the K best lists of a great number of random lists; (c) the K best lists among the solutions given by a great number of runs of the K-Medians.

The results are depicted in figure 1, for lists of $N = 200$ speakers taken among $M = 1000$ speakers, and with $\mathcal{L}_K$ gathering $K = 500$ lists taken from $400\,000$ initial lists. The number of occurrences of each speaker (black dots) is compared to the expected number $K \times N/M = 100$, corresponding to a uniform draw of 200 speakers among 1000. The figure shows that some speakers appear more often than the average across the series of lists characterized by their locally good quality. This suggests that there is a correlation between the quality of the lists and the fact that they contain some particular reference speakers.

Reverting this idea, we have studied if the N most frequent speakers in a set of lists characterized by their good quality would correspond to a good selection of reference speakers. Let $\mathcal{L}_K = \{L_k; \ k = 1, .., K\}$ be a set of speaker lists L_k, and $\delta_k(i) = \{1 \ \text{if speaker } x_i \in L_k; 0 \ \text{else.}\}$. The number of times the speaker x_i appears in $\mathcal{L}_K$ is therefore defined as:

$$\mathrm{Freq}(\,i\,|\mathcal{L}_K) = \sum_{L_k \in \mathcal{L}_K} \delta_k(i) \tag{20}$$

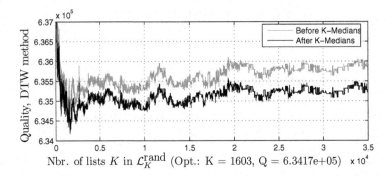

Fig. 2. Quality of the lists of Focal Speakers as a function of the number K of best random lists in $\mathcal{L}_K^{\mathrm{rand}}$, in the DTW modeling framework

Then, the speakers corresponding to the N highest values of $\mathrm{Freq}(\,i\,|\mathcal{L}_K)$ can be selected to constitute a list of so called Focal Speakers. This list will be noted $L_{\mathrm{foc}}(\mathcal{L}_K)$. Its quality $Q_A(L_{\mathrm{foc}}(\mathcal{L}_K))$ can be computed from various sets $\mathcal{L}_K$ of "good lists", and, in particular, the set $\mathcal{L}_K^{\mathrm{rand}}$ containing the K best of 400 000 random lists or the set $\mathcal{L}_K^{\mathrm{kmed}}$ containing the K best of 400 000 K-Medians results.

Figure 2 illustrates the evolution of $Q_{\mathrm{DTW}}(L_{\mathrm{foc}}(\mathcal{L}_K^{\mathrm{rand}}))$ versus the number K of lists in $\mathcal{L}_K^{\mathrm{rand}}$ (gray curve). The quality of the best list in $\mathcal{L}_K$ corresponds to $K = 1$. Lower quality values mean better lists: for every value of K, $L_{\mathrm{foc}}(\mathcal{L}_K)$ has a better quality than the best list in $\mathcal{L}_K$. Similar results have been observed in the other modeling frameworks than DTW, as well as with $\mathcal{L}_K^{\mathrm{kmed}}$. The most frequent speakers have been called *Focal Speakers* because they seem to concentrate the quality of the lists gathered in $\mathcal{L}_K$. The lists of focal speakers obtained for each K can be used to initialize additional runs of K-Medians. The resulting additional gain of quality is represented by the black curve.

The Focal Speakers approach naturally suggests a *joint optimization for the four speaker similarity modeling frameworks*. One can search the focal speakers among a set $\mathcal{L}_{K\times\mathrm{CV+DTW+GMM+HMM}}^{\mathrm{rand}}$ or a set $\mathcal{L}_{K\times\mathrm{CV+DTW+GMM+HMM}}^{\mathrm{kmed}}$ formed by gathering the K best lists obtained in CV, DTW, GMM and HMM. The corresponding results will be given in the next section.

5   Evaluation and Comparison of the Clustering Methods

Figure 3 compares the solutions of the various speaker selection algorithms for each of the four separate modeling frameworks:

– the gray Gaussian is the density of quality of 400 000 random lists. The black Gaussian is the density of quality for the related 400 000 K-Medians solutions. The short flag (at level 0.5) indicates the position of the best K-Medians solution (the lower the abscissa value, the better the quality);
– the medium sized flags (at level 1) indicate the solutions of the Hierarchical Clustering (HC) in the agglomerative case (solid); the divisive HC has been found

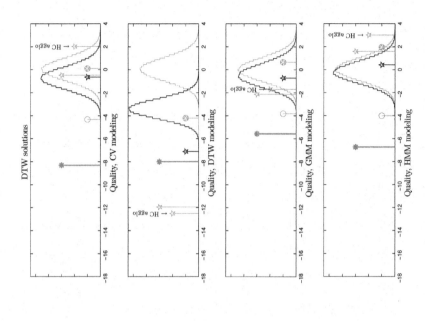

Fig. 4. Solutions optimized in DTW and evaluated in the context of CV, DTW, GMM and HMM, plus solutions corresponding to the Focal Speakers optimized in the joint frameworks

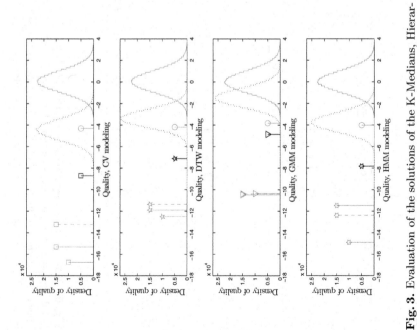

Fig. 3. Evaluation of the solutions of the K-Medians, Hierarchical Clustering and Focal Speakers methods, optimized in each separate frameworks

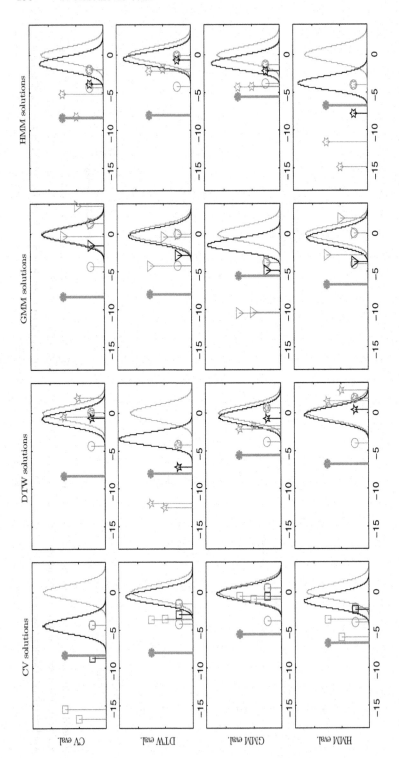

Fig. 5. Matrix of cross-evaluation of the K-Medians (small black flag & black histogram), agglomerative HC (medium) and Focal Speakers (tall) solutions obtained in the CV (□), DTW (✩), GMM (∇) and HMM (✩) frameworks. The quality of the Focal Speakers obtained from $\mathcal{L}^{kmed}_{500\times CV+DTW+GMM+HMM}$ is indicated as the bold gray line with the * marker. The reference random lists are indicated by the gray histogram and gray O marker.

to perform consistently worse than the agglomerative HC and its solutions have not been depicted here (see [6]);
– the taller flags (at level 1.5) indicate the solutions of the Focal Speakers method for the optimal K in $\mathcal{L}_K^{\mathrm{rand}}$ (dashed) and $\mathcal{L}_K^{\mathrm{kmed}}$ (solid).

The Focal Speakers method reaches qualities comparable to the agglomerative HC, both with the best random lists and with the best K-Medians. The optimal lists from FS and HC outperform the results of the K-Medians; however, they can be used to initialize a subsequent pass of K-medians which can push the quality a bit further (see [6]).

Figure 4 considers the solutions optimized in the DTW framework, and evaluates them in the context of the alternate similarity modeling methods. It shows that an optimal quality in a given modeling framework does not necessarily guarantee a good quality in the other ones. For example, the solution brought by the agglomerative HC applied in the DTW framework (distinctly marked in the figure) is the best in its framework of origin, but has a low quality with respect to CV, GMM and HMM modeling. The bold gray flags indicate the quality of $L_{\mathrm{foc}}(\mathcal{L}_{K \times \mathrm{CV+DTW+GMM+HMM}}^{\mathrm{kmed}})$ for $K = 500$ lists in each framework (2000 lists in total). As opposed to the previous cases, the quality of this optimal list of focal speakers is consistently good across all the frameworks. Besides, this list could be used to initialize an additional K-medians in each of the separate frameworks, giving 4 more lists with an even better (or same) quality in each framework (not depicted; see [6]). Using $K = 500$ lists for the FS method is, for the moment, an ad-hoc choice. We have observed that it does not influence so much the quality of the result: taking the 1000 best lists of each framework gave comparable results. Nevertheless, more elaborate ways to compose $\mathcal{L}_{\mathrm{CV+DTW+GMM+HMM}}$ could be studied.

Figure 5 generalizes figure 4 to all the cases of match or mismatch between the optimization context and the evaluation context. It shows that the Focal Speakers reach a consistently good solution across all the combinations.

6   Summary, Conclusions and Perspectives

In the context of the NEOLOGOS project, we have built a methodology for selecting a subset of reference speaker recordings, so as to keep a diversity of voices with respect to a variety of similarity criteria.

After formulating a FOM measure for quantifying the loss of quality of a subcorpus with respect to a reference corpus, we have defined and experimented a number of similarity metrics between speakers, that can be used in a clustering stage to partition the reference recordings into a predetermined number of classes. The concept of focal speaker has been introduced and experimented as a means to obtain a consistent and potentially robust set of selected speakers, making it easy to combine the results of several clustering processes (with different similarity metrics).

Given the constraints of the NEOLOGOS project, a validation of the method in terms of its impact on speech recognition error rates was not feasible. Nevertheless, this work contributes to the important area of corpus design by presenting a general framework, making explicit a well-defined methodology and associated tools, together with results that suggest its relevance and may encourage the investigation of similar approaches in other contexts.

References

1. Nagorski, A., Boves, L.: Steeneken: Optimal selection of speech data for automatic speech recognition systems. In: ICSLP, pp 2473–2476 (2002)
2. Lippmann, R.: Speech recognition by machines and humans. Speech Communication 22(1), 1–15 (1997)
3. Iskra, D., Toto, T.: Speecon - speech databases for consumer devices: Database specification and validation. In: LREC, pp. 329–333 (2002)
4. Nakamura, A., Matsunaga, S., Shimizu, T., Tonomura, M., Sagisaka, Y.: Japanese speech databases for robust speech recognition. In: Proc. ICSLP'96. Philadelphia, PA, vol. 4, pp. 2199–2202 (1996)
5. François, H., Boëffard, O.: Design of an optimal continuous speech database for text-to-speech synthesis considered as a set covering problem. In: Proc. Eurospeech'01 (2001)
6. Krstulović, S., Bimbot, F., Boëffard, O., Charlet, D., Fohr, D., Mella, O.: Optimizing the coverage of a speech database through a selection of representative speaker recordings. Speech Communication 48(10), 1319–1348 (2006)
7. Padmanabhan, M., Bahl, L., Nahamoo, D., Picheny, M.: Speaker clustering and transformation for speaker adaptation in speech recognition system. IEEE Transactions on Speech and Audio Processing 6(1), 71–77 (1998)
8. Johnson, S., Woodland, P.: Speaker clustering using direct maximisation of the MLLR-adapted likelihood. In: ICSLP. vol. 5(98), pp. 1775–1779
9. Naito, M., Deng, L., Sagisaka, Y.: Speaker clustering for speech recognition using vocal tract parameters. Speech Communication 36(3-4), 305–315 (2002)
10. Gauvain, J., Lee, C.: Maximum a posteriori estimation for multivariate gaussian mixture observations of markov chains. IEEE Transactions on Speech and Audio Processing 2(2), 291–299 (1994)
11. Reynolds, D.A., Quatieri, T., Dunn, R.: Speaker verification using adapted gaussian mixture models. Digital Signal Processing 10(1-3), 19–41 (2000)
12. Ben, M., Blouet, R., Bimbot, F.: A Monte-Carlo method for score normalization in Automatic Speaker Verification using Kullback-Leibler distances. In: Proc. ICASSP 2002 (2002)
13. Duda, R.O., Hart, P.E., Stork, D.G.: Pattern Classification. John Wiley and Sons, New York (2001)

Speaker Classification by Means of Orthographic and Broad Phonetic Transcriptions of Speech

Christophe Van Bael and Hans van Halteren

Centre for Language and Speech Technology,
Radboud University Nijmegen, The Netherlands
{c.v.bael, h.v.halteren}@let.ru.nl

Abstract. In this study we investigate whether a classification algorithm originally designed for authorship verification can be used to classify speakers according to their gender, age, regional background and level of education by investigating the lexical content and the pronunciation of their speech. Contrary to other speaker classification techniques, our algorithm does not base its decisions on direct measurements of the speech signal; rather it learns characteristic speech features of speaker classes by analysing the orthographic and broad phonetic transcription of speech from members of these classes. The resulting class profiles are subsequently used to verify whether unknown speakers belong to these classes.

Keywords: Speaker Classification, Linguistic Profiling, Orthographic Transcriptions, Broad Phonetic Transcriptions.

1 Introduction

Human listeners can rely on multiple modalities to determine a speaker's gender, age, regional background and -be it with less confidence- his or her level of education. Visual as well as auditory input can provide us with cues about a speaker's gender and age. In addition, auditory input can teach us a great deal about a speaker's regional background and his or her level of education.

The aim of our study was to investigate whether Linguistic Profiling, a supervised learning classification algorithm originally designed for authorship verification [1], can also be used to classify speakers according to their gender, age, regional background and level of education by investigating the lexical content and the pronunciation of their speech. Our procedure differs from conventional procedures for speaker classification in that our algorithm analysed written representations of speech rather than the speech signal proper; it analysed orthographic and broad phonetic transcriptions of speech to identify regularities in the use of words and the pronunciation of speakers of different genders, ages, regional backgrounds and levels of education. These regularities were subsequently combined into feature sets: one set of features describing the use of words as reflected in the orthographic transcription, and a second set of features describing the pronunciation characteristics as reflected in

C. Müller (Ed.): Speaker Classification II, LNAI 4441, pp. 293–307, 2007.
© Springer-Verlag Berlin Heidelberg 2007

the broad phonetic transcription. These feature sets were used to accept or reject unknown speakers as members of speaker classes that were defined in terms of the four aforementioned speaker characteristics. Since we wanted to study the performance of the algorithm with the individual features sets, the algorithm worked with one feature set at a time. The performance of the algorithm was evaluated through a comparison of its classification of unknown speakers with the information on the speakers as provided in the meta-data of the speech material.

This chapter is organised as follows. In Section 2, we describe the corpus material and the transcriptions. In Section 3, we describe the classification algorithm, the definition of speaker classes, the two sets of classification features and our general experimental setup. Subsequently, in Section 4, we present and discuss the results of the classification experiments, and in Section 5 we present our conclusions and our plans for future research.

2   Corpus Material and Transcriptions

We conducted our classification experiments with transcriptions of spontaneous telephone dialogues in the Spoken Dutch Corpus (Corpus Gesproken Nederlands, CGN), a 9-million word corpus comprising standard Dutch speech from native speakers in the Netherlands and in Flanders [2]. We considered recordings of telephone dialogues between speakers from the Netherlands only. These recordings were separated into two samples each (one sample per speaker). After excluding dialogues for which the meta-data were incomplete as far as relevant for our classification variables (see Section 3.2), and after excluding samples of which large parts were tagged as unintelligible in the orthographic transcription, we counted 663 samples from 340 different speakers. These samples ranged from 321 to 2221 words in length and comprised a total of 689,021 word tokens.

In addition to the words in the orthographic transcription, we also considered their part-of-speech tags. The orthographic transcription of the words in the CGN was created fully manually, the part-of-speech tags were generated automatically and manually corrected afterwards [2].

In order to study pronunciation characteristics we needed a canonical representation of the words in the orthographic transcription (i.e. the written representation of the standard pronunciation of the words in isolation from the context of neighbouring words [3]) and a broad phonetic transcription reflecting their actual pronunciation in the speech recordings. We generated a canonical representation of each recording by substituting every word in the orthographic transcription with its representation in a canonical pronunciation lexicon. The broad phonetic transcription was generated automatically because the CGN provides a manually verified phonetic transcription of only 115,574 out of the 689,021 words in our samples. We used an automatic transcription procedure which proved capable of closely approximating the manually verified phonetic transcription of the CGN [4]. In this procedure, the canonical representation of every utterance was first expanded into a network of alternative pronunciations. A continuous speech recogniser then chose the best matching phonetic transcription through forced recognition. In order to ensure the automatic generation of plausible phonetic transcriptions, we excluded speech utterances that, according to the orthographic transcription, contained non-speech, unintelligible speech, broken words and foreign speech. Samples containing overlapping speech were excluded as well. This resulted in automatic transcriptions for 252,274 out of 689,021 words, i.e. 136,700 words more than the 115,574 words for which the CGN could have provided a manually verified phonetic transcription.

3 Classification Methodology

3.1 Classification Algorithm

Linguistic Profiling is a supervised learning algorithm [1]. It first registers all classification features (e.g. pronunciation processes) that occur in at least N training samples (e.g. speech samples of individual speakers) of a corpus[1]. The algorithm then builds a 'profile' of each training sample by listing the number of standard deviations the count of each of the classification features deviates from the average count in the whole corpus. Subsequently, class-specific profiles are generated by averaging the profiles of all training samples from a specific speaker class (e.g. male speakers). The distance between the profile of a test sample and the profile of a given class of speakers is compared with a threshold value in order to determine whether the speaker of the sample should be attributed to that speaker class. The degree to which the distance does or does not exceed the threshold value indicates the confidence of the decision. We evaluated the algorithm's classification accuracy by comparing its decisions with the actual characteristics of the test speakers as provided in the metadata of the CGN. Since Linguistic Profiling is a verification algorithm, we measured its accuracy initially in terms of False Accept Rates (FARs) and False Reject Rates (FRRs). Since these values are threshold-dependent, however, we present a threshold-independent derivative instead, viz. the Equal Error Rate (EER), which is the value at which the FAR and the FRR are equal.

3.2 Classification Variables

We assigned the 663 selected samples to different classes according to the gender, age, regional background and level of education of the speakers.

The establishment of a male and a female speaker class was straightforward. We separated the samples into two classes: one class with 276 samples from 148 male speakers and another class with 387 samples from 192 female speakers.

All speakers were born between 1928 and 1981. We classified the speech samples age-wise according to two classification schemes. First, for every year, we generated a binary split of all speakers into those who were born in or before that year (e.g. $\leq$ 1955), and those who were born after that year (e.g. > 1955). This yielded classes with 24 to 639 samples from 11 to 329 speakers. In addition, for every year, we defined a class with subjects born within a symmetric eleven-year window around the target year (e.g. 1950 -1955- 1960). This yielded classes with 67 to 174 samples from 32 to 98 speakers.

---

[1] For each new classification task, the threshold (N) is empirically determined in order to keep the amount of information Linguistic Profiling has to deal with computable.

We retrieved the regional background of the speakers from the metadata of the CGN. We classified the speech samples in 16 classes according to the region speakers mainly lived in between the age of 4 and 16. Table 1 presents the distribution of samples per region. As a result of the large number of classes (we adhered to the original classification of the CGN), some classes contained only a few samples, in particular classes 2e (6), 3c (12) and 2f (13). However, since merging regional classes would probably have resulted in classes with more heterogeneous speech behaviour (which would probably have made speech from these classes harder to characterise and distinguish), we held on to the subdivision in 16 regional classes.

Table 1. Distribution of samples and speakers in terms of the speakers' regional backgrounds. From left to right: abbreviation, general geographical region in the Netherlands, specific geographical region, number of samples per class, number of individual speakers per class.

	general regions	specific geographical regions	sam	spk
1a	central	South Holland, excl. Goeree Overflakee	105	55
1b		North Holland, excl. West Friesland	112	50
1c		West Utrecht, incl. the city of Utrecht	21	12
2a	transitional	Zeeland, incl. Goeree Overflakee + Zeeland Flanders	42	21
2b		East Utrecht, excl. the city of Utrecht	42	19
2c		Gelderland river area, incl. Arnhem + Nijmegen	52	27
2d		Veluwe up to the river IJssel	19	14
2e		West Friesland	6	4
2f		Polders	13	4
3a	peripheral, North East	Achterhoek	18	10
3b		Overijssel	37	20
3c		Drenthe	12	7
3d		Groningen	17	11
3e		Friesland	20	10
4a	peripheral, South	North Brabant	113	60
4b		Limburg	34	16
			663	340

The metadata of the CGN also provided us with information on the level of education of the speakers. The speakers were tagged as having enjoyed higher education (university or polytechnic), secondary education or only primary education (no completed secondary education). In our samples, we counted 256 speakers who had enjoyed higher education, 75 speakers with secondary education and only 9 speakers with primary education. Because of the skewness of the distribution of speakers in these three classes, and because we didn't have reason to believe that the 9 subjects of the third class would heavily increase the heterogeneity in the large second class if we would merge these classes, we merged the 9 speakers of the third class with the 75 speakers of the second class. As a result, two speaker classes were established: highly educated subjects (256 speakers in 496 samples) and moderately educated subjects (84 speakers in 167 samples).

3.3 Classification Features

Per speaker class, the classification algorithm retrieved a set of lexical features from the orthographic transcription, and a set of pronunciation features from the broad phonetic transcription of the samples. The values of both feature sets were grouped into separate classification profiles modelling class-specific lexical use on the one hand and class-specific pronunciation characteristics on the other hand.

3.3.1 Lexical Features

The lexical features largely resembled the features that were used for the authorship verification experiments in [1]. This time, however, full syntactic analyses were not considered because the Amazon parser used in [1] has been developed for the analysis of written instead of spoken Dutch.[2] The lexical profiles represented the average utterance length in terms of number of word tokens, counts of uni-, bi-, and trigrams of words and the part-of-speech tags of the words. All counts were normalised for sample length by translating them to their frequency per 1000 tokens. In addition to these features, we tagged each utterance with information about the length, the linguistic status (declarative, interrogative and exclamatory, based on the punctuation marks) and the speaker (current speaker or interlocutor) of the preceding utterance. Only the features occurring in at least five samples were used. This led to a feature set of about 150.000 features potentially useful for classification.

3.3.2 Pronunciation Features

We characterised 'pronunciation features' in terms of the segmental differences between the canonical (standard) representation and the broad phonetic transcription of the words in the speech samples. We aligned the canonical and broad phonetic transcriptions with ADAPT, a dynamic programming algorithm designed to align strings of phonetic symbols according to their articulatory distance [5]. Figure 1 illustrates the alignment of a canonical (Can) and a broad phonetic transcription (PT), and the derivation of a pronunciation process: the deletion of schwa.

Can	l	d	@	l	A	p	@	l	l	v	A	l	t	l
PT	l	d	˜	l	A	p	@	l	l	f	A	l	t	l
Dutch	De				appel					valt				
English	The				apple					drops				

$$@ \rightarrow \emptyset \,/\, [\,l\,d\,] \underline{\quad} [\,l\,A\,p\,@\,l\,l\,v\,A\,l\,t\,l\,]$$

Fig. 1. Alignment of a canonical (Can) and a broad phonetic transcription (PT) and derivation of a pronunciation process (deletion of schwa). SAMPA symbols are used, word boundaries are marked as vertical bars.

---

[2] Part of the CGN is annotated for syntactic structure, but the amount of annotated data would have been insufficient to be of use for our experiments.

The segmental differences between a canonical and a broad phonetic transcription can be influenced by (at least) four main variables: the socio-situational setting (transcriptions of spontaneous speech typically yield more mismatches than transcriptions of prepared speech [6]), the consistency of the transcriptions (human transcribers may not always transcribe in a consistent manner [7]), the use of words (the use of specific words triggers specific pronunciation processes), and the pronunciation habits of the speakers (the focus of our study). Since we aimed at classifying speakers according to genuine speaker-specific pronunciation features only, we tried to filter out all pronunciation features that were due to the other three variables.

The first two variables (the socio-situational setting and the consistency of the transcriptions) were irrelevant for our study since we considered the transcriptions of speech uttered in one socio-situational setting only, and since the transcriptions were generated by a consistent automatic transcription procedure. This left us with one more variable to control: the lexical context in which the pronunciation processes occurred, which we modelled by means of the frequency of the current word and information about its context (its co-occurrence with surrounding words and the position of the word in the utterance).

We controlled for lexical context by means of a three-step procedure. First, we set up a ten-fold cross-validation training in which we consecutively built ten models for the impact of the lexical context on pronunciation, each time on the basis of 90% of the samples. The models represented the counts of all pronunciation processes observed in their canonical contexts. Next, we used each of these models in turn to predict the pronunciation in the left-out samples. For every canonical phone the pronunciation model predicted the probability of different phones being actually pronounced, considering all canonical contexts seen in the training material. As a final step, we compared the predicted pronunciation processes with the pronunciation processes observed in the automatic phonetic transcription. To this end, we counted the actual occurrences of all pronunciation processes in every sample, and for each process we calculated the difference between the predicted and the observed probability. These differences were considered to mainly indicate speaker-specific pronunciation processes, since these pronunciation processes were present *in addition to* the pronunciation processes that were predicted on the basis of the lexical context of the pronunciation processes. This additional variation was numerically represented as a feature vector of 94 numerical values, one for each of the 94 different pronunciation processes that were encountered in our material.

To investigate to what extent our approach was successful in removing the influence of the speakers' use of words on the observed pronunciation processes, we computed the Kullback-Leibler distance between the predicted and the observed pronunciation processes. This distance halved when the predictions were based on the observations of pronunciation processes in their lexical context instead of on the observations of the pronunciation processes by themselves, without considering their context. This leads us to believe that a significant part of the influence of the lexical context was indeed modelled by our method.

3.4 Experimental Setup

We organised our classification experiments as ten-fold cross-validations. To this end, we divided the 663 samples in ten mutually exclusive sample sets of comparable size. Each speaker occurred in one set only. Per classification variable (gender, age, regional background and level of education) and per feature set (lexical and pronunciation), we consecutively used nine sample sets to train the algorithm, and the remaining set to test the algorithm. Each time, Linguistic Profiling was trained and tested with a range of parameter settings. Upon completion, we considered the algorithm's accuracy at the parameter settings yielding the best performance over all ten folds in order to determine its performance ceiling.

After running our experiments, it became clear that the use of this Oracle approach had a negative consequence, in particular when we assessed the algorithm's performance for speaker classes with a small number of samples. We found that the EERs at the best performing parameter settings were lower than 50%, even when we attempted the classification of speakers in classes with randomly selected speakers. This is not surprising: there will always be some degree of variance around the expected value of 50% accuracy, and by selecting the best performing settings we are likely to end up with a score better than 50%. This effect grows stronger as the number of samples in the classification profiles becomes smaller.

In order to determine whether the algorithm's classification was above or below chance rate, we experimentally determined the mean and standard deviation of the algorithm's EER for the classification of speakers in *randomly* selected speaker classes of various sizes. When 300 or more random samples were used, we found a mean *random group EER* of 44% with a standard deviation under 2%. When our algorithm considered 50 to 100 random samples, we found a mean random group EER of 40% with a standard deviation of 3%. When smaller groups of random samples were considered, the mean random group EER gradually decreased while its standard deviation increased.

To facilitate the interpretation of the classification results in the upcoming sections, we compare each EER with the expected distribution of the random group EERs. We mark each EER with one asterisk if the probability that it belongs to the distribution of the random group EERs is smaller than 0.05, with two asterisks when $p<0.01$ and with three asterisks when $p<0.001$. In all cases where $p<0.05$, we will call the classification *"effective"*. Since both the expected EER value and the variance depend on group size, all values reported below for different speaker classes can only be compared directly if the number of speech samples in the classes is comparable.

4 Classification Results

4.1 Classification in Terms of Gender

For both genders, we conducted a ten-fold cross-validation in which nine-tenths of the transcriptions were used to identify gender-specific lexical and pronunciation features, and in which the transcriptions of the remaining samples were used to test the classification algorithm. Table 2 presents the results of this experiment for both genders and feature sets. The results were obtained with the algorithm's optimal parameter settings for each of the two feature sets.

Table 2. Best possible speaker classification in terms of gender with lexical and pronunciation features

gender	# samples	lexical features	pronunciation features
		EER (%)	EER (%)
male	276	23 ***	41
female	387	24 ***	42

Whereas the use of the lexical features led to a highly effective classification with error rates of about 24%, the use of the pronunciation features did not. In other words, the pronunciation features could not help the algorithm distinguish between the phonetic transcriptions of male and female speakers. The frequent misclassification of speakers from their pronunciation features may be due to several reasons. The most obvious reason would be the absence of gender-specific pronunciation characteristics at the broad phonetic level. A more disturbing reason (disturbing because it would question the validity of our automatic phonetic transcriptions as a knowledge source for our experiments), would be an inadequate representation of gender-specific pronunciation features in the automatic phonetic transcriptions.

There are two reasons to assume that the mediocre classification performance of our algorithm was due to the absence of gender-specific pronunciation characteristics at the broad phonetic level rather than to inadequacies in the automatic phonetic transcriptions. First, the linguistic literature has not yet reported systematic gender-specific pronunciation differences at the broad phonetic level. The only systematic gender-specific pronunciation characteristics that have so far been reported were based on measurements of the overall speech rate [8-9], and on measurements at levels of finer phonetic detail (e.g. a structural difference between the dimensions of the vowel space of male and female speakers [10]). None of these gender-specific pronunciation characteristics can be reflected in a broad phonetic transcription of speech, e.g. in the form of systematic phone deletions or substitutions. Second, our results are in line with [11], who could not discover gender-specific pronunciation characteristics through the alignment of a canonical and a manually verified (instead of an automatic) transcription of male and female speech from the Spoken Dutch Corpus either.

4.2 Classification in Terms of Age

For every year of birth between 1928 and 1981, we tried to classify the speakers in terms of them being born before that year (24 to 631 samples per class), after that year (32 to 639 samples per class), or in an eleven-year window around that year (67 to 174 samples per class - see Section 3.2). Since these classification experiments yielded many data points, we confine ourselves to a description of the general tendencies.

Despite the fact that the algorithm was able to retrieve and successfully use age-specific pronunciation features for most of these classes, it still performed better with the lexical profiles than with the pronunciation profiles. The binary before/after classifications showed relatively stable classification accuracies: ignoring three outliers at each side of the time scale, the EERs ranged between 18% and 23% for the lexical profiles (with a mean EER over all age classes of 20.5%) and between 26% and 36% for the pronunciation profiles (mean EER over all classes: 32.4%). The use of the lexical profiles consistently led to effective classification ($p < 0.001$), the use of the pronunciation profiles as well ($p < 0.01$, and in 90% of the tests even $p < 0.001$).

The classification of speakers according to the eleven-year intervals showed more variation: ignoring the same three outliers at each side of the time scale, we obtained error rates between 19% and 41% with the lexical profiles (mean EER over all classes: 32.0%), and between 28% and 46% with the pronunciation profiles (mean EER over all classes: 38.5%). The use of the lexical profiles led to effective classification for the years at the outskirts of the time scale ($p < 0.001$ for the years between 1928 and 1942, and between 1973 and 1981) whereas there was hardly any effective classification noticeable for the years between 1942 and 1973. The use of the pronunciation features showed a similar pattern, although fewer effective classifications were found.

4.3 Classification in Terms of Regional Background

Table 3 presents the results of the classification of our speakers according to the regional background they lived in between the age of 4 and 16. We classified the speakers in terms of 16 geographical regions (see Table 1 in Section 3.2) and by means of the two feature sets.

Table 3. Best possible speaker classification in terms of regional background with lexical and pronunciation features

region	# samples	lexical features EER (%)	pronunciation features EER (%)
1a	105	35 *	38
1b	112	36 *	40
1c	21	26 *	29
2a	42	23 ***	34
2b	42	36	37
2c	52	32 *	40
2d	19	30	32
2e	6	21	21
2f	13	14 **	30
3a	18	19 **	33
3b	37	27 *	37
3c	12	23	24
3d	17	36	26
3e	20	38	31
4a	113	32 ***	40
4b	34	27 *	22 ***

Table 3 shows that the classification algorithm obtained effective classification for 10 out of 16 regions when using the lexical classification features. This indicates that the orthographic transcription of (at least part of) the investigated speech contained useful information with which our classification algorithm could classify unknown speakers. The EERs in Table 3 can only be compared for speaker classes comprising a comparable number of speech samples, because the EERs decreased when speaker classes with fewer samples were considered (e.g. compare the 36% EER with class 1b, which was made up of 112 samples, with the 26% EER with class 1c with only 21 samples). This means that we cannot draw conclusions about specific regions being more easily recognised than other regions.

The results in Table 3 also show the inability of the algorithm to retrieve and use geographically determined pronunciation features from the broad phonetic transcription. The classification algorithm was only able to effectively classify speakers of one region (Limburg, a peripheral region in the South East of the Netherlands).

The poor performance of the classification algorithm with the pronunciation features can be due to several reasons. First of all, some of the above mentioned regions may have characteristic features, but of a kind that are usually not represented at the broad phonetic level. For example, the Dutch phoneme /r/ has many allophonic variants, some of which have been reported characteristic for specific regions in the Netherlands [12]. However, in our study these different realisations could not be used for classification because the broad phonetic transcription did not distinguish allophonic variants of the phoneme /r/. A second possible explanation for the disappointing performance of our algorithm is the absence of distinguishing pronunciation features in our automatic phonetic transcription. To verify whether this could indeed be so, we examined the pronunciation of word-final /n/ preceded by schwa in plural nouns and verbs, since this pronunciation process is known to be typical for speakers in specific regions of the Netherlands, notably 2d and 3a [13]. A comparison between the speech samples for which both an automatic and a manually verified phonetic transcription were available showed that the automatic phonetic transcription did not represent the pronunciation of such word-final /n/s, whereas the manually verified phonetic transcription of the CGN did at least in some cases. A third possible explanation for the mostly ineffective classification performance is the potential mismatch between the geographical boundaries of the 16 regions defined in the CGN and the regions that can actually be characterised by means of outspoken pronunciation features. A fourth potential explanation is the heterogeneity of the speaker populations in the regional classes, either because the pronunciation features in these classes are inherently heterogeneous or because some speakers in the CGN are not particularly representative of their region. Finally, of course, we should also consider potential limitations of Linguistic Profiling for the purpose of classifying speakers on the basis of pronunciation features. Perhaps its capabilities were hampered by the fact that it could only use 94 pronunciation features, while there were some 150.000 lexical features, comparable to the number of features used in [1].

Further research is needed to clarify the way in which the above mentioned factors affect the classification of speakers on the basis of manual or automatic broad phonetic transcriptions.

4.4   Classification in Terms of Education Level

Finally, we investigated whether our classification algorithm was able to classify speakers in terms of their level of education. Table 4 presents the classification results for the two speaker classes (highly educated and moderately educated) with both the lexical and the pronunciation features. Again, we show the Equal Error Rates at the algorithm's optimal parameter settings.

Table 4. Best possible speaker classification in terms of education level with lexical and pronunciation features

level of education	# samples	lexical features	pronunciation features
		EER (%)	EER (%)
highly educated	496	41	46
moderately educated	167	41	44

The results in Table 4 show that the algorithm was not able to classify speakers effectively in terms of their level of education. The classification results with the pronunciation features reflect the inconclusive results reported in [14]. While they found significant differences between the reductions of phones in 14 frequent words ending in –lijk spoken by highly versus moderately educated Flemish speakers (the moderately educated speakers reduced more phones), there was no significant difference between the phone reductions of highly and moderately educated speakers from the Netherlands.

Although our results do not imply that speakers cannot be categorised according to the influence of their education on their speech, the high EERs do imply that the lexical features as well as the pronunciation features were unsuitable for classifying speakers according to their education level. Future research should clarify whether a further division of the speakers into smaller, more specific classes can improve classification accuracy.

4.5   More Specific Speaker Classes

In the previous sections, we classified speakers in classes that were defined by one speaker characteristic (gender, age…) at a time. However, someone's speech is likely to be influenced by the interplay of all four aforementioned speaker characteristics. This implies that, when we classify speakers in broad classes of which all members have only one characteristic in common, the 'class-specific' speech features may show a great deal of dispersion. Evidently, speaker classification with very broad and therefore perhaps partially overlapping classification profiles for different speaker classes is more difficult than speaker classification with well defined and more exclusive classification profiles.

Therefore, we attempted an integrated classification of our speakers according to all four speaker characteristics by using classes of speakers for which all four characteristics were fixed. In order to have sufficient training data for each combined class, we restricted this experiment to the classification of highly educated women who were born before or in 1975 and who were raised in region 1a (South Holland) or 4a (North Brabant), and the classification of highly educated women who were born after 1975 and who were raised in North Brabant. Class profiles were created for each of these classes. Table 5 presents the results of this classification experiment.

Table 5. Best possible speaker classification in terms of three specific speaker classes according to a joint assessment of four speaker characteristics: gender, age, education, regional background

highly educated women		# samples	lexical features	pronunciation features
born	raised in		EER (%)	EER (%)
≤1975	1a	29	24 **	36
≤ 1975	4a	28	30	30
> 1975	4a	23	31	28

In order to evaluate the possible benefit of classifying speakers in more specific speaker classes rather than in general classes, one would ideally want to compare the EERs in Table 5 with the EERs reported in Sections 4.1 to 4.4. However, as was explained in Section 3.4, such a direct comparison is impossible because of the different number of samples in the speaker classes in the previous sections. It is possible, however, to compare the EERs obtained with the lexical and the pronunciation features for the three specific speaker classes in Table 5. These comparisons (24-36%, 30-30%, 31-28%) show that per speaker class, the EERs obtained with both feature types were much more similar than in the previous sections.

We hypothesise that in the previous sections, where we classified speakers in general speaker classes that were defined by only one common speaker characteristic (e.g. gender), classification was affected by an influence from the interplay of the remaining speaker characteristics (age, regional background and level of education) on the speech features in the classes. It may well be that in these circumstances, the classification algorithm could still benefit from the abundance of lexical classification features (around 150.000) to use the most distinguishing features for classification and to ignore less characteristic features. At the same time, the algorithm may have had more difficulties to select and use features out of the much smaller set of 94 pronunciation features which were characteristic of the classes and which were not influenced by an interplay of speaker characteristics.

5   Conclusions and Plans for Future Research

We investigated whether Linguistic Profiling, a supervised learning algorithm originally designed for authorship verification, can be used to classify speakers according to their gender, age, regional background and level of education on the basis of the lexical content and the pronunciation of their speech. Our approach differed from conventional speaker classification procedures in that our algorithm analysed written representations of speech rather than the speech signal proper; it analysed orthographic and broad phonetic transcriptions of speech in order to identify regularities in lexical content and pronunciation.

We conducted experiments to determine the performance of our algorithm for speaker classification with the aforementioned lexical and pronunciation features. These experiments showed that the algorithm was often able to retrieve and use characteristic lexical features from the orthographic transcriptions. The lexical features enabled the classification algorithm to distinguish between male and female speakers, to classify speakers in terms of their age, and to determine the region speakers spent most of their childhood in (this held for 10 out of 16 investigated regions). Despite these encouraging results, however, the use of the lexical features proved insufficient to effectively classify speech from moderately or highly educated speakers and from people who spent their childhood in specific (6 out of 16 investigated) regions in the Netherlands. Moreover, the algorithm's performance is probably not good enough for operational speaker classification: in general, we found equal error rates between 20% and 40%.

When the classification algorithm had access only to the pronunciation features as reflected in our automatic broad phonetic transcription, it was hardly ever able to classify speakers effectively. We have argued that this may be explained by 1) the absence in the material of distinguishing pronunciation features at the broad phonetic level, 2) the failure of the automatic phonetic transcription procedure to capture distinguishing pronunciation features, 3) a mismatch between our speaker classes and groups of speakers that possibly show distinguishing speech features, 4) the heterogeneity of our speaker classes (either because they are inherently heterogeneous or because the speakers were not representative of their classes), and 5) the limitations of our algorithm for classification with a small number of classification features.

In future research, some of these potential explanations may be further investigated. As for 1), we had hoped that the relatively large amounts of broad phonetic transcriptions would enable our algorithm to identify class-specific pronunciation features at the broad phonetic level. However, our approach to defining potentially useful pronunciation processes resulted in fewer than 100 such features, which appeared insufficient to distinguish speaker classes effectively. It remains to be seen if and how the number and the distinctiveness of the pronunciation features can be increased. One option might be to move towards more detailed phonetic transcriptions. This would increase the number of possible mappings between canonical representations and actual realisations, and hence potentially also the number of different pronunciation processes that can be used for classification. This approach may seem counterproductive because it might reduce the number of pronunciation processes that occur at least five times (the criterion used in this study). However, if more detailed transcriptions can be made reliably, we might gain after all, since the use of more diverse phonetic symbols can result in the definition of more diverse but also more systematic phone mappings representing characteristic pronunciation features. At the same time it is clear that the further we would move away from a broad phonetic transcription of speech, the closer we would come to the signal-based classification procedures reported elsewhere in this book. As for 2), we have identified at least one regional pronunciation phenomenon, viz. the presence of word-final /n/ preceded by schwa in plural nouns and verbs, which was not systematically represented in the automatic transcriptions. It may well be that the same holds for other pronunciation phenomena that are conventionally considered as characteristic for some geographical region; the automatic transcription procedure, which was based exclusively on local properties of the speech signal may have selected its symbols less 'systematically' than the human transcribers who may have been biased towards conventional regional characteristics on the basis of subtle cues in the signal. Again, this seems to suggest that we should try and move towards more detailed phonetic transcriptions. As for 3) and 4), we may attempt to classify speakers in more specific classes, hopefully with more homogeneous speech behaviour. In most cases, this is likely to mean a subdivision of the classes used in this study. Recall that we classified our speakers in just 16 predefined geographical regions, and that we attempted the classification of speakers in just two classes defined by their level of education. The training and use of more specific speaker classes may increase the homogeneity of speech characteristics in these classes, but it would inevitably also introduce a data sparseness problem. Finally, as for 5), we may consider increasing the number of classification features for our algorithm, but we have already argued that it is not obvious how this can be accomplished. Alternatively, we may consider investigating classification techniques that are designed to operate with smaller numbers of features.

Finally, for a real application rather than for a scientific investigation like this study, it will probably be suboptimal to base classifications on a single type of classification features. For the best possible classification, we should give the classifier access to as many and as large a variety of features as possible. This means combining both the lexical and pronunciation features presented here, and probably also other features which have proven useful for speaker classification, e.g. acoustic features that can be directly retrieved from the speech signal as illustrated in the other chapters of this book.

Acknowledgement

The work of Christophe Van Bael was funded by the Speech Technology Foundation (Stichting Spraaktechnologie), Utrecht, the Netherlands.

References

[1] van Halteren, H.: Author Verification by Linguistic Profiling: An exploration of the parameter space. ACM Transactions on Speech and Language Processing 4(1) (2007)
[2] Oostdijk, N.: The Design of the Spoken Dutch Corpus. In: Peters, P., Collins, P., Smith, A. (eds.) New Frontiers of Corpus Research, pp. 105–112. Rodopi, Amsterdam (2002)
[3] Laver, J.: Principles of phonetics. Cambridge University Press, Cambridge (1995)
[4] Van Bael, C., Boves, L., Strik, H., van den Heuvel, H.: Automatic Phonetic Transcription of Large Speech Corpora: a Comparative Study. In: Proceedings of ICSLP-Interspeech 2006, Pittsburgh PA, pp. 1085–1088 (2006)
[5] Elffers, B., Van Bael, C., Strik, H.: ADAPT: Algorithm for Dynamic Alignment of Phonetic Transcriptions. Internal report, Department of Language & Speech, Radboud University Nijmegen, the Netherlands. Electronically (2005), available from http://lands.let.ru.nl/literature/elffers.2005.1.pdf

[6] Binnenpoorte, D.: Phonetic Transcriptions of Large Speech Corpora. Ph.D. Dissertation. Radboud University Nijmegen, the Netherlands (2006)

[7] Cucchiarini, C.: Phonetic Transcription: a Methodological and Empirical Study. Ph.D. Dissertation. University of Nijmegen, the Netherlands (1993)

[8] Verhoeven, J., De Pauw, G., Kloots, H.: Speech rate in a pluricentric language: A comparison between Dutch in Belgium and the Netherlands. Language and Speech 47(3), 297–308 (2004)

[9] Byrd, D.: Relations of Sex and Dialect to Reduction. Speech Communiciation 15, 39–54 (1994)

[10] Henton, C.: Acoustic variability in the vowels of female and male speakers. The Journal of the Acoustical Society of America (JASA) 94(4), 2387 (1994)

[11] Binnenpoorte, C., Van Bael, C., den Os, E., Boves, L.: Gender in everyday speech and language: A corpus-based study. In: Proceedings of Interspeech 2005, Lisbon, Portugal, pp. 2213–2216 (2005)

[12] Verstraeten, B., Van de Velde, H.: Socio-geographical variation of /r/ in standard Dutch. In: Van de Velde, H., van Hout, R. (eds.) r-atics, sociolinguistic, phonetic and phonological characteristics of /r/. Etudes & Travaux - ILVP/ULB. No 4. Brussels, pp.45–61 (2001)

[13] Hol, A.R.: Dialectgrenzen in Gelderland. In: Wingens, M.F.M., Demoed, H.B., Scholten, F.W.J. (eds.) Gelders Erfgoed, Gelders cultuurhistorisch kwartaalblad, 2006-2, pp. 11–13 (2006)

[14] Keune, K., Ernestus, M., van Hout, R., Baayen, R.H.: Variation in Dutch: From Written MOGELIJK to Spoken MOK. Corpus Linguistics and Linguistic Theory 1(2), 183–223 (2005)

Author Index

Lecture Notes in Artificial Intelligence (LNAI)

Vol. 4411: R.H. Bordini, M. Dastani, J. Dix, A.E.F. Seghrouchni (Eds.), Programming Multi-Agent Systems. XIV, 249 pages. 2007.

Vol. 4410: A. Branco (Ed.), Anaphora: Analysis, Algorithms and Applications. X, 191 pages. 2007.

Vol. 4399: T. Kovacs, X. Llorà, K. Takadama, P.L. Lanzi, W. Stolzmann, S.W. Wilson (Eds.), Learning Classifier Systems. XII, 345 pages. 2007.

Vol. 4390: S.O. Kuznetsov, S. Schmidt (Eds.), Formal Concept Analysis. X, 329 pages. 2007.

Vol. 4389: D. Weyns, H.V.D. Parunak, F. Michel (Eds.), Environments for Multi-Agent Systems III. X, 273 pages. 2007.

Vol. 4384: T. Washio, K. Satoh, H. Takeda, A. Inokuchi (Eds.), New Frontiers in Artificial Intelligence. IX, 401 pages. 2007.

Vol. 4371: K. Inoue, K. Satoh, F. Toni (Eds.), Computational Logic in Multi-Agent Systems. X, 315 pages. 2007.

Vol. 4369: M. Umeda, A. Wolf, O. Bartenstein, U. Geske, D. Seipel, O. Takata (Eds.), Declarative Programming for Knowledge Management. X, 229 pages. 2006.

Vol. 4343: C. Müller (Ed.), Speaker Classification. X, 355 pages. 2007.

Vol. 4342: H. de Swart, E. Orłowska, G. Schmidt, M. Roubens (Eds.), Theory and Applications of Relational Structures as Knowledge Instruments II. X, 373 pages. 2006.

Vol. 4335: S.A. Brueckner, S. Hassas, M. Jelasity, D. Yamins (Eds.), Engineering Self-Organising Systems. XII, 212 pages. 2007.

Vol. 4334: B. Beckert, R. Hähnle, P.H. Schmitt (Eds.), Verification of Object-Oriented Software. XXIX, 658 pages. 2007.

Vol. 4333: U. Reimer, D. Karagiannis (Eds.), Practical Aspects of Knowledge Management. XII, 338 pages. 2006.

Vol. 4327: M. Baldoni, U. Endriss (Eds.), Declarative Agent Languages and Technologies IV. VIII, 257 pages. 2006.

Vol. 4314: C. Freksa, M. Kohlhase, K. Schill (Eds.), KI 2006: Advances in Artificial Intelligence. XII, 458 pages. 2007.

Vol. 4304: A. Sattar, B.-h. Kang (Eds.), AI 2006: Advances in Artificial Intelligence. XXVII, 1303 pages. 2006.

Vol. 4303: A. Hoffmann, B.-h. Kang, D. Richards, S. Tsumoto (Eds.), Advances in Knowledge Acquisition and Management. XI, 259 pages. 2006.

Vol. 4293: A. Gelbukh, C.A. Reyes-Garcia (Eds.), MICAI 2006: Advances in Artificial Intelligence. XXVIII, 1232 pages. 2006.

Vol. 4289: M. Ackermann, B. Berendt, M. Grobelnik, A. Hotho, D. Mladenič, G. Semeraro, M. Spiliopoulou, G. Stumme, V. Svátek, M. van Someren (Eds.), Semantics, Web and Mining. X, 197 pages. 2006.

Vol. 4285: Y. Matsumoto, R.W. Sproat, K.-F. Wong, M. Zhang (Eds.), Computer Processing of Oriental Languages. XVII, 544 pages. 2006.

Vol. 4274: Q. Huo, B. Ma, E.-S. Chng, H. Li (Eds.), Chinese Spoken Language Processing. XXIV, 805 pages. 2006.

Vol. 4265: L. Todorovski, N. Lavrač, K.P. Jantke (Eds.), Discovery Science. XIV, 384 pages. 2006.

Vol. 4264: J.L. Balcázar, P.M. Long, F. Stephan (Eds.), Algorithmic Learning Theory. XIII, 393 pages. 2006.

Vol. 4259: S. Greco, Y. Hata, S. Hirano, M. Inuiguchi, S. Miyamoto, H.S. Nguyen, R. Słowiński (Eds.), Rough Sets and Current Trends in Computing. XXII, 951 pages. 2006.

Vol. 4253: B. Gabrys, R.J. Howlett, L.C. Jain (Eds.), Knowledge-Based Intelligent Information and Engineering Systems, Part III. XXXII, 1301 pages. 2006.

Vol. 4252: B. Gabrys, R.J. Howlett, L.C. Jain (Eds.), Knowledge-Based Intelligent Information and Engineering Systems, Part II. XXXIII, 1335 pages. 2006.

Vol. 4251: B. Gabrys, R.J. Howlett, L.C. Jain (Eds.), Knowledge-Based Intelligent Information and Engineering Systems, Part I. LXVI, 1297 pages. 2006.

Vol. 4248: S. Staab, V. Svátek (Eds.), Managing Knowledge in a World of Networks. XIV, 400 pages. 2006.

Vol. 4246: M. Hermann, A. Voronkov (Eds.), Logic for Programming, Artificial Intelligence, and Reasoning. XIII, 588 pages. 2006.

Vol. 4223: L. Wang, L. Jiao, G. Shi, X. Li, J. Liu (Eds.), Fuzzy Systems and Knowledge Discovery. XXVIII, 1335 pages. 2006.

Vol. 4213: J. Fürnkranz, T. Scheffer, M. Spiliopoulou (Eds.), Knowledge Discovery in Databases: PKDD 2006. XXII, 660 pages. 2006.

Vol. 4212: J. Fürnkranz, T. Scheffer, M. Spiliopoulou (Eds.), Machine Learning: ECML 2006. XXIII, 851 pages. 2006.

Vol. 4211: P. Vogt, Y. Sugita, E. Tuci, C.L. Nehaniv (Eds.), Symbol Grounding and Beyond. VIII, 237 pages. 2006.

Vol. 4203: F. Esposito, Z.W. Raś, D. Malerba, G. Semeraro (Eds.), Foundations of Intelligent Systems. XVIII, 767 pages. 2006.

Vol. 4201: Y. Sakakibara, S. Kobayashi, K. Sato, T. Nishino, E. Tomita (Eds.), Grammatical Inference: Algorithms and Applications. XII, 359 pages. 2006.

Vol. 4200: I.F.C. Smith (Ed.), Intelligent Computing in Engineering and Architecture. XIII, 692 pages. 2006.

Vol. 4198: O. Nasraoui, O. Zaïane, M. Spiliopoulou, B. Mobasher, B. Masand, P.S. Yu (Eds.), Advances in Web Mining and Web Usage Analysis. IX, 177 pages. 2006.

Vol. 4196: K. Fischer, I.J. Timm, E. André, N. Zhong (Eds.), Multiagent System Technologies. X, 185 pages. 2006.

Vol. 4188: P. Sojka, I. Kopeček, K. Pala (Eds.), Text, Speech and Dialogue. XV, 721 pages. 2006.

Vol. 4183: J. Euzenat, J. Domingue (Eds.), Artificial Intelligence: Methodology, Systems, and Applications. XIII, 291 pages. 2006.

Vol. 4180: M. Kohlhase, OMDoc – An Open Markup Format for Mathematical Documents [version 1.2]. XIX, 428 pages. 2006.